The LabVIEW Style Book

The LabVIEW Style Book

Peter A. Blume
President, Bloomy Controls, Inc.

PRENTICE HALL

Upper Saddle River, NJ • Boston • Indianapolis • San Francisco
New York • Toronto • Montreal • London • Munich • Paris • Madrid
Capetown • Sydney • Tokyo • Singapore • Mexico City

Many of the designations used by manufacturers and sellers to distinguish their products are claimed as trademarks. Where those designations appear in this book, and the publisher was aware of a trademark claim, the designations have been printed with initial capital letters or in all capitals.

The author and publisher have taken care in the preparation of this book, but make no expressed or implied warranty of any kind and assume no responsibility for errors or omissions. No liability is assumed for incidental or consequential damages in connection with or arising out of the use of the information or programs contained herein.

The publisher offers excellent discounts on this book when ordered in quantity for bulk purchases or special sales, which may include electronic versions and/or custom covers and content particular to your business, training goals, marketing focus, and branding interests. For more information, please contact:

U.S. Corporate and Government Sales
(800) 382-3419
corpsales@pearsontechgroup.com

For sales outside the United States, please contact:

International Sales
international@pearsoned.com

This Book Is Safari Enabled

The Safari® Enabled icon on the cover of your favorite technology book means the book is available through Safari Bookshelf. When you buy this book, you get free access to the online edition for 45 days. Safari Bookshelf is an electronic reference library that lets you easily search thousands of technical books, find code samples, download chapters, and access technical information whenever and wherever you need it.

To gain 45-day Safari Enabled access to this book:

- Go to http://www.awprofessional.com/safarienabled
- Complete the brief registration form
- Enter the coupon code 27I1-GHWH-WV5A-TBGJ-78DC

If you have difficulty registering on Safari Bookshelf or accessing the online edition, please e-mail customer-service@safaribooksonline.com.

Visit us on the Web: www.prenhallprofessional.com

Library of Congress Cataloging-in-Publication Data:

Blume, Peter A.
  The labVIEW style book / Peter A. Blume.
    p. cm.
  Includes bibliographical references and index.
  ISBN 0-13-145835-3 (hardback : alk. paper) 1. Scientific apparatus and instruments—Computer simulation. 2. LabVIEW. I. Title.
  Q185.B568 2007
  005.4'38—dc22
                          2006035871

Copyright © 2007 Pearson Education, Inc.

All rights reserved. Printed in the United States of America. This publication is protected by copyright, and permission must be obtained from the publisher prior to any prohibited reproduction, storage in a retrieval system, or transmission in any form or by any means, electronic, mechanical, photocopying, recording, or likewise. For information regarding permissions, write to:

Pearson Education, Inc.
Rights and Contracts Department
One Lake Street
Upper Saddle River, NJ 07458
Fax (201) 236-3290

ISBN 0-13-145835-3
Text printed in the United States at Courier in Kendallville, Indiana.
Second printing, July 2008

*To the employees of Bloomy Controls who contributed to the evolution of this topic within our organization.
To my wife, Phia, and daughter, Eva,
for their patience, love, and support.*

Contents

Foreword   xv
Preface   xvii
Acknowledgments   xx
About the Author   xxii

▼ 1

The Significance of Style   1

   1.1   Style Significance 2
           1.1.1   Ease of Use 6
           1.1.2   Efficiency 7
           1.1.3   Readability 9
           1.1.4   Maintainability 11
           1.1.5   Robustness 12
           1.1.6   Simplicity 15

	1.1.7 Performance 17
	1.1.8 Style Tools 18
1.2	Style Versus Time Tradeoff 18

▼2

Prepare for Good Style   21

2.1	Specifications 22
	2.1.1 Best Practices for Specifications Development 24
	2.1.2 LabVIEW Project Requirements Specification 25
2.2	Design 29
	2.2.1 Search for Useful Resources 30
	2.2.2 Develop a Proof of Concept 30
	2.2.3 Revise the Specification 31
2.3	Configure the LabVIEW Environment 32
	2.3.1 LabVIEW Options Dialog Box 32
	2.3.2 Code Reuse 34
2.4	Project Organization, File Naming, and Control 39
	2.4.1 Disk Organization 40
	2.4.2 The LabVIEW Project 42
	2.4.3 File-Naming Conventions 44
	2.4.4 Source Control 45

Endnotes 46

▼3

Front Panel Style   47

3.1	Layout 48
	3.1.1 General Rules 48
	3.1.2 GUI VI Panel Layout 51
	3.1.3 SubVI Panel Layout 59
3.2	Text 61
	3.2.1 General Rules 61

		3.2.2 Control Labels... 66

        3.2.2 Control Labels...66
        3.2.3 SubVI Panel Text..68
        3.2.4 Industrial GUI VI Text..69
  3.3 Color..69
  3.4 GUI Navigation...71
        3.4.1 Control Scope..71
        3.4.2 Consistency...74
  3.5 Examples..75
        3.5.1 SubVI from Selection..75
        3.5.2 Dialog Utility VI..77
        3.5.3 Capacitor Test & Sort...79
        3.5.4 Centrifuge DAQ...81
        3.5.5 Spectralyzer...82
        3.5.6 Parafoil Guidance Interface..83

Endnotes..85

▼4

Block Diagram   87

  4.1 Layout..88
        4.1.1 Layout Basics...88
        4.1.2 SubVI Modularization...90
  4.2 Wiring..93
        4.2.1 Clear Wiring Techniques..94
        4.2.2 Cluster Modularization...97
  4.3 Data Flow..101
        4.3.1 Data Flow Basics..101
        4.3.2 Practical Variables and Sequence Structures..........................104
        4.3.3 Impractical Variables and Sequence Structures......................108
        4.3.4 Optimizing Data Flow...111
  4.4 Examples..115
        4.4.1 SubVI from Selection..115
        4.4.2 Excessively Nested VI...117
        4.4.3 Haphazard VI..119

	4.4.4	Right to Left VI	120
	4.4.5	Left to Right VI	121
	4.4.6	Centrifuge DAQ VI	122
	4.4.7	Screw Inspection VI	124
	4.4.8	Optical Filter Test VI	127

Endnotes ... 128

▼5

Icon and Connector — 129

5.1	Icon		132
	5.1.1	Icon Basics	132
	5.1.2	Icon Shortcuts	135
	5.1.3	International Icons	139
5.2	Connector Pane		140
5.3	Examples		145
	5.3.1	Obnoxious Examples	145
	5.3.2	Instrument Drivers	148
	5.3.3	Miscellaneous Examples	151
	5.3.4	Clever Examples	153

Endnotes ... 155

▼6

Data Structures — 157

6.1	Data Structure Design Methodology		158
	6.1.1	Choose the Controls and Data Types	158
	6.1.2	Configure the Properties	169
	6.1.3	Create the Data Constructs	170
6.2	Simple Data Types		172
	6.2.1	Boolean	173
	6.2.2	Numeric	175
	6.2.3	Special Numeric	177
	6.2.4	String, Path, and Picture	179

Contents

6.3	Data Constructs	180
	6.3.1 Simple Arrays and Clusters	181
	6.3.2 Special Data Constructs	186
	6.3.3 Nested Data Structures	187
6.4	Examples	193
	6.4.1 Thermometer VI	193
	6.4.2 OpenG Variant	194
	6.4.3 Haphazard Data	197
	6.4.4 Centrifuge DAQ VI	199
Endnotes		201

▼7

Error Handling — 203

7.1	Error Handling Basics	204
	7.1.1 Trapping Errors	205
	7.1.2 Reporting Errors	210
	7.1.3 Error Codes	214
7.2	SubVI Error Handling	217
7.3	Prioritizing Errors	222
7.4	Error Handling Tips	226
	7.4.1 Structure Wiring	226
	7.4.2 Merging Errors	226
	7.4.3 Clearing Errors	228
	7.4.4 Automatic Error Handling	230
7.5	Examples	230
	7.5.1 Continuous Acquire To File	230
	7.5.2 Suss Interface Toolkit	231
	7.5.3 Merge Parallel Errors	232
	7.5.4 Screw Inspection VI	234
	7.5.5 Test Executive VI	235
Endnotes		238

Design Patterns — 239

- 8.1 Simple Design Patterns ... 241
  - 8.1.1 Immediate SubVI ... 241
  - 8.1.2 Functional Global ... 244
  - 8.1.3 Continuous Loop ... 246
  - 8.1.4 Event-Handling Loop ... 250
- 8.2 State Machines ... 254
  - 8.2.1 Classic State Machine ... 257
  - 8.2.2 Queued State Machine ... 260
  - 8.2.3 Event-Driven State Machine ... 262
  - 8.2.4 Event Machine ... 265
- 8.3 Compound Design Patterns ... 267
  - 8.3.1 Parallel Loops ... 269
- 8.4 Complex Application Frameworks ... 272
  - 8.4.1 Dynamic Framework ... 272
  - 8.4.2 Multiple-Loop Application Framework ... 278
  - 8.4.3 Modular Multiple-Loop Application Framework ... 283
- 8.5 Examples ... 287
  - 8.5.1 Elapsed Time VI ... 287
  - 8.5.2 Poll Instrument Response VI ... 288
  - 8.5.3 Unconventional State Machine ... 289
  - 8.5.4 Centrifuge DAQ VI ... 292
  - 8.5.5 Transducer Control Utility ... 293
  - 8.5.6 Distributed Control System ... 296

Endnotes ... 297

Documentation — 299

- 9.1 Front Panel Documentation ... 301
- 9.2 Block Diagram ... 304

9.3		Icon and VI Description	311
9.4		Online Documentation	311
9.5		Examples	315
	9.5.1	SubVI from Selection VI	315
	9.5.2	Filter Test VI	316
	9.5.3	Meticulous Control Descriptions	317
	9.5.4	Temperature Profile Illustration	318
Endnotes			318

▼ 10
Code Reviews — 319

10.1		Self-Reviews	320
	10.1.1	VI Analyzer Toolkit	320
	10.1.2	Manual Checklist	330
10.2		Peer Reviews	334
Endnotes			338

Appendix A
Glossary — 339

Appendix B
Style Rules Summary — 349
Index — 357

Foreword

I've been writing VIs for a long time as a member of the LabVIEW R&D team at National Instruments. During my first week of LabVIEW training many years ago, I decided to write a LabVIEW version of my favorite card game, called "Set." It took me a while, but I eventually finished it, and it was probably some of the ugliest LabVIEW code you have ever seen. As Peter Blume would say, my code was THE definition of "spaghetti." Worse yet, the front panel was full of neon-colored controls and indicators that were blindingly bright for no good reason. My code was functional, but it was not at all useable, and certainly not maintainable. I tried to add some new functionality a few years ago (after I was much more familiar with LabVIEW programming), but I quickly gave up because I had no idea how my own code worked.

Many years later, I can comfortably say that my programming style has improved leaps and bounds over those initial months with LabVIEW. However, I can also say without a doubt that my growing pains regarding LabVIEW programming style would have been all but eliminated had *The LabVIEW Style Book* been around back then. This book is a fully comprehensive resource, covering every aspect of VI style, from the highest level (project planning and organization) to the smallest detail (wires with too many bends). If I had read this book *while* I was learning LabVIEW, my code would have been many times more useable and maintainable from the start.

The LabVIEW Style Book is a must-read for every LabVIEW developer. Not only does it contain essential style rules for any new LabVIEW developer, it is also a necessary refresher for seasoned LabVIEW veterans. In particular, there are some chapters (like Chapter 6, "Data Structures," Chapter 7, "Error Handling," Chapter 8, "Design Patterns") that contain crucial programming tips and techniques not available in any other single resource. As one of the VI style advocates in LabVIEW R&D, I will definitely be recommending that new VI developers on my team read this book, and I'm confident that I'll be referencing it frequently in style discussions with my more experienced colleagues.

Another truly unique feature of the book is the impressive number of example VIs that Peter uses to illustrate good (and sometimes bad) style. With his decade-and-a-half of LabVIEW experience, Peter has a vast library of VIs that he, his employees, or his customers have written. He utilizes this library of code effectively to illustrate applicable points in every chapter. What's more, he will sometimes take the same example over the course of several chapters and refine it more and more with style rules picked up along the way. This gives us a "real-time" view of how good style can positively affect VIs as they are being developed.

So with LabVIEW celebrating 20+ years of inspiring engineers, scientists, and even children worldwide, I'm confident that readers of *the* definitive book on LabVIEW style will ultimately appreciate the time and effort savings that result from developing VIs with consistent, appropriate style. If you're a LabVIEW beginner, get ready for your VIs to become—as Peter would say—"awe-inspiring" ...and if you're a LabVIEW expert, you're going to learn a lot, too. I know I did!

Darren Nattinger
Staff Software Engineer, LabVIEW R&D
National Instruments Corporation

Darren Nattinger has worked at National Instruments for eight years. He is currently the primary developer for the VI Analyzer Toolkit and the Report Generation Toolkit for Microsoft Office. He was also a reviewer for The LabVIEW Style Book.

Preface

The LabVIEW Style Book is a comprehensive reference on recommended LabVIEW development practices. It contains style rules designed to optimize the ease of use, efficiency, readability, maintainability, robustness, simplicity, and performance of LabVIEW applications. The book provides thorough explanations of each rule, including examples and illustrations. The material leverages the work of the early pioneers of the LabVIEW community[1], has evolved from many years of use by Bloomy Controls[2], and has been reviewed by esteemed representatives of the LabVIEW community. I invite you to learn from the experiences of myself and the staff at Bloomy Controls, Inc., by reading *The LabVIEW Style Book*. I hope you enjoy reading it as much as I enjoyed writing it!

Intended Reader

Intended readers include developers, managers, and organizations that develop or use LabVIEW applications. You must have a working knowledge of fundamental LabVIEW principles and terminology, as instructed in a LabVIEW Basics I and II hands-on course[3], and experience developing and deploying applications. Experienced beginners can use this book to form good programming habits early in their LabVIEW careers. Intermediate developers, who have mastered the fundamentals and are ready to take their skills to the next level, will learn the most from this material. No doubt you have experienced the power and flexibility of LabVIEW and are ready to concentrate on style. Advanced developers will strongly identify with the contents, reinforce their knowledge and experience, and have a useful reference to share with colleagues. You might use *The LabVIEW Style Book* to help reduce the training and support burden you might have within your organization. **Managers** and **Organizations** that employ multiple developers and users can gain maximum benefit by standardizing on these style rules across the organization. This approach ensures quality and consistency throughout an organization and helps satisfy industry quality standards.

Organization

The chapters of *The LabVIEW Style Book* present style rules and examples organized by topic. Chapter 1, "The Significance of Style," discusses the relationship between style and ease of use, efficiency, readability, maintainability, robustness, simplicity, and performance. Chapter 2, "Prepare for Good Style," presents considerations that influence style before you begin programming, including specifications, configuration of the LabVIEW environment, and project and file organization. Additionally, it presents a specialized standard for LabVIEW project specifications. Chapter 3, "Front Panel Style"; Chapter 4, "Block Diagram"; and Chapter 5, "Icon and Connector," present the basics for VI layout and development. Chapter 3 provides rules for layout, text, color, and navigation. It distinguishes separate rules for the front panels of GUI VIs and subVIs, where appropriate. Chapter 4 presents rules for layout, wiring, and data flow, along with techniques for optimizing data flow. Chapter 5 discusses good icon development practices and editing shortcuts, and covers standard connector terminal patterns, assignments, and conventions.

Chapter 6, "Data Structures," provides rules on data type selection and array and cluster development. A methodology is integrated with several useful reference tables for simplifying data type selection and configuration. Rules and examples for optimizing VIs involving complex data structures also are presented in this chapter. Chapter 7, "Error Handling"; Chapter 8, "Design Patterns"; and Chapter 9, "Documentation," expand upon the basics. Chapter 7 presents comprehensive rules for thorough error handling, along with special considerations for error handling within subVIs. Chapter 8 discusses common VI architectures that promote good style, beginning with simple subVI design patterns and progressing to single and multiple loop design patterns. It also describes several variations of the LabVIEW state machine. Additionally, Chapter 8 presents three complex application frameworks, including a dynamic framework that uses plug-ins, a multiple-loop framework, and a modular multiple-loop framework that uses loop-subVIs. Chapter 9 presents rules for documenting your source code, including the front panel, block diagram, and icon and VI description. Additionally, the generation and integration of online documents is discussed. Chapter 10, "Code Reviews," presents several methods of reviewing source code and enforcing style rules, including self-reviews utilizing a manual checklist, automated self-reviews utilizing the LabVIEW VI Analyzer Toolkit, and peer reviews. An application is evaluated using each of these techniques.

Appendixes include a glossary and a style rules summary. Appendix A, "Glossary," provides a list of terms and definitions; many LabVIEW and software industry terms are evolutionary and context sensitive. Any term that seems specialized or ambiguous is defined where it first appears within the book and used consistently in successive chapters. The definitions are repeated in the glossary for ease of reference. Appendix B, "Style Rules Summary," lists the style rules presented in each chapter.

Style Rules Priority Convention: Throughout *The LabVIEW Style Book*, two priority levels are applied to the rules and distinguished as bold or plain italic. High priority rules are laws that should almost always be followed with very few exceptions. They are denoted by bold italics as follows:

 Rule 2.2 Write a requirements specification document

Normal priority rules are recommendations that are generally considered as good practices, but are either not as critical as the high-priority rules, or more exceptions exist. They are denoted by plain italic as follows:

 Rule 2.3 Maintain good LabVIEW style throughout the proof of concepts

Endnotes

1. See the "Acknowledgments" section for a list of reviewers, contributors, and people who have helped advance the science of LabVIEW Style.
2. Bloomy Controls is a National Instruments Select Integration Partner with offices in Windsor, Connecticut; Milford, Massachusetts; and Fort Lee, New Jersey. Information is available at www.bloomy.com.
3. LabVIEW Basics I and II is a one-week hands-on course offered by NI Certified Training Centers. More information is available from http://www.ni.com/training.

Acknowledgments

I would like to thank the following reviewers and contributors of source code for illustrations:

Darren Nattinger, National Instruments
Greg Burroughs, Bloomy Controls, Inc.
John Compton-Smith, Dover Technology Limited
Crystal Drumheller, National Instruments
Greg McKaskle, National Instruments
Brian Powell, National Instruments
Heather Eisenbraun, National Instruments
Bob Hamburger, Bloomy Controls, Inc.
James Fowler, Bloomy Controls, Inc.
Bart Craft, National Instruments
Alex Khazanov, Harris Corporation
Robert Cornwell, Bloomy Controls, Inc.
Keith Brainard
Anthony Conaci, CiDRA
Robert Gough, JDS Uniphase
Ernie St. Louis, Ciencia
Jim Kring, James Kring, Inc.

Acknowledgments

Ken Tumidajski, Testand Corporation
Alan Blankman, LeCroy Corporation
Daniel L. Press, Prime Test Corporation
Danny Allard, Videotron
Brian Gangloff, DataAct Incorporated
Dave Galanis, UTC Power
Roger Emerick, Pioneer Aerospace Corp.
Robert Greene, NSK
Robert Breidenthal, Precision Optics

Additionally, I would like to acknowledge the early pioneers of LabVIEW style. Meg Kay and Gary Johnson wrote a LabVIEW style guide that has subsequently evolved into National Instruments *LabVIEW Development Guidelines*, part of the LabVIEW shipping documentation. Randy Johnson wrote *Rules to Wire By,* which was printed in the *LabVIEW Technical Resource* newsletter ("LTR") in 1999. More recently, Darren Nattinger of NI developed the LabVIEW VI Analyzer Toolkit, an add-on toolkit that automatically analyzes VI style. Finally, Noel Adorno wrote NI's *Instrument Driver Development Guidelines*, which are available for free download from www.ni.com.

About the Author

Peter Blume is the founder and president of Bloomy Controls, Inc., a National Instruments Select Integration Partner that specializes in LabVIEW-based systems development. Since LabVIEW Version 2.5, Blume and his staff of engineers have solved more than a thousand industrial applications for customers throughout the northeastern United States. To promote consistent quality among multiple developers in multiple offices, Blume established and evolved the company's LabVIEW development practices.

Blume has written and presented multiple LabVIEW style-related presentations, including *Bloomy Controls' Professional LabVIEW Development Guidelines* at NIWeek 2002 and *Five Techniques for Better LabVIEW Code* at NIWeek 2003. He also has published technical articles in various trade publications, including *Test & Measurement World, Evaluation Engineering, Electronic Design,* and *Desktop Engineering.*

Blume holds a Bachelor of Science degree in electrical engineering from the University of Connecticut. He is a National Instruments Certified LabVIEW Developer and Certified Professional Instructor. The company has offices in Connecticut, Massachusetts, and New Jersey. For more information, visit www.bloomy.com.

Readers who want to contact Blume regarding style-related suggestions, questions, or comments may do so at the following email address: lvstyle@bloomy.com. Readers interested in contracting Bloomy Controls for a LabVIEW development project should call us directly or contact us through our website at www.bloomy.com/quote.

The Significance of Style

1

LabVIEW is a graphical programming language for developing diverse applications in a multitude of industries. The block diagram provides a unique form of source code expression that is dissimilar to most programming languages and development environments. The data flow paradigm represents the program as wires, terminals, structures, and nodes rich with functionality and innovation. LabVIEW extends this innovation to the developer, providing tremendous freedom of expression and creativity. As such, there are many means to an end, or possible development styles, with LabVIEW.

1.1 Style Significance

A given software task might have numerous possible implementations that appear functionally equivalent. What on the surface seems a matter of personal development preference, creative license, or style has significant implications. Developer tendencies have direct effects on the outcome of applications. Throughout my career as a professional LabVIEW consultant, manager, and trainer, I have observed many different development styles and have enjoyed debating the pros and cons of each. I can tell you that style is a sensitive issue for many developers. However, it is extremely important to recognize that good development style is not merely a matter of personal preference. Some styles lend themselves to better performance, source code that is easier to read and maintain, and applications that are more reliable and robust. Hence, I present Theorem 1.1:

> **Theorem 1.1:** A direct relationship exists between LabVIEW development style and the ease of use, efficiency, readability, maintainability, robustness, simplicity, and performance of the completed application.

Theorem 1.1 is the foundation upon which this book is based. LabVIEW development style really *is* significant. Few would argue that clean and neat diagrams are easier to read than sloppy ones. Organized user interfaces are more intuitive and easier to operate. But did you know that VIs

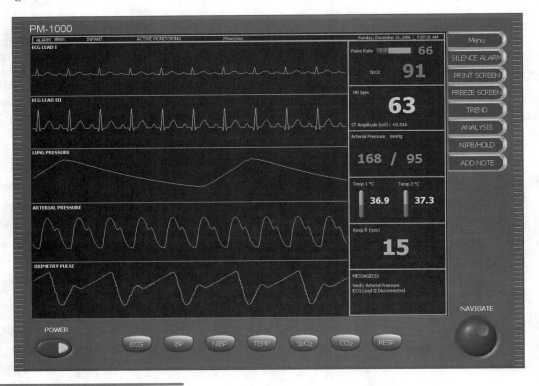

Figure 1-1
Meticulous VI—the front panel and block diagram for a complex application, developed with meticulous attention to detail

containing neat panels and diagrams normally execute more efficiently and with fewer bugs? Do you consider that your applications might need to be operated and maintained by people who are unfamiliar with your coding style? Ironically, this person might even be you 6 months, a year, or several years from now, after you have forgotten your own work. Indeed, I have seen developers confess their inability to explain source code that they developed last week, never mind last year. By contrast, many multideveloper teams that reside in remote locations can productively work together, share, and apply LabVIEW code seamlessly and without explanation. The difference is style.

Consider a few examples. Figures 1-1, 1-2, and 1-3 illustrate top-level VIs created by different developers for very dissimilar application types and complexity levels. However, each reflects the developer's distinctive programming style. Which front panel would you prefer to operate? Which diagram would you rather modify and maintain? Which VI executes efficiently? Which VI is more likely to have bugs, race conditions, or memory leaks?

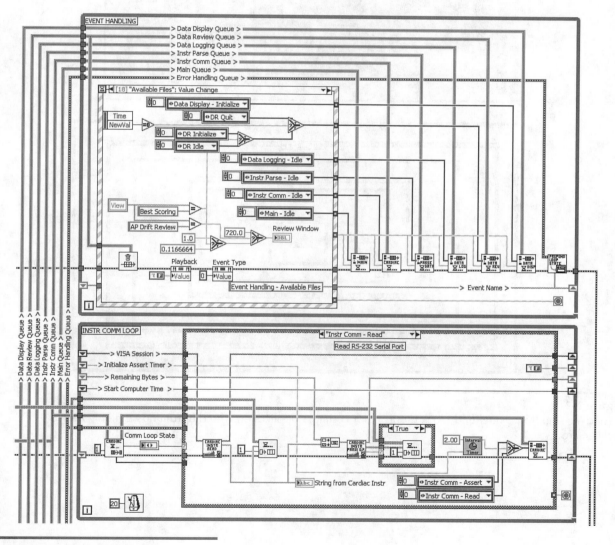

Figure 1-1   continued

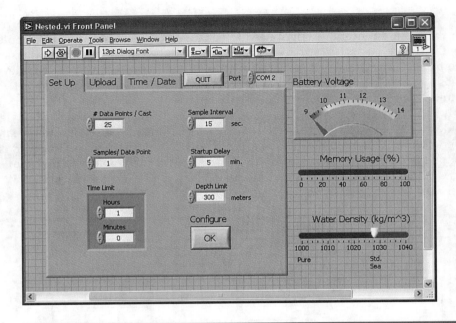

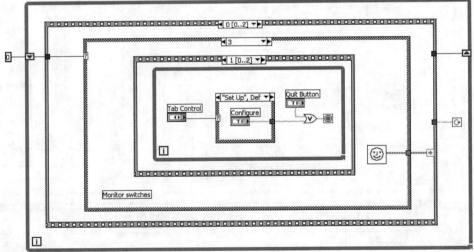

Figure 1-2
Nested VI—the front panel and block diagram for a simple application. The diagram has a highly nested architecture.

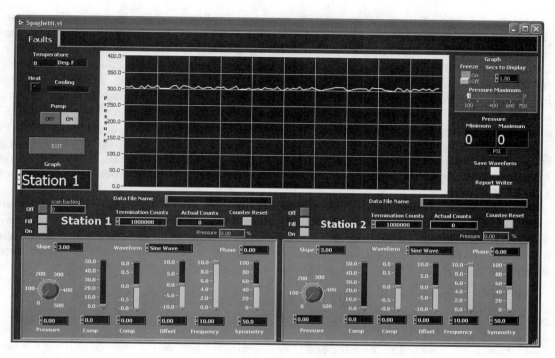
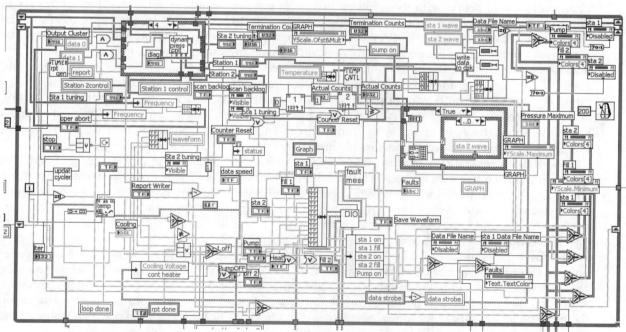

Figure 1-3
Spaghetti VI—the front panel and block diagram (which resembles spaghetti) for a medium-complexity application.

Meticulous VI in Figure 1-1 is a highly complex application used in the medical industry for patient physiologic data monitoring. It has a very professional and highly organized appearance. The front panel is densely functional yet appears neat and intuitive. The diagram is impressive, containing multiple parallel loops, networks of labeled wires, advanced constructs, and liberal documentation. Nested VI in Figure 1-2 is a relatively simple application that controls a scientific instrument. It has a clean, uncluttered front panel and block diagram. The most noteworthy characteristic is the diagram's multiple layers of structures nested within structures. Spaghetti VI in Figure 1-3 is an automated test application with medium complexity. The front panel contains many different colors and font styles, which tend to clash and distract the user. The diagram is a labyrinth of wires and nodes compacted together within a single loop. So dissimilar are these three programming styles that it is worth noting that all three examples are commercial applications developed by experienced LabVIEW professionals. Style is the main differentiator.

Let us break down Theorem 1.1 into seven different parts and evaluate each one individually: ease of use, efficiency, readability, maintainability, robustness, simplicity, and performance. Several of these terms are general, subjective, ambiguous, and overlapping. Therefore, each section begins with a definition of how the term is used here and then discusses how that term relates to development style.

1.1.1 Ease of Use

Ease of use is the ease with which the end user operates the software and accomplishes her objectives. This involves interacting with the application's graphical user interface (GUI). Ease of use ranges in importance from *less* important, for a one-time experiment to be run only by the application's developer, who is intimately familiar with the software; to *very* important, for a production application to be used by a variety of semiskilled operators within a single organization; to *extremely* important, for a mission- or safety-critical application, or a commercial application intended for distribution and resale. If you are designing a commercial application, read a GUI style reference for your application's target operating system. If your application is mission or safety critical, read an appropriate text on human factors. *The LabVIEW Style Book* contains general style rules that can be applied to most LabVIEW applications, from the lab to production, on most operating systems, from desktop PCs running Windows to embedded processors running a real-time or hand-held operating system.

Ease of use is related to readability and responsiveness of the GUI. In Figure 1-1, the front panel for Meticulous VI is intuitively laid out. The principal physiologic data is displayed in stacked waveform charts and large numeric indicators in the center of the panel. Ancillary data is contained on separate display screens. Boolean controls facilitate navigation in a conventional manner. The overall appearance resembles and behaves as a virtual instrument. The application is fast and responds quickly to operations that the user performs.

The front panel of Nested VI in Figure 1-2 uses a tab control to logically organize the front panel objects. The controls are clearly labeled and evenly spaced. However, the purpose of each tab and the order by which the user navigates the GUI are not immediately clear. The diagram contains Sequence structures that execute each frame to completion, regardless of any operations the user performs on the front panel. Its capability to respond to the user clicking **Quit** depends upon which frame of the Sequence structure the application is executing when the button's value changes.

The front panel of Spaghetti VI, shown in Figure 1-3, organizes controls and indicators using clusters. It is logical and organized, but it contains too many colors and font types, and not enough empty areas or "white space." More important, the performance is not reliable. No matter how attractive a GUI appears, if it does not perform its intended function, it will not provide a positive user experience.

Chapter 3, "Front Panel Style," contains style rules for front panel design that promote consistency and ease of use for your GUIs. Chapter 6, "Data Structures," and Chapter 8, "Design Patterns," provide additional style rules that help optimize the responsiveness of the GUI.

1.1.2 Efficiency

Efficiency pertains to the use of processor, memory, and input/output (I/O) resources. An efficient LabVIEW application executes quickly, without performing unnecessary operations, particularly ones that are performed repeatedly within looping structures. An efficient application also conserves memory by limiting the size of the four LabVIEW memory components: the front panel, block diagram, data space, and code. **Front panel** and **block diagram** memory store the graphical objects and images that comprise the front panel and block diagram, respectively. **Data space memory** contains all the data that flows through the diagram, as well as the diagram constants, default values for front panel controls, and the data that is copied when written to variables and front panel indicators. **Code** is the portion of memory that contains the compiled source code. Finally, efficient applications minimize I/O operations, such as GUI updates, instrument and network communications, and data acquisition (DAQ) calls. Execution speed and memory use are related. Memory and disk operations are a principal source of latencies in all modern computing devices. As memory consumption grows during the execution of an application, LabVIEW's memory manager is called upon to allocate new and larger memory blocks. This causes a delay while the memory manager runs and results in fragmented memory, for which some of the previous blocks are not efficiently used. The memory manager is a wonderful aspect of LabVIEW that handles memory allocation automatically. However, developers should be aware of the types of operations that might cause it to run and avoid these situations from occurring unnecessarily.

LabVIEW contains a tool called the **Profile Performance and Memory** window that directly measures the execution speed and data memory of all VIs loaded in memory. This tool, shown in Figure 1-4, is accessed by selecting **Tools»Profile»Performance and Memory**. This is an excellent tool for helping to improve the efficiency of an application. Use the Profile Performance and Memory window to determine which VIs consume the most time and memory, and examine those more closely. You can optimize your VI's efficiency by iteratively making improvements and checking the profile metrics.

Note that the efficiency of distinctly different applications, such as Meticulous VI, Nested VI, and Spaghetti VI, cannot be compared in any meaningful way. This is because execution time, memory consumption, and I/O depend on the application's requirements as well as the developer's programming style. For example, metrics from the Profile Performance and Memory window might indicate that Nested VI executes the fastest and uses the least memory. However, Nested VI is the least resource-demanding of the three applications and might even have plenty of room for further optimization. The Profile Performance and Memory window is best used for measuring the change in efficiency of a single application when incremental modifications have been made.

An alternate method of evaluating efficiency is simply to inspect the application for sources of inefficiency. Look for unnecessary operations within loops and for operations that create new data buffers. The diagram of Spaghetti VI (see Figure 1-3), for example, contains many nodes within the main While Loop. This includes read local variables, which make copies of their data when read from, and Property Nodes that are written within every iteration, regardless of whether their value has changed.

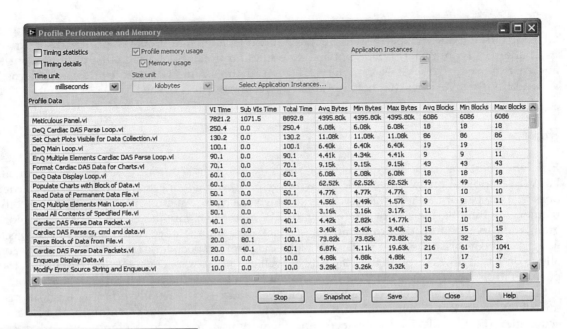

Figure 1-4
The LabVIEW Profile Performance and Memory window displays the data memory of Meticulous VI.

Inspecting the diagram of Nested VI (see Figure 1-2) reveals a **tight loop**, in which several control terminals are polled within the innermost While Loop as fast as the loop can run. High-speed GUI polling is not an efficient use of the processor. Most humans cannot distinguish between a 1 millisecond (ms) and a 100ms GUI response. More efficient alternatives include adding a delay using the Wait (ms) function or using an Event structure to service all user interface activity. These alternatives free the processor to work on parallel tasks and applications.

Meticulous VI (see Figure 1-1) uses multiple parallel loops on one diagram, many shift registers, and an Event structure. The following tasks execute in their own separate parallel loop: GUI event handling, instrument communications, data parsing, data logging, data review, data display, and error handling. A total of eight parallel loops are used. Each loop is finely tuned for optimum performance. However, the top-level diagram is extremely large, resulting in very large data and diagram memory components. The four components of memory use are provided in the VI Properties window, shown in Figure 1-5. You access the VI Properties window by selecting **File»VI Properties** and then selecting **Memory Usage** for the category.

Event structures are the most efficient method of capturing user interface activity. The GUI event handling loop sleeps, using no processor time, until it receives a registered event from the operating system. This is both more efficient and much more responsive than polling the control terminals in a loop. Hence, Meticulous VI is maximally responsive to GUI events. Finally, Meticulous VI makes extensive use of shift registers for passing data between loop iterations, and queues for passing data between parallel loops. Shift registers are much more efficient alternatives to variables, and queues are much more functional.

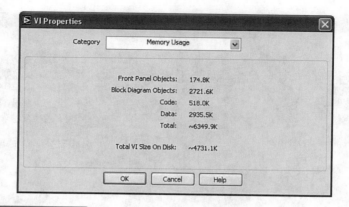

Figure 1-5
The VI Properties window indicates the four categories of memory use for Meticulous VI.

Nested VI makes good use of subVIs, which increase memory efficiency, but uses Sequence structures, which can decrease processing efficiency. SubVIs reduce the quantity of nodes on the diagram, which reduces block diagram memory. In addition, LabVIEW can reclaim memory buffers used by subVIs when the subVIs are not executing, thereby reducing data memory use. However, the nested structures, particularly the Sequence structures, are a potential source of inefficiency. When a Sequence structure is called, all frames must execute through completion, in consecutive order. In many situations, it is preferable to reorder or abort the sequence, to improve its efficiency.

The style rules presented throughout this book promote maximum efficiency. Chapter 4, "Block Diagram," and Chapter 8, "Design Patterns," discuss rules that influence processing efficiency. Chapter 6, "Data Structures," presents rules for efficient memory use.

1.1.3 Readability

Readability refers to how easily the developer can comprehend the source code. This includes both the front panel and the block diagram. The objects on the front panel should be clearly labeled and easily identified. The diagram should be neat, orderly, and easy to follow. The application should be well documented throughout.

The front panel of Meticulous VI appears both simple and perhaps a little mysterious. Figure 1-6 shows the front panel in edit mode. The menu controls along the bottom and right perimeters are Boolean controls with customized appearance and embedded text. A knob is used for rapid menu and waveform navigation. The main display area resembles a collage of indicators suspended on a black background. It is divided into two sections, a large subpanel control on the left and multiple overlapping indicators on the right. Twelve preset display configurations can be programmed by dynamically loading any one of twelve subVIs into the subpanel and making a subset of the overlapping indicators visible. The indicators are customized with a transparent background overlaid on a black decoration, giving them the appearance of floating in space. The diagram (see Figure 1-1) is easy to follow because all data flows through wires from left to right, with no more than one bend to wrap around structures and objects. The diagram is extensively documented, including visible terminal labels, free labels on long wires and within each structure, and enumerated case selectors.

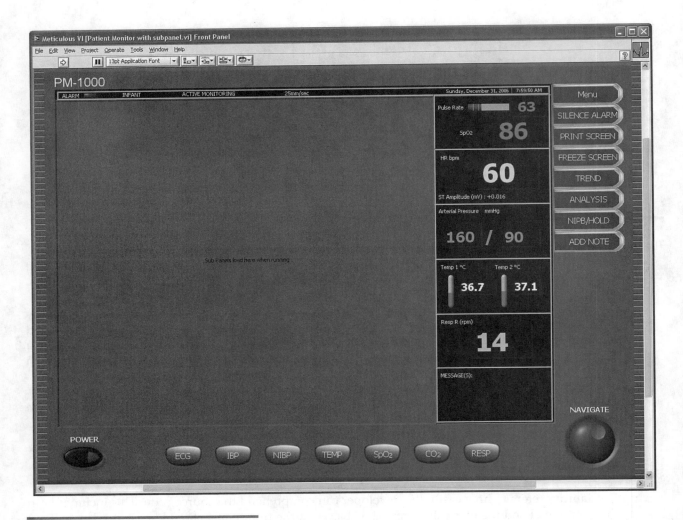

Figure 1-6
The front panel of Meticulous VI in edit mode. It contains customized Boolean controls along the bottom and right perimeters, a knob, a subpanel control, and multiple overlapping indicators.

The front panel of Nested VI (see Figure 1-2) is very simple and readable. The objects on the front panel and diagram are intuitively labeled and evenly spaced. The diagram contains free labels in every frame of every structure. However, the VI, control, and indicator descriptions are absent throughout the application. Descriptions are an important source of documentation, as discussed in Chapter 9, "Documentation." Figure 1-7 shows the Context Help window appearance for a VI and control, without descriptions.

Figure 1-7
The Context Help window reveals missing VI and control descriptions on the panel of Nested VI.

The front panel of Spaghetti VI (see Figure 1-3) contains simple controls, indicators, clusters, and decorations. Intuitive labels exist for most controls, but not all of them. For example, each station cluster contains two vertical fill slides that have the same abbreviated name, **Comp.** These controls are ambiguous. Each control should be uniquely named, and the abbreviation should be replaced with a more intuitive and meaningful term. Chapter 3 presents rules regarding front panel text. The diagram looks like, well, spaghetti. This is because the wiring scheme is haphazard, with data flowing in all directions, and the architecture is not adequate for the application's complexity. It is extremely difficult to visually trace most wires from source terminal to destination terminal. Chapter 4 presents rules for proper wiring and data flow that facilitate readability.

1.1.4 Maintainability

A LabVIEW application is **maintainable** if other LabVIEW developers besides the author understand the source code and if it can be easily modified and expanded to change or add new functionality. Hence, the source code must be readable to be maintainable. Additionally, the source code must use constructs that are modular and scalable, thereby allowing for future expansion of functionality.

The front panel of Meticulous VI contains the subpanel control and multiple overlapping indicators, as shown in Figure 1-6. The subpanel control promotes maintenance because new displays are created and existing displays are modified via component subVIs that are loaded into the subpanel control. This provides tremendous flexibility in the display appearance, along with the capability to modularize different display configurations as subVI panels. Hence, editing a display configuration involves editing a specific component subVI panel. However, changes to the overlapping indicators on the right require showing, hiding, and careful sizing and positioning of the indicators. Maintaining this portion of the front panel is tedious.

The diagram of Meticulous VI (see Figure 1-1) uses a complex application framework that consists of multiple parallel loops. Most loops, including the loop labeled **INSTR COMM LOOP,** use a variation of the State Machine design pattern. This consists of a Case structure with a separate case for each state of the application, and an enumeration wired to the case selector. The State Machine design pattern is readable because each frame has an intuitive label in the selector area that corresponds to the labels of the enumerated type definition and describes the function of each state. It is also scalable because states can be added and removed simply by adding and removing cases to the Case structure, and items to the enumerated type definition. Chapter 8 discusses the State Machine design pattern and Multiple Loop Application Framework.

The front panel of Nested VI (see Figure 1-2) uses a simple tab control that can be readily expanded by adding new tabs. Each tab contains a different display page and can be selected by the user when needed. The diagram however, contains Sequence and Case structures nested together, which obstructs the developer's view of the source code by forcing her to navigate each frame of each structure to understand how it works. Moreover, it is not a standard design pattern. Rather, it is confusing compared to the State Machine design pattern, which is more recognizable and less nested.

Spaghetti VI (see Figure 1-3) is not maintainable because the architecture is not scalable, the diagram is not modular, and the wiring is sloppy. The architecture consists of a single While Loop with all nodes that execute more than once inside the loop, and wires, structures, and nodes intermixed and overlapping. Known as the Continuous Loop design pattern, this is discussed in Chapter 8. If the scope of the application expands, the user either increases the size of the While Loop, which is already larger than what can be displayed on one screen in high resolution, or adds to the confusion within the existing area. Creating new wires and nodes within this framework would be tortuous. The front panel also has maintenance considerations. Adding a new station, for example, involves replicating the station cluster and the station-specific controls located outside the cluster. However, the front panel cannot support an additional station in its current form. Fortunately, a tab control can be integrated to provide substantial improvements with minimal effort. You can dedicate a separate tab for each station and then add and remove tabs as needed. In addition, you can incorporate the station-specific controls outside the cluster into the station cluster and save the cluster as a type definition. This way, any changes to the type definition are applied to all stations that use the type definition. Chapter 3 discusses tab controls, and Chapter 6 covers type definitions. The style rules presented throughout this book help ensure maintainability.

1.1.5 Robustness

A LabVIEW application is **robust** if it is bug free and never crashes. LabVIEW provides fundamental constructs that promote robust applications, including subVIs and error handling, as well as fundamental elements such as local and global variables that can hinder applications when used improperly. Good LabVIEW developers always modularize their diagrams using subVIs. This involves identifying a portion of code that performs a specific task and creating a subVI to perform that task. Testing and debugging subVIs is relatively easy, as long as they are each dedicated to a specific task comprised of a limited number of nodes. Modular applications that use previously tested and debugged subVIs are generally higher quality from the outset. Any bugs that are discovered are easy to identify and isolate because of the modularity.

LabVIEW contains a tool called VI Metrics that reports how many user subVIs and write variables an application uses, as well as the overall number of nodes on the diagram. The VI Metrics utility is accessed by selecting **Tools»Profile»VI Metrics**. Figure 1-8 shows the VI Metrics window with Nested VI loaded.

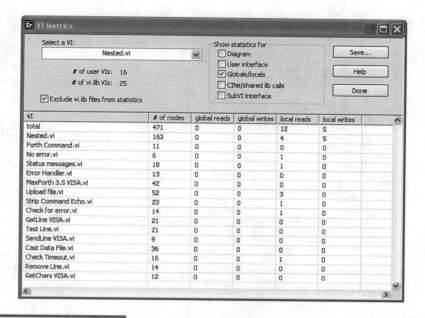

Figure 1-8
The number of nodes and the number of user VIs reported by the VI Metrics window help approximate the application's modularity.

For comparison purposes, we can define a modularity index as follows:

Equation 1.1
Modularity Index = (# user VIs / total # of nodes) × 100

I recommend a modularity index greater than 3.0. In our examples, Meticulous VI has 111 user VIs and 4,947 nodes, for a modularity index of 2.2. Nested VI has 16 user VIs and 471 nodes, for a modularity index of 3.3. Spaghetti VI contains 56 user VIs and 1,944 nodes, for a modularity index of 2.9. Chapter 4 discusses subVI modularization in detail.

Error handling consists of trapping errors by propagating the error cluster and reporting errors using a dialog or log file. Error handling is critical for robust application performance. Meticulous VI contains very thorough error handling, as can be seen by the error clusters that are propagated among all nodes that have error terminals. Any errors that occur within a loop are passed to a dedicated error handling loop via queue, where they are evaluated, reported, and logged. Figure 1-9 shows the **Error Handling Loop**.

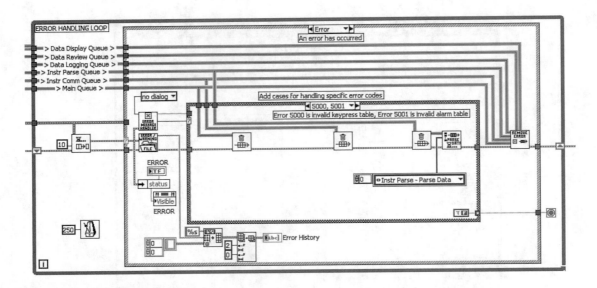

Figure 1-9
Meticulous VI contains a dedicated error handling loop that receives error information from the parallel loops via queue and evaluates, reports, and logs the errors.

The error handling scheme in Nested VI is less thorough and sophisticated than Meticulous VI, yet it is simple and functional. Referring to Figure 1-10, the first subVI called in each case of the inner Case structure has an unwired **error in** terminal, whereas the **error out** terminal is propagated between the nodes of its case, through the Case structure's output tunnel, to a Sequence Local. The Sequence Local passes the error information to the error handling subVI in frame 1 of the outermost Sequence structure. You can improve error handling style by propagating the error cluster between loop iterations and wiring it to the **error in** terminals of all nodes. Efficiency also can be improved if the error handling code is not called unless an error occurs.

Finally, error handling is both incomplete and ineffective in Spaghetti VI. As you can see in Figure 1-3, no apparent error handling scheme is used. Overall, Meticulous VI is the most robust in terms of error handling. Proper error handling is an essential ingredient for reliable and robust applications. Chapter 7, "Error Handling," covers this topic in detail.

Writing to local and global variables can cause unexpected results in an application. For example, a race condition occurs when two copies of a write variable are written to at the same time. As a general rule, the more write variables an application has, the greater the potential for unexpected and undesirable behavior. The quantity of local and global write variables is quickly determined from the VI Metrics window. Meticulous VI contains 92 write variables, Nested VI contains only 5, and Spaghetti VI contains 46. Hence, Nested VI is the most robust in terms of fewest write variables. Chapter 4 discusses variables in greater detail. The rules presented throughout this book ensure robust applications.

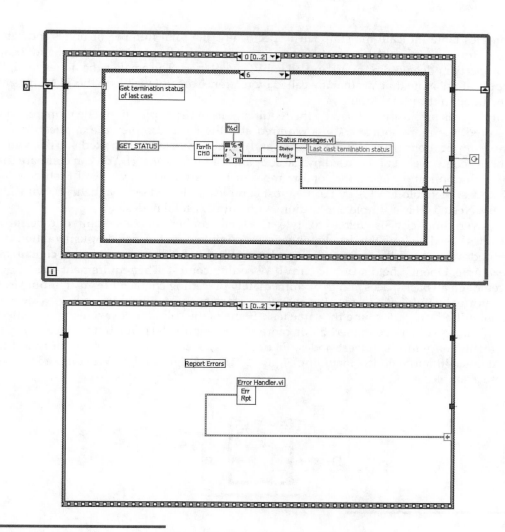

Figure 1-10
The error handling scheme of Nested VI is functional but not optimal. The error cluster propagates between subVIs, through the Case structure, to the sequence local. The error is processed using the Error Handler VI in frame 1 of the Sequence structure.

1.1.6 Simplicity

Simplicity inversely relates to the quantity of nodes and terminals that comprise an application. Simplicity is the opposite of complexity. The fewer front panel objects and block diagram nodes, the simpler the application. Simplicity affects readability and performance. It is generally desirable to minimize complexity by choosing the simplest implementations that require the fewest nodes.

The VI Metrics window provides a useful interface for evaluating simplicity. Choose a VI in the **Select a VI** control, such as the top-level VI for an application. The totals displayed in the top row of the metrics table indicate the simplicity. Specifically, the total number of nodes indicates the total number of functions, subVIs, structures, front panel terminals, constants, global and local variables, and property nodes within the VI and all subVIs combined. The smaller the nodes total, the simpler the application. Note that VIs that are called by reference are not visible to the VI Metrics tool and are not included in the nodes total.

Simplicity is primarily dictated by the requirements of the application. The more functionality that is required, the more source code is required and the more complex the application is. The three example applications from Figures 1-1, 1-2, and 1-3 all have statically linked subVIs, with the exception of the component subVIs used by the subpanel of Meticulous VI. You can compare the complexity of each application using the total number of nodes reported by the VI Metrics window. With 4,947 nodes, Meticulous VI is by far the most complex, followed by Spaghetti VI, with 1,944. On the contrary, Nested VI is a simple application, with only 471 total nodes.

However, simplicity also relates to style. The total number of nodes required to perform a specific task varies based on how the developer chooses to implement the task. Implementations that contain fewer nodes are generally more efficient and readable than implementations that contain more nodes. For example, OpenG[1] held a public LabVIEW coding contest. The requirement was to develop a VI that removed all backspaces and their immediately preceding character from an input string. The VIs were informally judged in terms of icon style, performance, and diagram style. A wide variety of VIs were submitted, all performing the same task. Some of the submissions were implemented with only 12 or 13 nodes; others contained 25 or more nodes. Figure 1-11 illustrates three submissions. The implementation with the fewest nodes, Figure 1-11A, is easier to read and understand than implementations with more nodes, including Figure 1-11B and Figure 1-11C. We compare the performance of these examples next.

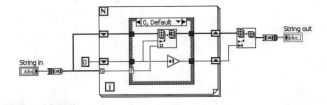

Figure 1-11A
This subVI implementation uses only 13 nodes to remove all backspaces from the input string.

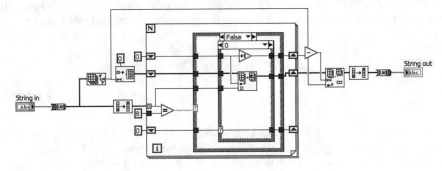

Figure 1-11B
This implementation uses 22 nodes, including nested Case structures, to achieve the same functionality.

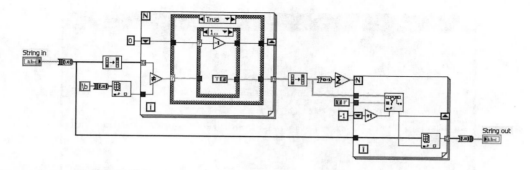

Figure 1-11C
This implementation uses 25 nodes including two For Loops and nested Case structures.

1.1.7 Performance

Performance is a particularly broad term that has many definitions. In this discussion, we consider application and subVI performance. **Application performance** refers to how well the application or VI completes its intended mission. Therefore, application performance measures are related to the requirements. For example, if an application is developed to reduce test time, test time is the primary performance measure. If improving product quality is the primary objective, a reduction in the number of defective products or customer returns might be the performance measure. Most applications have multiple objectives. The most common objectives include increasing the speed and efficiency by which a task is accomplished while improving ease of use. Therefore, application performance can be considered a combination of efficiency (see Section 1.1.2) and ease of use (see Section 1.1.1). Robustness (see Section 1.1.5) also is required. If the application is not reliable and robust, performance is meaningless.

Style affects application performance. Referring to the application examples, the primary objective of Meticulous VI (see Figure 1-1) is to perform reliable acquisition, logging, and display of physiologic data. This objective is achieved using the Multiple Loop Application Framework, which prioritizes the important tasks in dedicated loops. Nested VI (see Figure 1-2) performs configuration of a scientific instrument and then uploads data acquired from a variety of remote sample sites. In this case, the simplistic GUI aids the user in error-free operation. Spaghetti VI (see Figure 1-3), however, suffers from unreliable and sluggish performance because of the inefficiency of the single-loop architecture and other style-related factors previously discussed.

Consider **subVI performance** as its execution speed. Section 1.1.6 mentioned the relationship between simplicity and performance. Often the fewer nodes that are used to implement a subVI, the faster the subVI executes. For example, the three implementations of the Remove Backspace VI in Figure 1-11 have different execution speeds. Figure 1-12 shows the Profile Performance and Memory window containing the timing statistics after each VI was called 1,000 times within a loop, with a long string passed to the **string in** terminal. Thirteen Nodes VI consumed only 5.6ms, Twenty-Two Nodes VI consumed an average of 5.9ms per run, and Twenty-Five Nodes VI consumed 12.8ms. Although 5.6ms, 5.9ms, and 12.8ms might seem insignificant, if the three development styles are extended to every subVI throughout a large application, the performance difference becomes substantial.

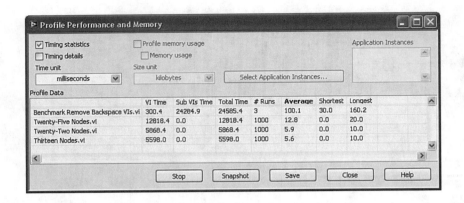

Figure 1-12
The Profile Performance and Memory window displays the timing statistics after each Remove Backspace subVI runs 1,000 times.

1.1.8 Style Tools

LabVIEW contains several built-in tools that help evaluate style. In the previous sections, you observed the Profile Performance and Memory window used to evaluate data memory use and execution speed. The Memory Usage VI properties page determines the memory allocated to each of the four memory components of a single VI. These tools provide metrics for efficiency (see Section 1.1.2) and performance (see Section 1.1.7). The VI Metrics window determines simplicity (see Section 1.1.6) and is used to calculate the modularity index to help evaluate robustness (see Section 1.1.5). Additionally, code reviews provide a more direct method of evaluating style. Code reviews are performed utilizing a combination of self reviews, automated inspection, and peer reviews. The Style Rules Summary in Appendix B can be used as a style checklist for performing self reviews. Additionally, many of the rules presented in *The LabVIEW Style Book* can be evaluated using the LabVIEW VI Analyzer Toolkit from National Instruments. Chapter 10, "Code Reviews," discusses these tools in greater detail.

1.2 Style Versus Time Tradeoff

Good LabVIEW developers are often in high demand within our organizations. The applications we develop are critical to our companies' research, development, or production endeavors. Intense time-to-market pressures encourage fast application development cycles. As such, LabVIEW developers often gravitate toward shortcut and fast development techniques for speeding their development time. After all, that is why we all chose LabVIEW in the first place—it substantially accelerates our development time versus traditional text-based programming languages.

> **Definition 1.1:** *Development time includes the hours required to develop, document, test, modify, and maintain an application* **throughout its entire life cycle***.*

On the surface, the pursuit of fast development time appears at odds with good programming style. The former might entail shortcuts that bypass the latter. At Bloomy Controls, we are familiar with this apparent paradox. Many of our customers contract us as a means of achieving tight deadlines. Any style rules that increase development time without an offsetting productivity benefit must be eliminated. However, it is important to look at the entire life cycle of the application when considering development time. Many developers mistakenly consider the time from conception until the first version of the software is released as the development time and strive to minimize this time. In actuality, development time includes the hours consumed for all testing, debugging, documentation, revisions, and upgrades that occur throughout the entire life cycle of the application. When taken in this context, it behooves us to use good style.

> **Theorem 1.2:** Good style **reduces** development time and effort.

Good development style might require more up-front time and effort, but the quality of the software is higher from the start, and the software is more useful throughout the life cycle of the application. I cannot overemphasize this concept. In numerous instances, Bloomy Controls has been contracted to make minor functional modifications and bug fixes to customers' preexisting applications that required disproportional effort. In some instances, we were forced to rework the application as if starting over from scratch because bad style prevented modifications. Using proper style from the outset reduces development time throughout the application's life cycle and makes modifications easy.

Furthermore, this book provides only rules that are practical to implement in real-world situations, including projects with tight deadlines. I have eliminated rules that are not practical. One such rule that we had many years ago was a requirement for developers to edit the VI History *every* time *any* VI was saved, regardless of the project, development model, or stage of development. As a rule, all the Bloomy engineers had LabVIEW configured to automatically prompt for revision history comment. It immediately became clear that this was impractical and a nuisance for many applications during the early stages of development, prior to the first release. This rule was abandoned less than two weeks after it began.

An example of a rule that some consider a hindrance but that I adamantly require among all of my staff is writing VI descriptions for every VI. VI descriptions are a critical source of documentation. I have seen employees, customers, and colleagues try to save time by saving this simple snippet of documentation for *later* because they're trying to get a piece of production machinery online or meet some critical deadline. However, more often than not, *later* is really *never*. Even with the best intentions, the same time pressures that cause one to skip the descriptions in the first place usually linger around later. Alternatively, if we do return to our code and write the descriptions later, it is much more time-consuming because we have to locate the undocumented VIs one at a time and refresh our memories about their purpose. It is much less time-consuming and more reliable and efficient to write the description immediately after the VI is written or revised and it is fresh in our minds. This practice ensures that the descriptions are written promptly, accurately, and efficiently. Chapter 9 covers documentation, including VI descriptions. The style rules presented throughout this book are practical to implement after they are learned and applied.

Prepare for Good Style

2

LabVIEW style is positively influenced by preparing to develop your applications before you begin coding. First and foremost, write a requirements specification. It is much easier to develop good code when working from a specification instead of deriving the requirements on-the-fly. Second, design the software. Combine standard software design principles, such as the State Machine Diagram, with LabVIEW best practices, such as the State Machine design pattern. Third, configure the LabVIEW environment. Make use of desired preferences and previously developed code, such as reusable templates, instrument drivers, and utilities. Finally, establish some conventions for organizing, naming, and controlling your project files. An organized project spawns organized source code.

This chapter describes preparation techniques that are conducive to good development style. It is noteworthy that several of these techniques relate equally to configuration management (CM) as well as style. Indeed, many of the Rules in this chapter, as well as throughout the book, may complement or extend a CM process. All topics are presented in the context of style, which is one element of CM. Due to the wide diversity of LabVIEW applications and target industries, there is no universal CM system that can work for most organizations. Therefore, the reader may evaluate the relevance of these materials to their own CM system.

2.1 Specifications

Most LabVIEW applications begin with a business objective that involves an industrial measurement and automation challenge. Business objectives include accelerating research and development, increasing production throughput, improving quality, and reducing manual labor. Most people use LabVIEW because it is the fastest method of developing applications for achieving their desired business objective. There are many other benefits people consider, such as the graphical paradigm, open support for thousands of instruments, platform portability, and network connectivity. However, most industrial decisions are justified by considering the savings of time and money. As such, most industrial users choose LabVIEW because of the rapid software development cycle.

LabVIEW provides the opportunity to develop applications very quickly. High-level tools such as Express VIs, instrument drivers, templates, and examples empower the user to develop software with very little planning or preparation. The speed with which you can connect to one or more instruments and begin making measurements is extraordinary. Additionally, rapid code development provides developers and managers, often under significant time pressures, instant gratification. As a result, many LabVIEW developers adopt fast programming habits. Shortcuts are routinely taken. Any development steps that are deemed unnecessary are frequently skipped.

Requirements specifications development constitutes a critical phase of the application development cycle. Absence of a specification leads to significant time and effort expended developing software that may not satisfy the intended objective. Alternatively, the requirements that are understood today may not be the same as what is required tomorrow—a phenomenon known as *scope creep*. Moreover, how do we know when our application is complete without documented specifications?

In practice, I would estimate that fewer than 25% of LabVIEW developers write formal requirements specifications. Most projects are initiated in meetings in which hardware and overall system requirements dominate the agenda. Hardware seems to take priority over software because the products, features, part numbers, prices, and lead times seem more tangible and inflexible. Since LabVIEW provides unlimited flexibility with respect to software functionality, the features, prices, and lead times of a LabVIEW application appear virtual or even self-imposed. In fact, every software feature has a lead time and price in terms of development hours. This simple fact is often overlooked or underestimated. As a result, misunderstood requirements and scope creep abound throughout the LabVIEW community.

Scope creep is not purely evil and forbidden. In R&D environments in particular, measurement and automation requirements often change naturally as the research emphasis, or product or process design evolves. LabVIEW's flexibility and rapid development capability are advantageous in such dynamic industrial settings. Alternatively, if the original requirements are satisfied ahead of schedule, then the project scope may expand to include greater levels of automation. This correlates to greater benefits. However, if the specifications are not documented, then it becomes very difficult to verify what the requirements are at any given point in time throughout the project life cycle, whether you are succeeding, whether the application scope has expanded, and most importantly, when the project is complete.

> **Theorem 2.1:** *Written specifications positively influence LabVIEW style.*

The presence or absence of a written specification has a direct effect on LabVIEW programming style, as well as the overall outcome of the application. A specification is a detailed description of the project's requirements. Writing a specification entails significantly more planning and attention to detail than verbal communications. It is a much different thought process than source code development. It forces the writers and readers to envision the entire scope of the application before the software development begins. The specification may be reviewed and refined by the contributors as much as necessary to build consensus. Once the coding begins, the risk for misunderstandings, oversights, and unforeseen scope change is dramatically reduced.

Reviewing or contributing to the specification is an excellent exercise for a developer preparing to design the application. During all meetings I attend, I always strive to understand the project's requirements as thoroughly as possible. No matter how confident I may feel at meeting's end, I have never proceeded to sit down and write a specification without new questions and ideas arising in the process. Following up on these ideas translates to better understanding and agreement of project scope, which translates to a better LabVIEW application, and more satisfied end users.

> **Theorem 2.2:** *Unforeseen scope changes hinder good style.*

Unanticipated expansion of scope can hinder good style, and in the most severe instances may even diminish an application into spaghetti. A developer may choose to begin with the simplest architecture that satisfies the application's initial requirements, in order to expedite development. For example, a simple top-level architecture consisting of a single While Loop may be chosen. This architecture is often not amenable to an application's expanded requirements. When the scope changes substantially, the While Loop is expanded to accommodate more nodes and wires and may become large and unwieldy. Also, as the diagram's node quantity increases, there are more obstacles preventing neat placement of additional nodes, wires, and structures, resulting in a sloppy diagram. Indeed, this may have happened in the Spaghetti VI example from Chapter 1, "The Significance of Style." If you take a second look at Figure 1-3, you will see a single While Loop containing many functions, subVIs, wires, terminals, and Case structures. This is an application that *might* have started out neat at one time and evolved, as the work scope evolved, into spaghetti. The architecture might have been appropriate for the original requirements but was inappropriate for the eventual requirements. Realistically, if this were the case, the developer certainly should have changed the architecture, among many other possible improvements, after the work scope changed. It is also likely that the application was developed from the outset without a fundamental understanding of the requirements and without a style convention. In any event, the process of writing and reviewing a specification significantly reduces the potential for large unexpected changes of scope, reducing the extent of corresponding source code revisions, thereby improving style.

If a specification document does not exist, either the developer does not fully understand the requirements or they reside in the developer's memory. The two situations are very similar. In the former situation, the software will require substantial modifications as the developer's understanding evolves. This is also known as the Code and Fix software life cycle development model. The more overall edits there are, including changes because of misunderstood requirements, the messier the source code can become, per Theorem 2.2. This is generally true with all programming languages.

In the latter situation, in which the requirements are understood but not documented, several problems can arise. Skipping the documentation reduces the amount of planning and consideration, which can result in an inferior initial release that requires many revisions. Hence, the Code and Fix life cycle model still applies. Additionally, the memory-resident specification might become the developer's exclusive intellectual property. Inevitably, the developer will forget some of the details, and her ability to maintain the application decays over time. More concerning is that the developer might not be available to support the application throughout its life cycle. Most people change responsibilities, jobs, companies, and careers, and LabVIEW developers are no exception.

Speaking from experience, Bloomy Controls routinely receives inquiries from companies that need to maintain, fix, or refactor their legacy LabVIEW applications. When the specifications and documentation are lacking, the source code is often sloppy. These applications are generally rewritten from scratch, beginning with a specification. Conversely, when good specifications exist, the source code is predictably high quality—not always, but most of the time.

2.1.1 Best Practices for Specifications Development

Good specifications originate as manually recorded notes. Many of the meetings in which projects are specified resemble fast-paced brainstorming sessions, and high-speed diligent note taking is essential. Bound notepads are a standard method for professional engineers and scientists to track their daily activities. They are also an ideal method of initially capturing raw specifications.

Rule 2.1   Maintain a LabVIEW project journal

The bound notepad serves as a LabVIEW project journal. Record the date of each activity, and maintain your project-related notes in chronological order. As notes are appended throughout the project's life cycle, they need not be limited to requirements; notes are commonly extended to include design notes and maintenance logs. The notes can also include bug descriptions, wish lists, data, manual calculations, notes from support inquiries, and artistic doodles. The function of a project journal for a LabVIEW developer is similar to a laboratory journal for a scientist or inventor. Laboratory journals are used to preserve the experimental data and observations that are part of any scientific investigation. They are the basis for further analysis, discussion, evaluation, and interpretation[1]. Similarly, project journals contain any mixture of specifications and data that form the basis of the software design. In many situations, scientists and inventors use LabVIEW to conduct their experiments. Indeed, many scientists and inventors are also LabVIEW developers. Conversely, LabVIEW developers support the scientific method when we record our activities in a project journal. In some cases, the data that is recorded might even supplement the proof of discovery or invention when a company applies for a patent.

Recording the raw specifications in a project journal constitutes a good specification-development practice. Per Theorem 2.1, specifications positively influence LabVIEW style. The developer can refer to her notes as needed throughout the development and maintenance of the application, as long as she remains available to do so and does not lose or destroy the notepad. Moreover, project journals form the basis of a more formal requirements specification.

Simplicity and speed are the primary advantages of project journals, yet there are several limitations:

1. Bound notepads provide minimal organization. Project notes are entered chronologically. Inactive requirements, ideas, and notes are often mixed with active ones. It is often difficult to distinguish active requirements from inactive ones.
2. Many people do not have clear, legible handwriting. Indeed, it is very difficult for even the neatest writers to maintain legibility throughout a rapid brainstorming session, the kind of environment in which many LabVIEW projects are specified.

3. Notepads are not transferable. It is inefficient, at best, for multiple project team members to share someone's handwritten notes as project reference materials. Instead, notepads function more like a personal hard copy of one individual's daily activities related to a project.
4. If the raw notes are not formalized and expounded upon, the composer does not gain new insights and perspectives. Moreover, it is not possible to receive feedback, build consensus, and confirm the requirements among multiple project team members if the project team does not share and review the requirements.

 Rule 2.2 Write a requirements specification document

Compose a requirements specification document for each application, using the raw notes from the project journal as a reference. A requirements specification is a statement of the project's primary objectives and a prioritized listing of the required features. It contains sufficient detail to clearly describe all important requirements that contribute to the objectives. It should contain very few design and implementation details. Hence, the requirements specification should not place constraints on *how* the project is developed, except where these constraints affect the project's outcome.

The Institute of Electrical and Electronics Engineers (IEEE) offers a Recommended Practice for Software Requirements Specifications, standard 830-1998, and a Guide for Developing System Requirements Specifications, standard 1233-1998. These are excellent references for writing software and system specifications that apply to any programming language, industry, and application type. Specifications written according to these standards are very thorough, high-quality documents.

In some evolutionary industries, the rapid pace of changes or time-to-market pressures pose challenges to the process of writing formal requirements specification documents. In many situations, it is desirable to have the software developed in parallel with product development. Many of the performance specifications that affect the application's requirements are unknown or change frequently during application development. These situations do not justify abandoning the requirements specification process. Instead, a very aggressive specification-development process should be pursued in parallel with product and software development. Specifically, Agile project-management methods such as Extreme Programming are best suited to evolutionary projects. At Bloomy Controls, we dedicate a resource to the specifications-development and -maintenance effort, separate from the software, to gather and maintain the requirements in an iterative manner. The requirements specification need not finalize requirements that are evolving. Instead, try to determine most-probable and worst-case scenarios for the undefined requirements, and include them in the specification. If this is not feasible, simply use "TBD" and come back to it later. In many government-regulated industries, however, TBDs are expressly forbidden. New projects are formed exclusively to resolve the unknown requirements before the software development commences. However, government-regulated industries are generally not rapidly evolving.

In summary, the LabVIEW project journal contains an informal history of the project requirements and evolution, and the requirements specification document establishes the active project scope. These elements are prerequisites for good programming style.

2.1.2 LabVIEW Project Requirements Specification

LabVIEW projects represent a specialized subset of the vast software industry for which the IEEE standards were created. Specifically, LabVIEW is primarily applied to measurement and automation applications, and a relatively fast development cycle is expected. Most applications perform acquisition, analysis, and presentation. These factors can be used to form a specialized standard for writing requirements specifications for LabVIEW projects. Specifically, I recommend the following elements:

- Statement of high-level objectives
- Budget
- Timetable
- Detailed requirements for acquisition, analysis, and presentation
- Priorities for each requirement
- Test methodology

Figure 2-1 shows a LabVIEW Project Requirements Specification template.

The high-level objectives should clearly describe why the project is being undertaken. It is particularly useful to describe how the objectives relate to the company's core products and services. This is also known as the business objective. For example, a high-level objective might be stated as follows:

> Evaluate a low-emissions exhaust system that is being considered by the automotive industry for 2010 model automobiles. The system must comprehensively test the first 100 prototypes in January 2008. The system must be scaled into high-volume production in July. The market potential of the product is estimated to be $1 billion over a 5-year span.

Budget and timetable are very important considerations that not only directly affect the system design, but might also determine the overall feasibility. It is extremely important to get an idea of how much funding is available and the required completion date at the outset of the project. It is common to identify significant obstacles at this point. For example, you might find that the lead time or expense of some of the required elements exceeds the requirements. The earlier in the project that you identify such discrepancies, the more time and effort you will save.

I like to organize the requirements into four primary categories: acquisition, analysis, presentation, and test methodology. The **acquisition** section describes the required measurement and control hardware interfaces. I find it useful to create a spreadsheet that itemizes every physical parameter or input and output (I/O) point that must be measured or controlled, including the engineering units, range, accuracy, and rate at which the parameters must be acquired or generated. This information can then be used to specify transducers, signal conditioning, digitizers, instruments, and control devices. These choices are then entered right into the same spreadsheet, providing further definition of the system. Note that the hardware interfaces and I/O requirements help determine the scope of the LabVIEW development required to access these interfaces, such as readily available instrument drivers, lower-level development using VISA or DAQmx, or simple interactive configuration using Express VIs.

The **analysis** section describes any mathematical processing that is required, either online as data is acquired or offline using data files and a separate routine. Some types of analysis include data reduction or decimation, corrections, digital filters, peak searches, limit evaluations, statistics, and control algorithms. For example, statistical process control (SPC) is very common in automotive manufacturing applications. SPC entails computing such parameters as mean, range, control limits, standard deviation, process capability, and more, and graphically displaying the results as Xbar & R and other charts. The analysis section of the requirements specification describes such calculations.

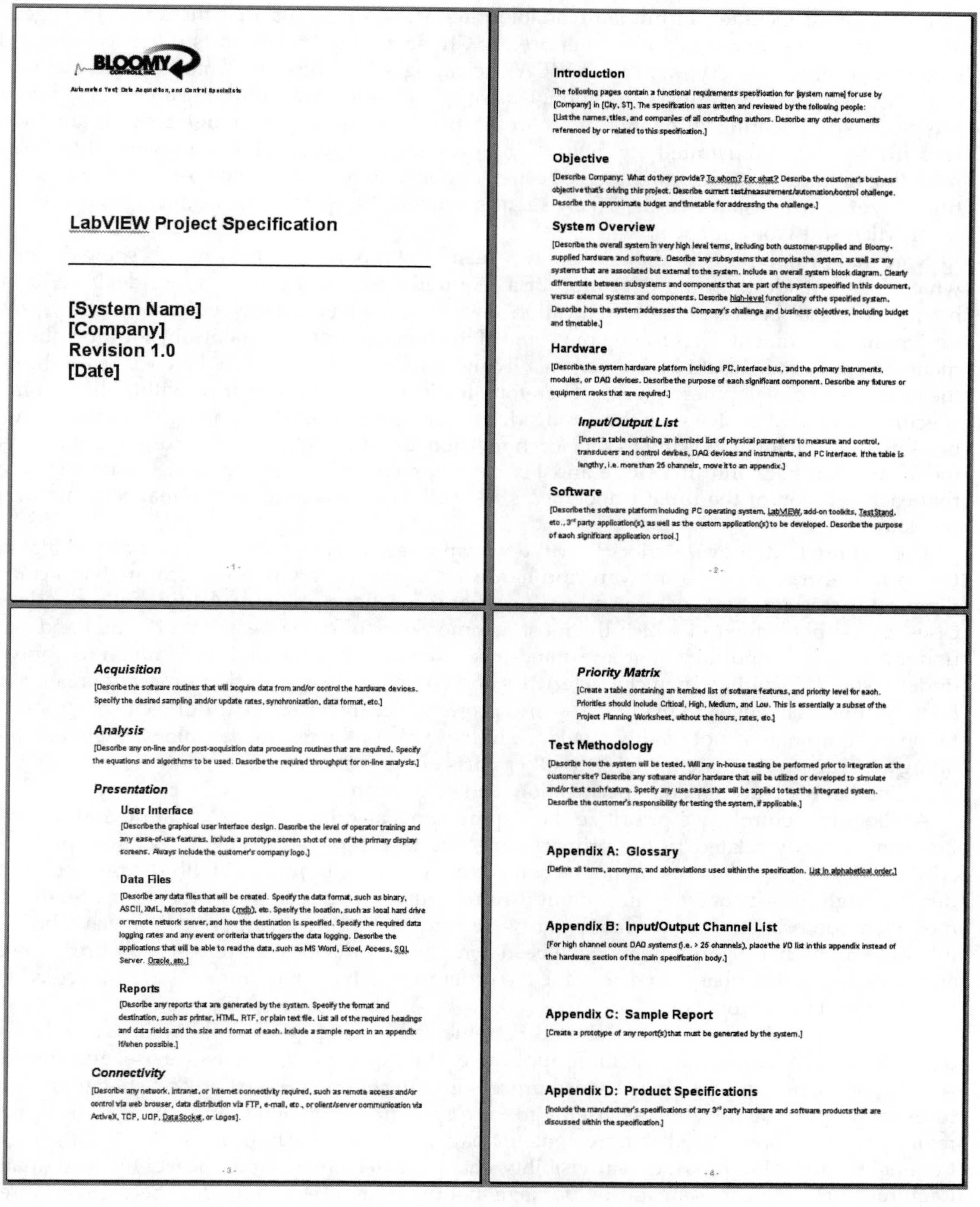

BLOOMY CONTROLS, INC.

Automated Test, Data Acquisition, and Control Specialists

LabVIEW Project Specification

[System Name]
[Company]
Revision 1.0
[Date]

- 1 -

Introduction

The following pages contain a functional requirements specification for [system name] for use by [Company] in [City, ST]. The specification was written and reviewed by the following people: [List the names, titles, and companies of all contributing authors. Describe any other documents referenced by or related to this specification.]

Objective

[Describe Company: What do they provide? To whom? For what? Describe the customer's business objective that's driving this project. Describe current test/measurement/automation/control challenge. Describe the approximate budget and timetable for addressing the challenge.]

System Overview

[Describe the overall system in very high level terms, including both customer-supplied and Bloomy-supplied hardware and software. Describe any subsystems that comprise the system, as well as any systems that are associated but external to the system. Include an overall system block diagram. Clearly differentiate between subsystems and components that are part of the system specified in this document, versus external systems and components. Describe high-level functionality of the specified system. Describe how the system addresses the Company's challenge and business objectives, including budget and timetable.]

Hardware

[Describe the system hardware platform including PC, interface bus, and the primary instruments, modules, or DAQ devices. Describe the purpose of each significant component. Describe any fixtures or equipment racks that are required.]

Input/Output List

[Insert a table containing an itemized list of physical parameters to measure and control, transducers and control devices, DAQ devices and instruments, and PC interface. If the table is lengthy, i.e. more than 25 channels, move it to an appendix.]

Software

[Describe the software platform including PC operating system, LabVIEW, add-on toolkits, TestStand, etc., 3rd party application(s), as well as the custom application(s) to be developed. Describe the purpose of each significant application or tool.]

- 2 -

Acquisition

[Describe the software routines that will acquire data from and/or control the hardware devices. Specify the desired sampling and/or update rates, synchronization, data format, etc.]

Analysis

[Describe any on-line and/or post-acquisition data processing routines that are required. Specify the equations and algorithms to be used. Describe the required throughput for on-line analysis.]

Presentation

User Interface

[Describe the graphical user interface design. Describe the level of operator training and any ease-of-use features. Include a prototype screen shot of one of the primary display screens. Always include the customer's company logo.]

Data Files

[Describe any data files that will be created. Specify the data format, such as binary, ASCII, XML, Microsoft database (.mdb), etc. Specify the location, such as local hard drive or remote network server, and how the destination is specified. Specify the required data logging rates and any event or criteria that triggers the data logging. Describe the applications that will be able to read the data, such as MS Word, Excel, Access, SQL Server, Oracle, etc.]

Reports

[Describe any reports that are generated by the system. Specify the format and destination, such as printer, HTML, RTF, or plain text file. List all of the required headings and data fields and the size and format of each. Include a sample report in an appendix if/when possible.]

Connectivity

[Describe any network, intranet, or Internet connectivity required, such as remote access and/or control via web browser, data distribution via FTP, e-mail, etc., or client/server communication via ActiveX, TCP, UDP, DataSocket, or Logos].

- 3 -

Priority Matrix

[Create a table containing an itemized list of software features, and priority level for each. Priorities should include Critical, High, Medium, and Low. This is essentially a subset of the Project Planning Worksheet, without the hours, rates, etc.]

Test Methodology

[Describe how the system will be tested. Will any in-house testing be performed prior to integration at the customer site? Describe any software and/or hardware that will be utilized or developed to simulate and/or test each feature. Specify any use cases that will be applied to test the integrated system. Describe the customer's responsibility for testing the system, if applicable.]

Appendix A: Glossary

[Define all terms, acronyms, and abbreviations used within the specification. List in alphabetical order.]

Appendix B: Input/Output Channel List

[For high channel count DAQ systems (i.e. > 25 channels), place the I/O list in this appendix instead of the hardware section of the main specification body.]

Appendix C: Sample Report

[Create a prototype of any report(s) that must be generated by the system.]

Appendix D: Product Specifications

[Include the manufacturer's specifications of any 3rd party hardware and software products that are discussed within the specification.]

- 4 -

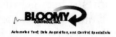

Figure 2-1
A template for a LabVIEW Project Requirements Specification[2]

Presentation includes all human-readable interfaces to the data that the application generates. This includes the graphical user interface, ASCII data files, and reports. One technique that is extremely useful and very easy in LabVIEW is creating a prototype user interface and including one or more screenshots in your specification. Do not spend a lot of time building an actual working prototype or experimenting with fonts and colors at this point; just apply enough controls and indicators on a front panel to illustrate how the user interface might appear. This is a powerful technique for revealing requirements that might have been overlooked or misunderstood. As the saying goes, a picture is worth a thousand words. Likewise, it is beneficial to prototype a sample report or data file using Microsoft Word or Excel.

Note that prototyping candidate display screens and reports generally entails some design effort, which falls outside the pure definition of a requirements specification. In an ideal world, a solid boundary exists between the specification and design phases. However, in practice, prototype designs are instrumental in fleshing out some of the functional requirements. In addition, the requirements specification should be viewed as a living document that is revised as necessary throughout the design and development phases. Therefore, the level of detail contained within the requirements specification might evolve with time. Indeed, a requirements specification might eventually form the basis of a user manual. Be sure to save each revision, or you might lose sight of the original requirements and fall back into the Code and Fix development model. Source code control tools ensure that each revision of the project documents, as well as source code, is automatically archived in a repository.

The **test methodology** describes how the system is tested. A test scenario is conceived that verifies the correct operation of the software and hardware. Ideally, every required feature has a corresponding test that isolates the feature and verifies that it functions properly. Additionally, a series of use cases can be performed in which the most common uses of the software are defined and executed under controlled conditions. For a test and measurement application, it is common to apply a unit under test (UUT) with known characteristics and compare the application's measurements with the known values. However, in many real-world projects, a characterized UUT or some other instrumentation or equipment is not readily available. In these cases, the test methodology might entail developing software to simulate the UUT or other parts of the system. In any event, the test methodology is outlined in the requirements specification, and details can be added as the project evolves.

At Bloomy Controls, we prioritize the requirements as critical, high, medium, and low. **Critical** items are directly related to the primary objective of the project. They are showstoppers. If any critical-priority requirements cannot be completed, we cease to proceed with the project as defined. Items of **high** priority are very important features, but not necessarily showstoppers. **Medium** is the most common priority. Items of **low** priority are features that we incorporate if time and budget permit, after all higher-priority items have been completed. Priorities are referenced during the design and implementation phases to determine the order in which the requirements are addressed and the time that we budget for each.

The requirements specification should be reviewed by multiple people involved in developing, using, and maintaining the system, if applicable. It is common to discuss, revise, and improve the specification before proceeding. A word processor with comment-insertion and change-tracking features is simple but nonetheless useful for incorporating comments and changes from multiple reviewers. More specialized requirements-management tools such as Telelogic DOORs and IBM's Rational RequisitePro provide clear visibility, change management, and traceability. Additionally, NI Requirements Gateway is an add-on package that provides a traceability link between requirements specifications formed with any of the previously mentioned tools and your LabVIEW source code. This enables you to analyze and report the compliance of your software to the project requirements.

2.2 Design

Software is designed using traditional techniques such as flow charts and pseudo code, or using Unified Modeling Language (UML), including class, sequence, collaboration, state, and component diagrams. Many excellent references describe these standard design techniques;[3] I do not describe them here. Instead, I would like to note that Theorem 2.1 can be extended to include design documentation. The more overall planning there is, including design documents as well as specifications, the better the LabVIEW style will be.

Many developers are tempted to skip straight from the requirements specification to the software development. If fewer than 25% of LabVIEW developers write specifications, far fewer are apt to create design documents. Again, I have observed a discrepancy between hardware- and software-development practices. For hardware, the design phase is a given. Schematics always precede the soldering of circuits. Block diagrams and CAD drawings precede the fabrication of mechanical assemblies, fixtures, and equipment racks. Design documentation is essential for hardware. Why should software be any different?

A few explanations, albeit excuses, might exist. In some applications, the software design is discovered by experimenting with an implementation. For example, one or more prototype LabVIEW VIs are developed to help characterize the application that the software is intended to solve. Specifically, prototypes are often developed to determine how fast a data acquisition and online analysis routine can run on a given computer. The working prototype VIs might eventually evolve into an integral portion of the software. Many LabVIEW developers prefer to expand the prototype into the finished application, following the Code and Fix development model. This can lead to trouble, as you saw in the Spaghetti VI example in Chapter 1. Discipline is required to stop with a working prototype and to go back and complete the specification and design phases before continuing with software development.

Another explanation is that a design phase is not necessary for very simple applications. Indeed, LabVIEW applications range in complexity from very simple to very complex. If a given application can be solved by configuring a few Express VIs or examples, the design phase is not necessary. Also, if you can quickly implement an application using a combination of your favorite design pattern templates, drivers, and utilities—all of which you are intimately familiar with—and the application is strictly for your own use, do not bother creating a design document. At what point does the application's complexity become worthy of a separate design phase? The requirements specification should provide some insight—assuming that you wrote one.

Finally, sometimes a conventional design does not translate well to a LabVIEW diagram or might even adversely affect your application's style. For example, the literal translation of a flow chart to a LabVIEW diagram might entail networks of Sequence and Case structures that do not constitute good programming style. In fact, the result might appear like a cross between Nested VI and Spaghetti VI from the examples in Chapter 1, while occupying the space of Meticulous VI. This would be the worst of all worlds! Instead, it is best to understand the precepts of good LabVIEW style and to apply modeling and design techniques in a manner to complement your diagrams. For example, the State Machine Diagram specifies a design that gracefully translates into the most popular LabVIEW design patterns.

2.2.1 Search for Useful Resources

Search for resources that will expedite design and development while positively influencing style. Search online for products, references, examples, drivers, toolkits, colleagues, and consultants. Start with www.ni.com. **NI Developer Zone,** in particular, is a forum for developers in need of LabVIEW resources, such as articles, example code, and support. The **LabVIEW Tools Network** is an NI-sponsored site within the NI Developer Zone that lists add-on products, books, tutorials, and more. Additionally, visit OpenG.org, and the LAVA user group. **OpenG** is an organized community committed to the development and use of open source LabVIEW tools; the URL is www.openg.org. The LAVA website is dedicated to the open and unbiased exchange of ideas on intermediate to advanced topics in LabVIEW; the URL is www.lavausergroup.org.

Do not overlook the many offline resources and media available as well. The LabVIEW environment contains hundreds of example VIs that are searchable using NI Example Finder (**Help»Find Examples**). The *LabVIEW Help* contains application notes and white papers. You can purchase many LabVIEW books from your favorite bookstore or online reseller. Additionally, NI offers hands-on instructor-led training through NI Certified Training Centers, and customized solutions through the NI Alliance Program. The NI Alliance program is a worldwide network of more than 600 consultants, systems integrators, developers, channel partners, and industry experts. For example, Bloomy Controls is a Select Alliance Partner and hosts two NI Certified Training Centers.[4]

Now back to the low-emissions exhaust system application example. Our requirements include control of several specialized gas-analyzing instruments and SPC analysis of the data. Hypothetically, we looked on NI Developer Zone and did not locate drivers for our specific instruments. We also inquired with the instrument manufacturers but came up empty. We did find the SPC Toolkit for LabVIEW, part of the LabVIEW Enterprise Connectivity Toolset from National Instruments. It contains VIs and charts for all the SPC analysis that we need. Hence, we can elect to use the SPC Toolkit and save ourselves substantial development effort. Also, our LabVIEW style is positively influenced by incorporating the high-quality VIs that are provided with an NI toolkit, and reducing the scope and complexity of the source code.

2.2.2 Develop a Proof of Concept

In many cases, it is useful to develop a proof of concept to evaluate a specific instrument, hardware component, or software design. For example, you might have a critical requirement for high-speed acquisition and online data processing. You might be uncertain that a given computer, instrument or DAQ module, or communications bus, combined with a data processing algorithm, can execute within a specific time interval. Therefore, you should design a test that will perform or simulate the critical acquisition and analysis algorithms, and benchmark the execution speed. You then could use the results of your benchmark tests to validate your design; seek alternatives, such as higher-performing instrumentation or faster data processing algorithms; or relax the requirements in your specification, if necessary.

In regard to the exhaust system, we need to be able to send set points to mass flow controllers, acquire data from several gas analyzers synchronized within 10ms of each other, and perform SPC analysis algorithms and chart updates in 100ms intervals. These are critical requirements. There is no use working on any of the other features until we are certain that these requirements can be satisfied. Therefore, we develop a proof of concept consisting of VIs that perform these functions on a small scale, and we set up some experiments to monitor the performance.

When the proof of concept is complete, make sure you save it. You might be able to reuse all or portions of it in the final application. Alternatively, you might find that the performance of your completed application differs from the proof of concept. In this case, you should return to your proof of concept as a troubleshooting tool. If it still functions properly, you can begin to isolate performance issues in other areas. If the proof of concept suddenly stops working, the hardware, system configuration or other conditions might have changed since the last successful run. Reverting to a previously tested software and hardware configuration is sometimes referred to as a **sanity check**. If nothing has changed but the system no longer works properly, perhaps the only reasonable explanation left is that we, the developers, have lost our sanity.

As you might have guessed, the proof of concept positively influences LabVIEW style. Because the proof of concept normally addresses the most critical requirements of the application, the software will likely either incorporate or be modeled after the proof of concept. Implementing a routine that has already been tested and debugged reduces the risk of significant changes. Additionally, the proof of concept might help define some of the application's primary constructs, such as the design pattern and data structures. When these elements are proven, the corresponding application requires fewer changes, and fewer changes equates to neater code, per Theorem 2.2.

 Rule 2.3   Maintain good LabVIEW style throughout the proof of concepts

My personal pet peeve is when developers are sloppy with their proof of concepts. For example, we have all met developers who prototype VIs using default control labels such as **numeric** and **numeric 2,** and VI names such as **Untitled 1.** They expect to show us their prototype and have us understand what it does, as if we have some type of telepathy. As discussed, a good proof of concept should be archived and should function as a project reference. This makes sense only if it is developed using good style. Ideally, developers should apply the same good style conventions on the proof of concepts that they apply to the finished application. Indeed, good style allows the proof of concept to scale gracefully into an application. Good style is important all the time.

2.2.3   Revise the Specification

When the design phase is complete and the critical requirements have been evaluated via proof of concept, revisit the specification. Changes are inevitable. Conflicts might exist between several of the requirements, and tradeoffs and adjustments might be necessary. The specification is a living document. Be sure to revisit the specification on a periodic basis and make revisions. Remove any obsolete requirements. Add any new requirements. Evaluate the impact on schedule. Elaborate on any aspects of the application that are better understood than when the project began. Maintain the specification in a controlled repository where your project team members can access it easily. Manage changes and revisions according to your organization's applicable standards.

Considering our exhaust system, we might need to update the acquisition and analysis requirements based on our finding with the resource search and proof of concept development phases. Specifically, expand the SPC requirements to include more than simple Xbar & R charts, due to the VIs available in the SPC Toolkit. Also update the acquisition and data processing rates based on the benchmarks acquired using the proof of concept. Finally, we might reduce the risk level and time budgeted for each task as a result of these shortcuts.

2.3 Configure the LabVIEW Environment

Developers, do you

- Begin most applications with a blank VI?
- Use LabVIEW's default settings for the size, color, and fonts of GUI controls and indicators?
- Create user interfaces that are functional but bland?

Alternatively, do you spend excessive time

- Reconstructing your favorite design patterns?
- Searching previous projects for reusable source code?
- Deselecting **View As Icon** or disabling auto **wire routing**?
- Making similar repetitive edits throughout most applications?

If you answered "yes" to any of these questions, you need to customize the LabVIEW environment. If you start with blank VIs and default control properties, you might have decided that attractive panels and ease of use are not important for your particular application, or you are not artistically inclined to design a better scheme. More likely, you have discovered and applied good combinations of colors, fonts, and properties that go very well together. In this case, you might be starting with the defaults because you do not have a good system of tracking and reusing elements of your previous applications. Your favorite design patterns and user interface themes should be readily accessible as templates. The palette views and front panel and block diagram options should be customized to your personal preferences. This section covers how to configure the LabVIEW environment to save time and improve your programming style.

2.3.1 LabVIEW Options Dialog Box

The LabVIEW Options dialog box, accessed from the menu bar by selecting **Tools»Options,** is used to set preferences related to the work environment. Many of these options are related to style. For example, some block diagram options I avoid include **Show dots at wire junctions, Show subVI names when dropped,** and **Place front panel terminals as icons.** Diagrams tend to look neater without the dots, labels, and icons. Dots at wire junctions help developers distinguish between the junction of multiple wire segments and overlapping wires. However, many LabVIEW developers search for dots that represent coercions, which have negative connotations. Triple-clicking any wire can reveal its branches, and dots at junctions can be avoided. Terminals as icons and subVI labels are simply a waste of precious real estate. I recommend disabling these options. Figure 2-2 shows my own personal preferences for the block diagram options.

On the front panel, I prefer most of the default options. Seasoned LabVIEW developers really appreciate transparent labels and modern style controls. In previous versions, many of us repetitively colored our labels transparent on each and every control of every panel we created. Transparent labels look much nicer than the raised labels of the past. Modern style controls, also known as 3D controls, were one of the great innovations of LabVIEW 6, enabling our VI front panels to resemble (or transcend) their physical counterparts: the panels of traditional boxed instruments. Also related to front panel style is the Alignment Grid option: **Show front panel grid** and other associated properties. The grid appears while in edit mode and aids the developer in visually aligning and spacing front panel objects.

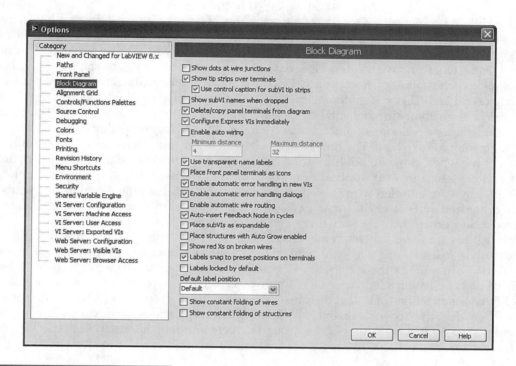

Figure 2-2
LabVIEW options dialog box, with block diagram preferences shown

 Rule 2.4   Document your LabVIEW options and back up the LabVIEW.ini file

In an ideal world, all developers within an organization would standardize on the same LabVIEW environment settings. In addition to promoting consistent style, this improves the interchangeability of development machines and is a good CM practice. The LabVIEW options are maintained in the LabVIEW.ini configuration file located in the root of the LabVIEW installation directory. I recommend backing up this file or perhaps even saving screen shots of the options dialog box. This enables you to reconfigure a new or existing machine quickly. It is also possible to distribute the LabVIEW.ini file to maintain consistent options across multiple machines. However, beware that the contents of the LabVIEW.ini file are version, platform, and system sensitive. In addition to environment preferences, the LabVIEW.ini file maintains paths to the most recently used files and directories, which are relevant only to a specific machine.

Most LabVIEW developers reorganize their palettes frequently, to suit their particular editing needs at any given time. This consists of arranging the top-to-bottom order of the categories, as well as expanding, minimizing, showing, and hiding them. Additionally, custom palettes are configured using **Tools»Advanced»Edit Palette Set**. Custom palettes are strategically important with respect to code reuse.

2.3.2 Code Reuse

Code reuse is an essential technique in all modern programming environments. The LabVIEW VI is a self-contained software module that can be called from higher-level VIs as a statically linked subVI, or dynamically by configuring **Call Setup**, or using the VI Server Call by Reference Node. SubVIs lend themselves very well to code reuse. Additionally, controls can be customized and saved as custom controls or type definitions. Design patterns and application frameworks can be saved as VI or project templates. These tools can save the developer days of development time versus starting with blank VIs and developing each application from scratch. Every developer should maintain a software reuse library containing useful LabVIEW code. Moreover, every organization that employs multiple developers should maintain a reuse library in a networked repository maintained under source control. The potential benefits include faster development time, greater commonality, and higher-quality applications.

We *all* want to save time! The time savings element of reusing source code is very important. Why reinvent the wheel if we do not have to? Always look for existing design patterns, VIs, and controls that you can reuse in your applications. Many developers use NI Example Finder to locate useful examples that are bundled with LabVIEW. This is a very well-integrated tool that presents examples in an organized and searchable manner. Far fewer developers integrate and use their own LabVIEW source code, perhaps because it might not be obvious how this is best accomplished.

We begin with the most commonly shared LabVIEW construct: the instrument driver. Although thousands of instrument drivers are available from NI Developer Zone's Instrument Driver Network (www.ni.com/idnet), there are tens of thousands of programmable instruments out there. Most of us have had to develop a few drivers on our own, as well as customize some of the downloaded drivers. Many organizations maintain their instrument drivers in a repository. The key to using them is to make them directly accessible on the LabVIEW **Functions** palette of the developer's computer. This is accomplished by placing copies of the desired drivers into the LabVIEW\instr.lib folder on the developer's machine. LabVIEW reads the contents of this folder each time it launches and populates the **Instrument Drivers** palette with subpalettes for each driver it finds. This feature facilitates seamless integration of instrument drivers, copied from anywhere, with the LabVIEW environment. Figure 2-3 shows a **Functions** palette with the **Instrument Drivers** subpalette expanded.

Figure 2-3
Functions palette with the **Instrument Drivers** subpalette expanded. Each icon corresponds to an instrument driver contained within the LabVIEW\instr.lib folder.

Rule 2.5   Develop reusable SubVIs

- Perform specific, well-defined tasks
- Use controls instead of constants
- Apply good style

Likewise, most of us have developed subVIs for manipulating strings and arrays, reading and writing files, performing computations, and more. Any such subVIs are reusable if they satisfy certain criteria. First, they should perform a specific useful task in a manner that is flexible and general purpose. In some cases, this involves using controls assigned to connector terminals instead of constants on the diagram. Second, the subVI must be developed employing good style, per the recommendations in this book.

For example, suppose we are developing an application in which we need a routine to parse multiple data packets returned from an instrument that are delimited by carriage return and line feed character combinations. We can write a subVI that calls Match Pattern within a While Loop. If my regular expression is a string formed by concatenating the carriage return and line feed constants on the diagram, the subVI will not be very general purpose and reusable. Instead, I should create a string control on the front panel, label it **Search String**, and assign it to a prominent connector terminal. Now the subVI performs a useful and specific task: It parses data packets with variable delimiter and is general purpose and reusable.

 Rule 2.6   Make reusable libraries accessible from the LabVIEW palettes

- Place reusable libraries in `LabVIEW\user.lib`
- Customize the LabVIEW **Functions** and **Controls** palettes

The final step in making the subVIs reusable is to make them readily accessible to the developer. Searching through archives that contain source code from previous projects can be cumbersome and unwieldy. Instead, identify reusable components as you code and place them in a dedicated repository. Then you can upload copies of the reusable components to locations in your LabVIEW installation directory that will integrate with the LabVIEW environment. Specifically, the `LabVIEW\user.lib` folder behaves similarly to `LabVIEW\instr.lib`. LabVIEW maps any folders, project libraries, or LLBs that it finds in this directory to subpalettes underneath the **User Libraries Functions** palette.

Figure 2-4 illustrates an example of a **Functions** palette with the **User Libraries** category expanded. In this case, the `user.lib` folder contains five libraries: OpenG, Bloomy Utilities, Bloomy Library, Win Utilities, and an unidentified library containing a generic icon. LabVIEW automatically assigns the generic icon, shown on the far right, to any folders, project libraries, or LLBs that do not have an associated menu file. The palette properties, including the folder or library mapping, as well as the arrangement of subpalettes, menus, icons, and VIs, are specified by selecting **Tools»Advanced»Edit Palette Set**.

Figure 2-4
The **Functions** palette with the **User Libraries** category expanded reveals five user libraries.

SubVIs that perform low-level tasks that complement or extend the capabilities of the built-in LabVIEW functions are known as **utility VIs**. It is desirable to access the utility VIs directly from the associated LabVIEW palettes, in addition to or instead of the **User Libraries** palette. For example, the OpenG and Bloomy reuse libraries both contain File I/O libraries. These VIs are accessed by navigating the **User Libraries** palette, as shown in Figures 2-5A and 2-5B. However, it might be faster and more intuitive if these libraries were accessible from the **File I/O** palette. Fortunately, LabVIEW provides complete flexibility in customizing the palettes. In Figure 2-5C, redundant subpalettes have been created for the File I/O utility libraries on the **File I/O** palette. This is accomplished from the palette set editor by inserting new subpalettes on the associated LabVIEW palettes that are linked to the MNU files, folders, project libraries, or LLBs that comprise the utility libraries. It is important to recognize that the source files uniquely remain in the `LabVIEW\user.lib` folder. The MNU files that define the palette contents are edited by the palette set editor and maintained at the following path: `<default data directory>\<LabVIEW version number>\palettes`.

Most of us customize the controls and indicators available from LabVIEW's **Controls** palette. Some of the most common edits include changing the data type or representation, resizing, coloring the various foreground and background elements, and editing fonts of the labels or embedded text. Alternatively, the control editor can be applied to dramatically modify the appearance and create a brand new type of control. Customized controls can be saved as custom controls, type definitions, or strict type definitions, and can be reused throughout your applications. Any such controls can be accessed from the **User Controls** palette simply by placing them within the `LabVIEW\user.lib` folder. LabVIEW uses the file extensions to determine whether each item appears on the **Controls** palette or **Functions** palette. Files with the `.vi`, or `.vit` extension appear in the **Functions** palette; files with the `.ctl` extension appear in the **Controls** palette.

Templates are VIs that contain commonly used design patterns or constructs, saved with the `.vit` extension. Some common templates include the SubVI with Error Handling, Dialog Using Events, and Standard State Machine. These templates are all available from LabVIEW's **New** dialog box, which appears every time you select **VI or Project from Template** from the **Getting Started** window, or **New** from the File menu. The **New** dialog box is shown in Figure 2-6.

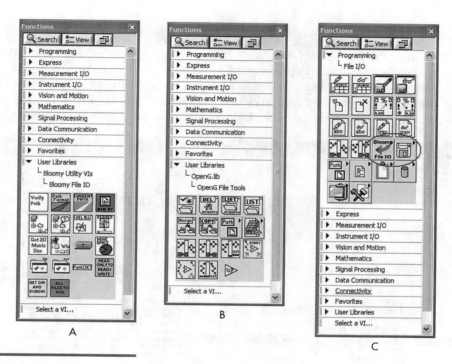

Figure 2-5
The OpenG and Bloomy libraries contain File I/O utility VIs that reside within `LabVIEW\user.lib`. The LabVIEW palettes are customized to access these VIs equally from the LabVIEW **User Libraries** palette, as shown in palettes A and B, and the **File I/O** palette, as shown in palette C.

    The **SubVI with Error Handling** consists of a front panel with the **error in** and **error out** clusters assigned to the lower left and right connector terminals, and a block diagram containing a Case structure with **error in** wired to the selector terminal. The developer adds other controls and indicators to the front panel, assigns them to appropriate connector terminals, and creates and wires together diagram objects inside the **No Error** frame of the Case structure. This template might save only a couple minutes of development effort, but it can be applied repetitively, reducing tedium.

    Dialog Using Events is a VI that opens its front panel when called and prompts the user with any data or selections provided on the panel. Its Window Appearance property is configured as Dialog, and it contains two dialog box–style Boolean controls labeled **OK** and **Cancel**. The diagram consists of a While Loop that contains an Event structure that sleeps until the Value Change event fires on either control, which stops the VI. If all your dialog boxes originate from this template, your application's graphical user interface will exhibit a style and behavior that is consistent with the native operating system dialog boxes and is consistent throughout your application.

    The **Standard State Machine** is a common design pattern that consists of a Case structure within a While Loop, a shift register for passing the case selector value, and an enumerated type definition that contains the state names. State Machines are a popular type of design pattern that are described in much more detail in Chapter 8, "Design Patterns."

Rule 2.7   Place reusable templates in the `LabVIEW\templates` *folder*

Figure 2-6A shows the LabVIEW **New** dialog box with the aforementioned templates that ship with LabVIEW. You can expand upon the standard templates and also incorporate your own by placing them in the `LabVIEW\templates` folder. In addition to increasing productivity, templates will help improve your LabVIEW programming style, as long as your templates follow good style. Figure 2-6B shows some templates that are part of the Bloomy Controls software reuse library.

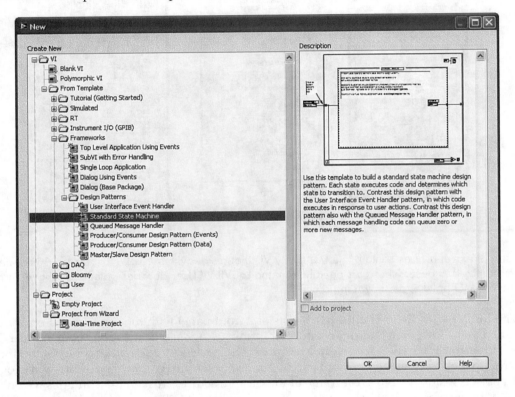

Figure 2-6A
The **New** dialog box provides access to templates that ship with LabVIEW, including the Standard State Machine.

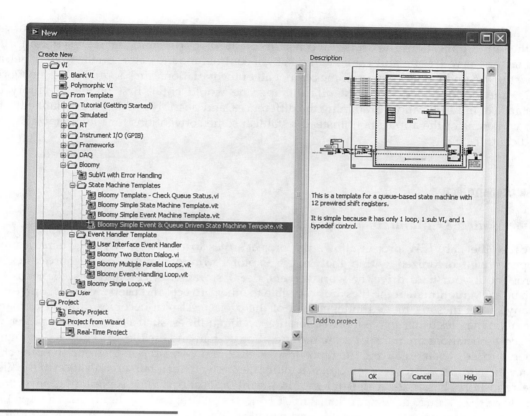

Figure 2-6B
An organization's templates are accessible from the **New** dialog box when placed in the LabVIEW\templates folder.

2.4 Project Organization, File Naming, and Control

LabVIEW 8.0 introduced Project Explorer, an interface for organizing project files within a project tree. The properties that specify the project are stored within an XML file called the LabVIEW project (.lvproj). The files referenced by the project can reside almost anywhere, can be named almost anything, and can target on a growing variety of computing devices. This presents new opportunities, as well as new challenges for developers and organizations. Conventions for organizing, naming, and controlling project files are desirable.

Many of us take it for granted that source files are organized, named, and controlled appropriately. From experience, I have learned that this assumption is entirely *not* valid. For example, it seems that some developers strive to minimize the length of their filenames. I have seen naming conventions that consist of very short acronyms followed by an underscore and number, as if there was an 8.3-character limitation reminiscent of DOS. Surprisingly, this is not uncommon. I have seen large applications that consist of hundreds of such VIs within a single folder or LLB, none of them marked as top level. In these situations, when the developer is long gone and documentation is scarce, *I wish I could buy a vowel!*

Furthermore, I have seen test labs that contain multiple PCs, each maintaining dozens of variations of the same application. The users edit constants on the block diagram that represent transducer scaling factors, as well as default values and ranges of controls and indicators on the front panel, and then save the VIs under a new name. Some but not all of the variations are backed up to a network server, CDs, and various media. Any significant upgrade would entail first taking inventory of every instance of the application, examining the differences, and possibly making redundant edits to multiple variations of the application. Instead, establish some conventions for file organization, naming, and control.

2.4.1 Disk Organization

Rule 2.8   Maintain an organized repository on disk

Because the LabVIEW project allows the project's files to reside almost anywhere, a project that appears well organized within the Project Explorer window might have source files scattered throughout your hard drive, network servers, peers, storage devices, and miscellaneous targets. It can be a CM nightmare if the files are accessible to other projects and users, without source control, or if any of the files are moved independent of the project. Also, it becomes difficult to manage and maintain the files using conventional operating system utilities, such as Windows Explorer, when the project associations are transparent to the operating system and the files are scattered in many places. Therefore, create an organized file repository on disk for every project, to group the project files and place them under source control. Create a folder hierarchy to maintain organization of the files within the repository. Figure 2-7A illustrates a project file repository, with separate top-level folders for `Application Build`, `Data`, `Documentation`, `Graphics`, and `LV Source`. The `Application Build` folder contains the executable and install utility. The `Data` folder contains sample data sets generated by the application for the developer's reference. The `Documentation` folder contains reference documents, such as project specifications and instrument user manuals, as well as documents that ship with the product, such as help files and user manuals. The `Graphics` folder contains image files such as screen shots, company logos, and icons. The `LV Source` folder contains LabVIEW source files that are uniquely developed for the project. Source files from a reuse library that are used by the project but not edited within the project are located in a separate repository for use in multiple projects. This includes reusable utilities and components such as instrument drivers.

Rule 2.9   Create a LabVIEW source folder hierarchy that reflects your application's architecture

The five folders shown in the `[Project Name]` folder of Figure 2-7A form a base set of file categories. If your project files are located in the roots of these five folders, all of which reside within a single project folder, you have successfully created a project repository. However, each of the folders can be further divided into subfolders for greater organization on disk. In particular, you might create a subfolder hierarchy to organize the LabVIEW source files in a manner that reflects the application's architecture. For example, some typical subfolder names include `Analysis`, `Configuration`, `DAQ`, `Data Manipulation`, `File IO`, and `User Interface`. I recommend placing the top-level VIs at the root of `LV Source` and placing all subVIs and controls in the appropriate subfolders. Likewise, the other base folders can be further organized into subfolders. Figure 2-7B provides an example of an expanded folder hierarchy for a complex application.

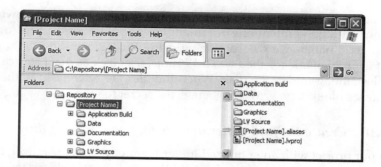

Figure 2-7A
Folder hierarchy for maintaining project files on disk, with a basic set of top-level folders

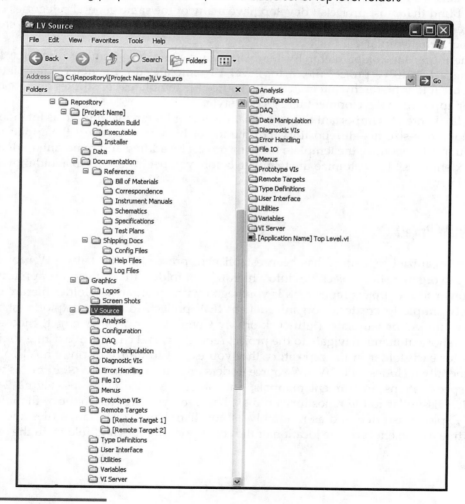

Figure 2-7B
Expanded folder hierarchy that provides detailed organization and an accelerated starting point for complex application development

Most developers agree with the usefulness of an organized repository. However, like many Rules, we often start out with good intentions but might lose focus under the pressures of tight deadlines. Unless we form good habits, we risk sacrificing many Rules, including the source folder hierarchy. Furthermore, it is problematic for the LabVIEW project, as well as most source code control tools, to move source files among folders on disk. When the files are referenced within the project, moving them on disk causes file-linking errors. So here is the secret to maintaining an organized repository:

 Rule 2.10 Create the folder hierarchy before you begin coding

If you initiate your project with an organized folder hierarchy from the outset, maintaining an organized repository is as simple as saving each file in the appropriate folders as you create and edit the files throughout the development cycle. I find that it is much more feasible if the folder hierarchy already exists. Furthermore, you need not create a new folder hierarchy for every project from scratch. I find that most projects I develop have many of the same general categories of files, similar to Figure 2-7B. After you have created one such folder hierarchy, use it as a model to create a reusable folder hierarchy template.

Furthermore, the folder hierarchy template can contain any number of source file templates, including a LabVIEW project file, a top-level VI template, documentation templates, and more. Indeed, the folder hierarchy integrated with source file templates constitutes an accelerated starting point for application development using good style.

On a final note, it is important to realize that the folder hierarchy template is intended as a starting point, not a one-size-fits-all repository. For the folder hierarchy to reflect the application's architecture, you must customize the template for your project by adding and removing folders as necessary. To the extent possible, customize the template before you begin coding, to avoid moving source files on disk later.

2.4.2 The LabVIEW Project

After an organized repository has been established, proceed to the LabVIEW project. Use Project Explorer to organize the project files into a hierarchy of folders that reflects the application's architecture, similar to the project repository. If you began with an organized folder hierarchy on disk, it is extremely simple to create an organized LabVIEW project. In Project Explorer, start with a new LabVIEW project or template, right-click on My Computer or another target, choose **Add Folder** from the shortcut menu, navigate to the project repository, and choose an existing folder. Repeat this process for each folder in the repository that you expect to utilize during LabVIEW development. This generally includes all LabVIEW source folders appearing under LV Source, as well as requirements specifications, instrument manuals, and other reference materials. Figure 2-8 provides an example. Dissimilar to the repository on disk, however, you are free to move file references within Project Explorer as often and as randomly as you like. How you arrange files and folders on the project tree has no effect on the location of the corresponding files and folders in the repository.

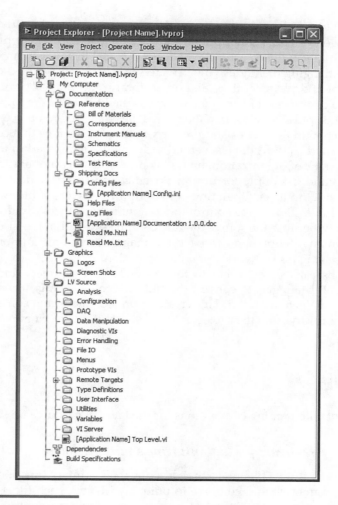

Figure 2-8
The LabVIEW project contains an organized folder hierarchy that reflects the application's architecture, similar to the project repository.

 Rule 2.11   Organize LabVIEW source files into cohesive project libraries, where appropriate

The LabVIEW project library is a file with the .lvlib extension that maintains properties shared by a collection of LabVIEW source files. The properties include a name prefix, version, access rights, password, palette icons, and menu names. Project libraries are intended to maintain cohesive collections of source files that work together to perform a specific set of functions, such as an instrument driver. Project libraries prevent namespace conflicts that arise when two files containing the same name are opened. Specifically, all the source files that comprise the project library inherit the library name as the prefix for the source filename. This allows project libraries to contain identically named source files, as long as the library names are unique. For example, multiple instrument drivers can contain an Init VI and a Close VI, which can be loaded in memory at the same time. The file prefix is provided by the `.lvlib` filename, which is unique to each instrument driver. Additionally, you can use the Library to define VIs as either public and accessible to all users or private and usable only by the developer. With the ability to control which VIs are usable, you can maintain a consistent public VI interface while changing the underlying private VIs without fear of breaking the user's code. Note that legacy instrument drivers, toolkits, and other libraries developed prior to version 8.0 are generally distributed in a compressed file format known as the LLB. LLBs are a different type of entity than a project library. LLBs physically contain the source files that they reference, whereas project libraries are XML files that contain references and property values for the files that comprise the library.

Libraries distributed as LLBs have several disadvantages compared to folders and project library files. These include lack of organization, limited compatibility with operating system utilities, longer load and save time, and risk of file corruption. Specifically, LLBs provide only two levels of organization: top level and not top level. Operating system utilities such as Windows Explorer cannot search for source files within LLBs. Source control tools cannot check the individual files in and out within an LLB. Additionally, because the source files are compressed within the LLB, they take longer to load and save. Finally, many people tend to pack too many source files into one LLB, further reducing organization and risking loss of work from file corruption.

Fortunately, LLBs can be converted to folders of VIs with project libraries. First, use the LLB Manager (**Tools»LLB Manager**) to convert the LLB to a folder of VIs. Then go into Project Explorer and add the folder of VIs to a project. This creates a project folder that contains the library VIs. Next, choose **Convert to Library** from the project folder's shortcut menu. Finally, name and save the project library.

2.4.3 File-Naming Conventions

This section presents file-naming conventions for the LabVIEW source files.

   Rule 2.12   Create unique and intuitive source filenames

It is important to use intuitive filenames that describe their primary function. As a rule, the source file and project library names should combine to uniquely identify each file. Consider the hypothetical situation that you might need to take an unexpected leave of absence during the middle of an important project. Your colleagues should be able to locate and identify your source files, understand your programming style, and resume project development in your absence. This is a true test of good style.

   Rule 2.13   Do not abbreviate filenames

Avoid conventions that rely on very short abbreviations, acronyms, and numbers. For example, the source filename `IPS_Osc_H_to_0.vi` might be meaningful to the original developer, but not to many others. Use as many characters and distinct words as necessary to uniquely identify the source file. Avoid using characters that not all file systems accept, such as slash (/), backslash (\), colon (:), and tilde (~). LLB Manager and VI Analyzer are tools that can be used to test the platform portability of your filenames.

   Rule 2.14   Never use LabVIEW's default filenames

Never rely on a default name and number to uniquely identify a source file. This contradicts LabVIEW's built-in default VI naming convention, in which LabVIEW automatically assigns the name **Untitled** <number> and **Control** <number> to each new VI and custom control or type definition, respectively. Avoid default names and numbers at all costs. In my opinion, this is a severe violation of good style.

Many developers append numbers to the filenames to indicate the version number. Note that this is not necessary if your source files are maintained within project libraries or if the project repository is under source control. Additionally, VI revision history (**Tools»Source Control»Show History**) is a built-in LabVIEW feature that maintains the revision number of your VIs. For example, it can be configured to automatically increment and prompt for comment each time you save or close a VI with changes. This is the only method in which the revision number is maintained directly within the source file.

Legacy LabVIEW Plug and Play instrument driver files developed prior to the LabVIEW project follow a specific file-naming convention. Each filename contains an instrument prefix, which is an abbreviation of the instrument vendor name and the instrument model. For example, the Keithley 2000 digital multimeter has the prefix **ke2000**. In addition, multiple required files are explicitly specified in the VXI*plug&play* Instrument Driver Functional Body specification. You can view the complete specification at www.vxipnp.org. However, modern project-style instrument drivers maintain the instrument prefix in the project library instead of in the source filenames. Hence, the instrument prefix is no longer needed within the individual filenames.

Rule 2.15   Identify the top-level VIs

The most important source files to identify in any LabVIEW project are the top-level VIs. Always distinguish these files by maintaining them near the root of the folder hierarchy on disk, as well as within the project, and apply an appropriate naming convention, such as *<project name>* Main VI, or *<project name>* Top Level VI. Figures 2-7B and 2-8 illustrate this convention. Because the top-level VI is not contained within any of the project's libraries, the project name is included within the filename. This ensures that the filename is unique and intuitive.

2.4.4   Source Control

Source control is a process by which source files are secured, shared, and maintained in a multideveloper environment. Source control is facilitated by a third-party source control application. LabVIEW integrates with multiple source control applications by enabling us to perform the most common source control operations from within the LabVIEW project. As of LabVIEW version 8.2, LabVIEW supports Microsoft Visual SourceSafe, Perforce, MKS Source Integrity, IBM Rational ClearCase, Serena Version Manager (PVCS), Seapine Surround SCM, Borland StarTeam, Telelogic Synergy, PushOK (CVS and SVN plug-ins), and ionForge Evolution. Under Windows, LabVIEW uses the Microsoft Source Code Control Interface. On non-Windows platforms, LabVIEW supports only Perforce via the command-line interface.

Source control is essential in a multideveloper environment. It allows a project team to collaborate on the same project at the same time by controlling access to the project folders and source files. This prevents multiple developers from editing the same source files at the same time. Also, most source control packages enable you to configure permissions for each user. For example, you could configure a project documentation repository to give write access only to managers, while you could configure your source code repository to give write access only to software developers. Alternatively, some source control tools give multiple developers unlimited access to the source files, and the source control tools resolve conflicts when changes are merged back into the repository.

Source control tools track changes and revisions to all types of project files and maintain a comprehensive backup. These features are as beneficial for a single-developer project as they are for a multi-developer project. Additionally, CM and source control are common requirements for different certifications, such as ISO 9000.

Rule 2.16   Follow your organization's CM Rules

If your organization has a CM process in place, follow all Rules that have been established, including source control. There could be a specific procedure that you must follow that includes specific tools and configuration settings. Placing a project under source control usually involves creating or moving your repository to a designated location, adding the files to source control, and configuring the access rights for the files within the source control software. You can then proceed with development from the LabVIEW environment, adding new files to source control as they are created, checking out existing files for editing, checking them back in, and so on. For more information, refer to the *LabVIEW Help*.

Rule 2.17   Avoid moving source files on disk

One point of emphasis is that after development has started, project files can be organized freely within Project Explorer but should *not* be moved on disk. Otherwise, broken links will result within the LabVIEW project and any source control tools that reference the dislocated files.

Now we are ready to begin coding!

Endnotes

1. Howard M. Kanare. *Writing the Laboratory Notebook*. Washington D.C.: American Chemical Society, 1985.
2. Free downloadable materials are available from `www.bloomy.com/resources`.
3. Software design references:

    Larman, Craig. *Applying UML and Patterns: An Introduction to Object-Oriented Analysis and Design and Iterative Development*, second edition. Upper Saddle River, NJ: Prentice Hall PTR, 2001.

    McConnell, Steven. *Code Complete*, second edition. Redmond, WA: Microsoft Press, 2004.

    McConnell, Steven. *Software Project Survival Guide*. Redmond, WA: Microsoft Press, 1997.
4. Bloomy Controls is an NI Select Integration Partner that provides systems development and training services throughout the Northeast United States, including offices in Windsor, Connecticut; Milford, Massachusetts; and Fort Lee, New Jersey.

Front Panel Style

3

The LabVIEW front panel is named to emphasize its capability to emulate the panels of traditional self-contained instruments. Indeed, when LabVIEW was first released in 1986, its primary purpose was to control laboratory instruments via GPIB and serial interface. *Oh, how times have changed!* Today LabVIEW is applied to many different types of applications, industries, and instrumentation. The complexity of applications has dramatically increased. The evolution of front panels is no exception. 3D controls, tab controls, tree controls, containers, Property Nodes, VI Server, Event structures, and subpanels are modern innovations that facilitate powerful LabVIEW applications.

Despite these innovations, many front panel design techniques used by early LabVIEW developers still apply today. Additionally, many of the modern innovations aid our ability to organize and simplify an otherwise complex graphical user interface (GUI). This chapter provides guidelines that help make your LabVIEW front panels intuitive, user friendly, and conforming to industry standards. We begin by defining some terms that I use to describe the scope of each guideline.

Top-level VIs are VIs that reside at the highest level of the application hierarchy, with front panels that comprise the primary display screens. Some applications have only one top-level VI, which developers often refer to as the Main VI. Other applications have a top-level VI that dynamically loads other high-level VIs that comprise the primary display screens. For our purposes, all high-level VIs that contain user-viewable panels, and a hierarchy of subVI calls are considered top-level VIs.

Dialog VIs are GUI-related subVIs that open their front panels and prompt the user for information. Dialog VIs have far fewer subVIs and much less overall functionality than top-level VIs. Their purpose is to perform a specific transaction with the user. It is also common for dialog VIs to be **modal**, for which the front panel window is locked on top of all open windows until the interaction is completed and the window closes. Modal behavior is a default property setting with LabVIEW's **Dialog** window appearance, as selected from **File»VI Properties»Window Appearance**. However, in the context of this book, dialog VIs need not contain exactly the same properties as the **Dialog** window appearance and are not necessarily modal.

Collectively, I refer to the set of all VIs that have user-viewable panels, including top-level VIs and dialog VIs, as **GUI VIs**. VIs that do not open their panels during program execution are **subVIs**. This is a simplification because many top-level and dialog VIs are called as subVIs. In this book, however, **GUI VIs** have panels that users can view and operate on, whereas only developers can view **subVI** panels. Therefore, the scope and function of GUI VI panels substantially differ from those of subVI panels, and separate guidelines apply to the front panels of each. These two VI categories are distinguished whenever separate guidelines are applicable, as each style topic is presented throughout the chapter.

GUI VIs intended for personal use in an office, private laboratory, or other individual computing environment are referred to as **desktop** GUI VIs. GUI VIs designed for multiple users to interface with industrial equipment are **industrial** GUI VIs. Desktop GUI VIs appear and behave similar to other native applications on the computer's operating system, whereas industrial GUI VIs complement the industrial application and associated equipment, regardless of the computing platform upon which they are installed. Hence, separate rules apply for desktop and industrial GUI VIs.

Some subVIs are application specific, and others are general purpose and reusable. In theory, one might apply a different level of development effort, depending on the intended scope of subVI reuse. In practice, it is beneficial to design *all* subVIs expecting them to be reusable, to help ensure consistent quality and promote reuse. In a few places, I differentiate rules for **commercial** subVIs that are part of a product, such as a developer toolkit or an instrument driver, and **standard** subVIs, which are assumed to be reusable from a stylistic perspective but are not commercial. If the subVI scope is not specified as standard or commercial, I am referring to all subVIs, both standard and commercial.

This chapter presents front panel style rules and examples that conform as well as violate the rules. However, I make no attempt to discuss how to design and implement a graphical user interface. I recommend *LabVIEW GUI Essential Techniques*, by David Ritter,[1] as a comprehensive resource on GUI development. Additionally, if you are developing a commercial application for a specific operating system, I recommend reading a style guide for the target operating system. The objective of this chapter is to provide rules to make your front panels intuitive, consistent, and conforming to industry standards.

3.1 Layout

Layout pertains to how objects are arranged on the front panel. Layout influences ease of use, both for users who operate GUI VI panels and for developers who access subVI panels. The following are some rules regarding front panel layout, organized into sections for general rules, GUI VI panels, and subVI panels.

3.1.1 General Rules

The general rules presented in this section apply to the layout of all types of VI front panels, including both GUI VIs and subVIs.

Rule 3.1   Group related controls using decorations, spacing, tabs, and clusters

Identify logically related controls, and group them in the same area of the front panel. Consider the example shown in Figure 3-1. One LabVIEW application controls two identical test fixtures, labeled **Station One** and **Station Two**. The GUI contains all controls and indicators for operating both test fixtures on one panel. Several distinct groupings exist. First, the panel is divided vertically into two halves, with objects associated with **Station One** on the left half and **Station Two** on the right half. In addition to the divisional grouping, decorations visually partition the groupings and provide the appearance of two separate and distinct panels. Within each decoration are two primary subgroupings of indicators: temperature and power. Thirty-two digital indicators are tightly spaced within clusters near the center of each decoration, displaying the temperature measurements, and six digital indicators are bundled within clusters at the bottom of each decoration, displaying the power measurements. The borders and light background shading within each cluster further delineate the contents of each.

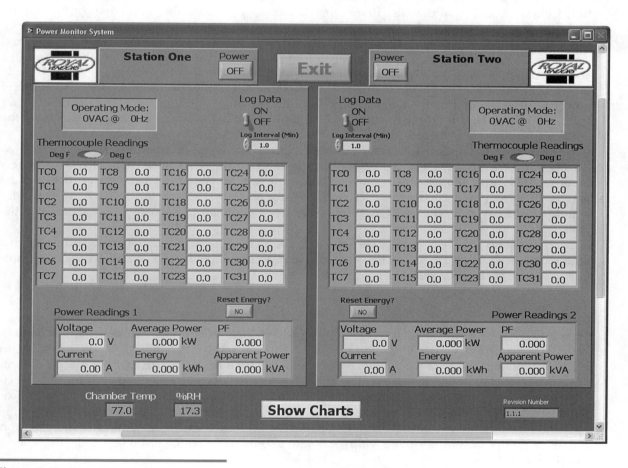

Figure 3-1
The GUI VI panel contains objects for two test fixtures grouped using spacing, decorations, and clusters.

Rule 3.2   Apply symmetry and spacing to front panel objects

My favorite LabVIEW editing tools are **Align** and **Distribute Objects**, accessible from the toolbar. I use these tools on *every* front panel I create, whether it is a GUI VI or a subVI. These tools are instrumental to effective and professional-looking front panel layouts. When viewing a panel for the first time, you can immediately see whether the developer uses these tools. Panels with aligned and evenly distributed objects are neat and symmetric, whereas panels without them appear more haphazard. The best thing about these tools is that they are quick and easy to use. There is no good reason not to use them.

The front panel of Figure 3-1 is symmetric, with the left and right halves of the screen appearing as mirror images of each other. The indicators within the temperature and power clusters are also symmetric and evenly spaced. The 32 temperature indicators are arranged in 4 columns of 8 indicators, tightly compressed so that each indicator is as close as possible without overlapping the next one. This is achieved by selecting a set of objects, such as one column of indicators, and then using the **Vertical Compress** tool from the **Distribute Objects** menu. Next, align the right edges of the column using the **Right Edges** tool from the **Align Objects** menu. When you are happy with the appearance and spacing of one column of objects, group them by selecting **Group** from the **Reorder** menu, and clone the group four times to create four even columns of indicators. Finally, select the four groups and space them evenly by selecting any of the horizontal distribution tools from the **Align Objects** menu. The same basic techniques are used to create three columns of two indicators for the cluster of power readings. Note that the temperature indicators are more tightly compressed than the power indicators. This is because the temperature labels are adjacent to the axis of compression, allowing the indicators to abut each other.

Just as grouping is used to associate related objects, spacing is used to disassociate unrelated objects. In Figure 3-1, a small amount of the panel background appears between the two primary decorations, delineating the separate stations. Additionally, a fair amount of space appears between the temperature and power clusters for each station. Space at the bottom and top of the panel resembles margins that help frame the panel and direct the user's attention to the objects in the center of the panel.

Symmetry and spacing need not be limited to GUI VIs. Apply the same rules to subVIs to give your source code an organized and professional appearance. Section 3.2.3, "SubVI Panel Text," provides an example and some additional rules that pertain to subVIs.

 Rule 3.3   Size similar objects the same

Objects that have similar function and priority should be identically sized. Sometimes it is difficult to know in advance exactly what size will work for all the related controls in a group. For example, in Figure 3-1, the indicators in the power clusters are related but have varying levels of precision, from one to three decimals. On the surface, it might seem that wider controls are required for three decimals instead of for one or two. Instead of having indicators of varying sizes, resize all of the indicators to the width of the widest indicator. This is easy using the tools in the **Resize Objects** ring. Specifically, select all the controls in the group and then choose **Maximum Width**. Additionally, you can resize groups of controls to the width or height of the smallest or largest object using the respective **Resize Objects** tools.

3.1.2 GUI VI Panel Layout

This section presents rules pertaining to the layout of GUI VI panels.

 Rule 3.4   Maximize the top-level VI panels for industrial applications

Many industrial applications serve as a dedicated interface to the industrial equipment that they monitor or control. It is usually not desirable to provide visible access to the operating system, tempting users to access resources unrelated to the application. Consequently, most industrial applications launch automatically during PC boot up, and the top-level VI panel covers the entire screen. This is accomplished by using the **Maximized** Window Run Time Position property, selected from **File»VI Properties**. Choose **Window Run-Time Position** from the **Category** ring, and select **Maximized** from the **Position** ring. Beware that extra front panel real estate becomes visible when the display resolution increases. This can reveal controls that were intentionally scrolled off the visible area during development. Be sure to make such controls invisible if you do not want the user to find them.

 Rule 3.5   Size dialog VI panels much less than full screen

 Rule 3.6   Center dialog VI panels

Center the front panels of dialog VIs and size them much smaller than full screen. This helps distinguish them as dialogs. Specifically, selecting the **Centered** Window Run Time Position property ensures that the panel appears in the center of the screen when opened, regardless of the host computer's display resolution. Without centering, your panels would open at the exact location where they were last saved and might require manual positioning by the user.

Additionally, your panel position and appearance is affected by variations in display resolution. If you expect the display resolution to change, select **Maintain proportions for different monitor resolutions** and **Scale all objects on the front panel as the panel resizes**. You can select these properties in the **Window Size** VI property category; they are also programmable using VI Server's VI Class properties. If your GUI is comprised of multiple panels that are designed to open adjacent to each other for simultaneous viewing, create a routine that programmatically adjusts the panel positions. Use VI Server functions to programmatically read the **Display.All Monitors** application property to determine the host computer's display resolution. Reposition and resize the panels and objects according to the resolution, using VI Server functions and front panel window VI properties.

 Rule 3.7   Use LabVIEW's dialogs for desktop applications; avoid them for industrial applications

LabVIEW's native dialogs are conveniently accessed from the **Dialog & User Interface** palette. They include the One, Two, and Three Button Dialog functions, as well as Prompt for User Input and Display Message to User Express VIs. The LabVIEW dialogs are simple to use and designed to resemble the dialogs native to the host computer's operating system. Hence, they are very useful for platform-portable desktop applications. However, their appearance cannot be customized. Also, the dialog windows and objects are relatively small compared to an industrial VI panel with 3D or classic style controls and larger panel size. The LabVIEW dialogs can be difficult to read unless you are sitting right in front of the monitor and are paying close attention. Consequently, the LabVIEW dialogs are *not* satisfactory for industrial GUI VIs. In many industrial settings where the users are multitasking while an application runs, a LabVIEW dialog might open unnoticed. Yet dialogs are often used for critical purposes, such as prompting for manual actions or displaying an unexpected error or warning. Fortunately, it is very simple to create more industrial alternatives that follow the style rules for industrial GUI VIs. Figure 3-2 compares desktop and industrial style dialog windows.

Figure 3-2
LabVIEW's Two Button Dialog function is well suited for desktop applications, as shown on the left. A larger Two Button Dialog VI is preferred for industrial applications, as shown on the right.

 Rule 3.8 Use system controls for desktop dialogs; use 3D controls for industrial dialogs

Custom dialog VIs can be developed for desktop as well as industrial applications. For desktop applications, simply set the **Window Appearance** property to **Dialog** and use only system style controls from the **System** palette. You set the **Dialog** window appearance from **File»VI Properties» Window Appearance**. The panel background color, modal behavior, and other properties are selected to resemble an operating system dialog. Likewise, the controls, indicators, and decorations available from the **System** palette resemble the style of objects used by the operating system. Use only the Dialog window appearance and System controls, to maximize consistency with operating system dialogs.

Although maintaining operating system consistency is desirable for desktop applications, it is generally *un*desirable for industrial applications. Instead, industrial GUI VIs, including dialogs, should complement the industrial application and equipment they are designed for. As discussed, this usually means larger panels, objects, and text. 3D style controls have a larger and more industrial appearance, and help improve visibility over dialog or classic style controls.

 Rule 3.9 Enlarge and center the objects of an industrial GUI VI in proportion to their importance

The default size of LabVIEW's controls and indicators is adequate for most desktop GUI VIs but is not sufficient for many industrial GUI VIs. Important objects on an industrial GUI should be readily visible from several feet away from the monitor. The more important the object is, the more prominent it should appear. The object's size and location help distinguish its importance. Larger and more centered objects are the most visible. Resize important controls and indicators larger than the default, and position them near the center of the screen.

Figure 3-3 shows separate desktop and industrial versions of a custom dialog VI used for stereo receiver measurements. The VI prompts the user to configure some test parameters and adjust the volume control of the receiver being tested. The VI controls a signal generator while monitoring the receiver's audio output level. The desktop version in Figure 3-3A is comprised of a panel with **Dialog Window Appearance** and controls from the **System** palette. All the controls, indicators, and text have default size and appearance, consistent with the GUI standards for the operating system. This forces the user to be physically close to the monitor and to examine all the controls. The desktop dialog VI is well suited for an R&D lab, where the user closely monitors the application, and can use any combination of control values with equal priority.

The industrial version in Figure 3-3B contains larger text, 3D controls, and a single **OK** button. The **Left** and **Right Audio Output** indicators are very large and vertically centered, drawing the user's attention. These features make it easy for the industrial user to operate the panel quickly. The user can skim the GUI; identify, view, and operate the most important data and controls; and complete the transaction quickly and efficiently. Industrial dialog VIs are preferred for a repetitious production environment.

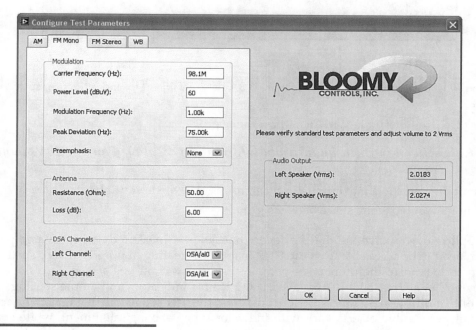

Figure 3-3A
This custom dialog VI is designed for desktop use, such as a typical R&D environment.

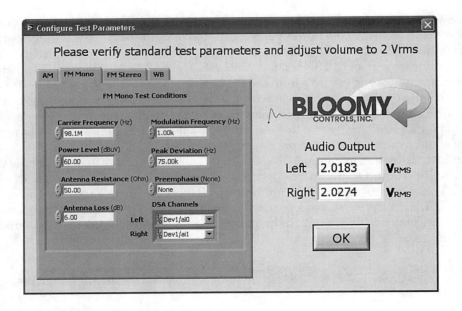

Figure 3-3B
This custom dialog VI is functionally equivalent to Figure 3-3A but is designed for industrial use, such as a production environment.

> *Rule 3.10 Limit the quantity of information displayed on a GUI VI panel*

- Limit the panel to a maximum of seven groups of seven unique objects
- Maintain ample white space between groups

Some applications involve hundreds or even thousands of measured and calculated parameters, all of which need to be accessible from the GUI. Instead of creating hundreds of very small indicators in an effort to maximize the number of parameters displayed on one screen, form logical groups of parameters that are reasonably displayed using legible indicators and ample spacing. Organize the groups onto multiple tabs, panels, or subpanels that the user can readily navigate. The actual size and number of permissible objects vary with indicator type, labeling, alignment, white space, monitor size, and resolution. One rule of thumb is seven groups of up to seven unique objects per group.

The Power Monitor System VI from Figure 3-1 contains seven groups of objects, as shown in Figure 3-4. These include one group for each station delineated by decorations (1, 4), groups of thermocouple indicators (2, 5), and power indicators (3, 6) for each station delineated as clusters, and the group of controls and indicators common to both stations located across the bottom (7). The clusters of power indicators contain six unique indicators within each. The clusters of thermocouples contain 32 indicators that are sequentially arranged and labeled. The user does not have to examine all 32 indicator labels to understand them. After she reads three or four of them, she recognizes the ordering, layout, and pattern. In terms of identification, the group of 32 indicators is analogous to a group of only a few different objects. Nonetheless, there are 32 unique data items in the group. Hence, this VI *approximately* satisfies the seven groups of seven unique objects criteria.

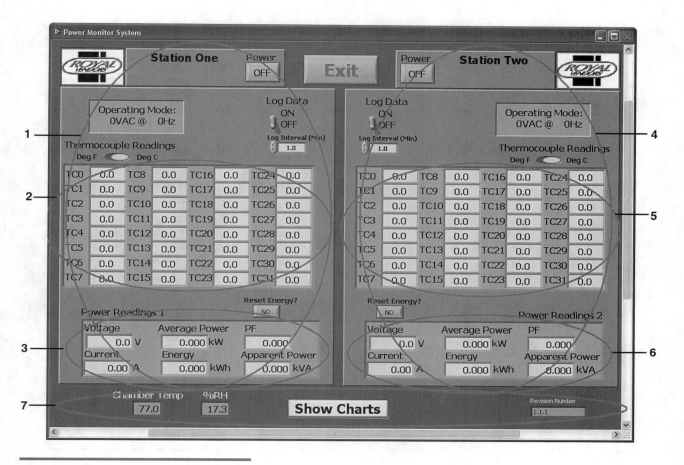

Figure 3-4
Power Monitor System VI contains seven groups of objects. The cluster containing 32 TCs is sequentially arranged and labeled for ease of identification.

Power Monitor System VI is quite dense. It contains 91 total controls and indicators on a panel that is sized for 1024×768 monitor resolution. The overall density is easily reduced using a tab control and dividing logical groups of objects onto separate tabs. Figure 3-5 contains a simple arrangement, with separate tabs for each station. The tab control is sized to fill the entire panel, providing the functional equivalent of multiple panels, with one panel visible at a time. The total number of visible controls and indicators is reduced to 48 per station. Each tab is divided into five logical groups, including three decorations, one cluster of temperature indicators, and one cluster of power indicators. The reduced object density increases white space and makes the existing objects easier to recognize and view.

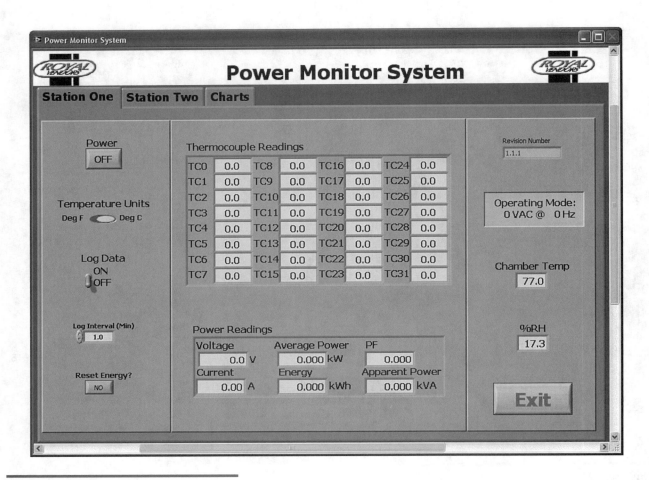

Figure 3-5
A tab control organizes the objects by station, reducing the total number of visible objects to 48 per station.

The organization of tabs and indicators in Figure 3-5 limits the user to viewing only one station at a time. Because both stations operate simultaneously, the user is forced to frequently flip back and forth between stations. An alternative is to include an additional tab that contains a high-level overview of both stations, using a limited number of controls and indicators. For example, the **Overview** tab in Figure 3-6 displays the maximum and average temperatures for each station, the average power, a bar plot of temperatures, and the **Power** ON/OFF Boolean controls. The bar plot is a very useful indicator for visualizing multiple data channels. As you can see, it is easier to notice the relative temperatures of each channel, as well as trends and outliers, than the cluster of numeric indicators. This illustrates the impact that display type has on readability. Hence, the **Overview** tab provides summary information for both stations, and the user can flip to the individual station tabs to view the details. This reduces the amount of navigation that is required when running both stations and reduces the overall density of each tab.

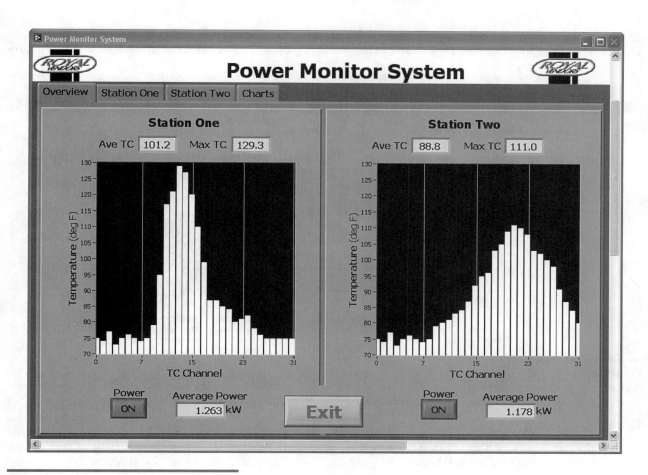

Figure 3-6
An overview tab provides a summary of both stations using a limited number of controls and indicators, including bar graphs.

 Rule 3.11   Avoid overlapping visible objects

LabVIEW performs less efficiently when front panel controls and indicators overlap. In particular, never overlap a graph indicator with another indicator because LabVIEW must redraw the entire graph each time *either* indicator's value changes. Graph updates are one of the most processor-intensive operations that LabVIEW performs, so avoid making them happen unnecessarily. Also, graph indicators contain numerous properties that are programmed to create a wide variety of effects, so overlapping other objects is usually not necessary.

In some circumstances, overlapping objects do not affect performance, and are permissible. In one example, only one of the objects is visible at a time. LabVIEW does not update invisible objects, so overlapping them does not cause extra processing. Another example involves importing a large static image and overlapping some controls or indicators on top of the image. For example, consider a large image that represents an industrial process or unit being tested and overlaying digital indicators that display live data at positions that depict the locations of the measurements. As long as the indicator backgrounds are not transparent, indicator updates do not cause the underlying image to be redrawn. Only the visible portion of the live indicator is redrawn with each update. Figure 3-7 provides an example.

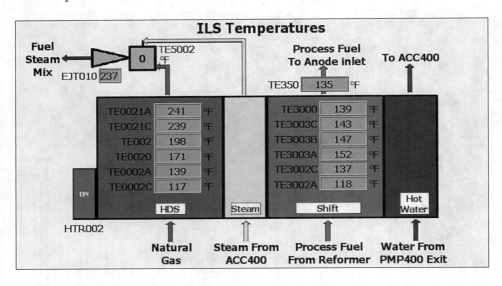

Figure 3-7
A static image of an industrial process with overlapping digital numeric indicators operates efficiently because only the digital indicators are redrawn with each update.

Several Window Appearance properties need to be configured for a GUI VI. Most of them fall into the categories of **Top level application window** and **Dialog**, which are preconfigured and conveniently selected via the **Window Appearance** property page. The most significant is to disable the **Show Abort button**, for several reasons. First, this button would allow users to abort the application immediately and abruptly, regardless of the state of your application and without notifying the executing source code. No programmatic method exists for detecting or responding to the abort operation. This is a potentially dangerous capability to provide your application's users. Most LabVIEW applications control some type of instrumentation, and many applications are networked or distributed across multiple targets. Abort causes LabVIEW to suddenly stop, regardless of the state of any connected instruments or applications. This can cause the external resources to hang up in some awkward state, possibly requiring a reboot. Additionally, associated memory buffers can become fragmented and open files can be corrupted. Therefore, always hide the **Abort Execution** button from your completed applications and provide a graceful method of closing your GUI VI panels and exiting the application.

 Rule 3.12 Hide the toolbar

Better yet, hide the entire toolbar. When the **Abort Execution** button is not visible, the only tools remaining with runtime functionality are **Run**, **Run Continuously**, and **Pause**. These tools are primarily useful for novice-level training purposes. If you are reading this book, you have probably mastered the While Loop and debugging tools. Also, most top-level VIs are configured with the **Run when opened** Execution property selected. Therefore, the LabVIEW toolbar is usually not needed.

Rule 3.13   Include your company logo for a professional appearance

A nice finishing touch is to apply your company logo to the panel of the top-level GUI VI. LabVIEW can import most common graphic files. Select **Import Picture to Clipboard** from the **Edit** menu and then paste the image onto the panel. I normally apply the logo in a margin near the top of the panel.

3.1.3   SubVI Panel Layout

This section presents rules for the layout of subVI panels, which remain closed during the execution of the application.

 Rule 3.14   Use default appearance for subVI panels, objects, and most text

Because only developers view the panels of subVIs, it is usually not necessary to expend a lot of effort to make them attractive. Instead, it is a convention that subVI panels appear bland. This helps identify them as subVIs. This way, when a developer opens a VI and sees default colors, control appearances, and fonts, she generally recognizes it as a subVI. I recommend expending just enough effort to make subVI panels organized and professional. Use the default appearance for the panels, objects, and most text. This means gray panels sized less than full screen containing controls and indicators that resemble their appearance on the palettes, with 13-point Application font for most text.

 Rule 3.15   Arrange controls to resemble the connector assignments

 Rule 3.16   Resize the panel for a snug fit

The rules for arranging the location of controls and indicators on a subVI panel are similar to and consistent with the rules for connector patterns and terminal assignments that are discussed in Chapter 5, "Icon and Connector." Place controls on the left side of subVI panels, and indicators on the right. I/O name and Refnum controls belong at the top, and error clusters at the bottom. For a dense subVI panel that contains more than eight controls and indicators, arrange the most important controls and indicators in the farthest left and right sides of the panel, respectively. Put the less frequently used controls and indicators in the horizontal center. Arrange the controls and indicators in these locations, resize the panel to fit the objects, and then use the alignment and distribution tools to snap them perfectly into place, for a professional grade subVI panel. When the panel layout is complete, the connector assignments are performed according to the layout.

Figure 3-8 shows two candidate panels for Interval Timer VI, a utility subVI that times an interval. The **Interval** control is the most important parameter, and its corresponding terminal priority is **required**. The **Pause?** and **Mode** controls are the least often used. The panel in Figure 3-8A uses the default panel size and position, and the controls are arranged haphazardly. The **Interval** control is located in the center of the panel and assigned to a middle terminal. This location is well suited for an important parameter on a GUI VI panel but is incorrect for a subVI. The panel in Figure 3-8B has the controls arranged neatly and the panel sized appropriately for the controls. It is organized with the most important controls on the left, starting with the **Interval** control at the top left, and the least important controls in the horizontal center. Error clusters have been added for proper subVI error handling. The connector assignments appropriately reflect the panel layout. Chapter 7, "Error Handling," discusses error handling, and Chapter 5 covers connector assignments and priorities.

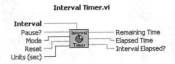

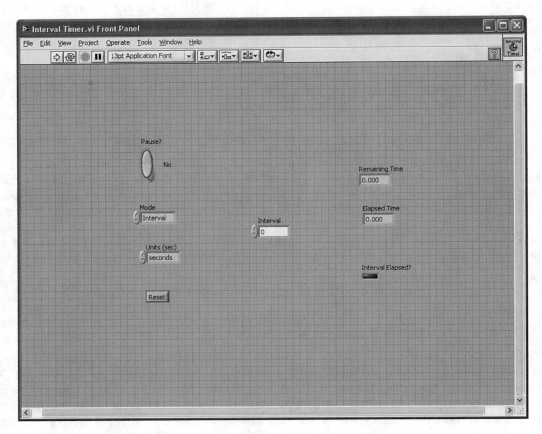

Figure 3-8A
The controls of Interval Timer VI are haphazardly arranged and the panel is oversized.

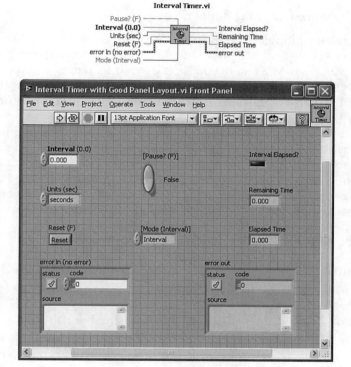

Figure 3-8B
The controls are aligned, distributed, and prioritized via locations and text style. The panel is resized for a snug appearance.

3.2 Text

Text has a substantial bearing on the perceived user-friendliness of software, including the LabVIEW applications that we develop. The following are some specific rules regarding front panel text. We begin with some general rules that apply to all varieties of GUI VI and subVI panels.

3.2.1 General Rules

Have you ever read an entire software license agreement that is displayed in a dialog when installing new software? Or do you instinctively press the **Accept** button? When you begin using new software, do you learn it by reading the documentation, online help, and readme files, or do you immediately operate the GUI by navigating the controls and menus? Have you ever been annoyed by lengthy text on a GUI?

Rule 3.17   Minimize front panel text

Intuitive front panels are graphical, not textual. Users and developers alike do not like to read a lot of text on the front panels. Instead, we prefer buttons, controls, objects, icons, and images that are universal and self-explanatory. As a result, you should avoid sentences, long phrases, and paragraphs of text within labels and controls. It either slows people down or is ignored entirely. The more text your front panels have, the more complicated and user-*un*friendly they appear. There are more appropriate places for lengthy text and documentation.

Sometimes, as a developer is preparing to bring a new application online, there is a tendency to embed instructions on the front panels of the GUI VIs to aid the user in operating the software. However, text that is applied primarily for this purpose is meaningful only for a new user running the software for the first time, at best. Thereafter, the text is probably ignored and a waste of space. A good question to consider is, how often will a new user operate the software, versus a more experienced user? Unless the application is explicitly designed for new or infrequent users, such as a wizard or installation utility, the text should reside elsewhere.

I recommend the following alternatives to lengthy front panel text:

- Create intuitive GUI VIs that are robust and self-explanatory, and do not require textual explanations.
- Provide a Boolean control labeled **Help** or **?**, and a menu labeled **Help** that launches help information on demand, such as an online document, the LabVIEW Context Help window, or a dialog VI with string indicator displaying the text.
- Create a user manual or help topic in an external document or compiled help file, and launch it from LabVIEW on demand.

 Rule 3.18   Delete template instructions immediately after edits are performed

Another situation in which developers use lengthy text is when leaving notes or instructions for other developers on the panels of reusable templates. For example, the LabVIEW instrument driver template VIs contain text labels that tell the developer exactly what changes are required to each VI. Always delete the template instructions immediately after the prescribed edits have been made. Otherwise, the application appears incomplete or raises uncertainty. It is very disappointing to open a VI that contains editing instructions within a completed application.

 Rule 3.19   Apply consistent fonts and capitalization

Most applications require more than one text style to distinguish different types of data and events. For example, on an industrial GUI, it is common to display important data using a large bold style to make it stand out prominently on the panel. Additionally, you might want a large, plain style for headings, or a high-contrast colored style for important messages such as alarms and warnings. However, too many text styles makes the panels appear unnecessarily busy and perhaps even confusing. Most applications require exactly three text styles: a standard style for most text, a large or bold style for emphasizing important data, and a heading or contrast style for special situations. It is important to choose the corresponding fonts and remain consistent throughout the GUI panels of an application.

LabVIEW supports many fonts, but not all are equal. Application, System, and Dialog fonts are not really fonts; they are *symbolic mappings* to fonts that are native to the host computer's operating system. Application font is the standard font the operating system uses for most text. Dialog font is used for dialog control labels and messages. System font is used for window titles and important messages intended to grab attention. The symbolic fonts maintain the look and feel of the operating system for each of these three contexts. However, the actual font used varies based on the operating system's supported fonts and the user's configured preferences. Moreover, Windows currently uses only one font for all three contexts. Specifically, in the United States all three symbolic fonts default to Tahoma under Windows XP, and Segoe UI under Windows Vista. The advantage of the symbolic fonts is that they maintain platform compatibility and portability. Hence, if you port your application to a new operating system, or even a new version of the same operating system, your application maintains the native operating system look and feel. We do not know which fonts Windows and other operating systems might utilize in the future, but the symbolic fonts will map accordingly.

Alternatively, most operating systems support universal fonts such as Arial, Times New Roman, and Tahoma. Choose a universal font instead of a symbolic font if you want to specify the exact font that is used. These fonts generally appear the same on each platform. Hence, a tradeoff exists between native operating system look and feel, and consistent application look and feel. Use the symbolic fonts for the former and universal fonts for the latter.

Sometimes a specialized font is required to achieve a special effect. For example, I use the Symbol font to apply Greek characters to labels indicating engineering units. I use monospace fonts, such as Courier New, to limit the visible characters in a string control or indicator. Beware that special fonts might behave differently on different platforms. Additional maintenance is often required to resize or reposition the text fields or labels. The Greek letter μ, which I use extensively, becomes the letter *m* on some machines. In fact, some of the specialized fonts do not even display consistently between the front panel and diagram windows of the same VI. Specialized fonts are not recommended if you are distributing applications for use on multiple platforms. An alternative is to create the specialized label in an external application, save it as an image file, and import the image into LabVIEW.

 Rule 3.20 Choose only one font, and vary the size, boldness, and color to obtain multiple styles

Application, System, and Dialog are named to connote three different contexts. Yet the world's most popular operating systems, Microsoft Windows XP and Vista, use one font for all three. My recommendation is to choose only one font, either Application font for portability or any universal sans serif font, such as Arial, for consistency. Vary the size, boldness, and color to obtain different styles for the different contexts. Choosing only one font prevents the clashing appearance caused by multiple dissimilar fonts.

Power Monitor System VI contains eight different font styles, as shown in Figure 3-9A. Specifically, the Application font is used with the following variations of point size and color: 42-point bold for the heading **Power Monitor System**, 24-point bold for tab labels, 20-point bold for all indicator labels and data, 13-point bold for **Log Interval** label and data, 13-point bold for **Reset Energy** Boolean text, 13-point plain for **Revision Number** label and data, and 37-point bold red for **Exit** Boolean text. The different variations might have been chosen to distinguish the importance of each control. However, it becomes indistinguishable and distracting to use more than three text styles. Instead, the size and locations of the controls can help distinguish their priorities. In Figure 3-9B, the same panel is revised using three text styles. A 20-point bold Application font is used for all owned labels, data, and

Boolean text. A 20-point plain font is used for minor labels, including units and toggle switch **ON/OFF** labels. The **Power Monitor System** heading remains 42-point bold. This front panel appears neater and more intuitive than the one in Figure 3-9A.

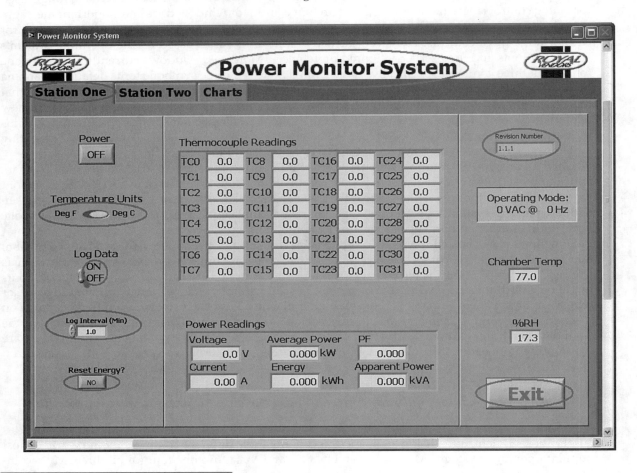

Figure 3-9A
Power Monitor System VI contains eight different styles of text based on variations of point size, boldness, and color.

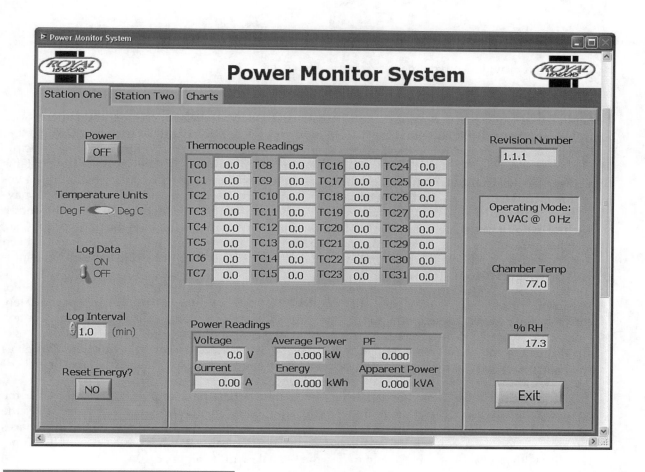

Figure 3-9B
Power Monitor System VI is revised with three fonts applied consistently. It appears neater and more intuitive than Figure 3-9A.

Apply capitalization consistently as well. If you have 12 indicators of similar context and significance on a GUI panel, naturally, if you capitalize the first letter of the label on one indicator, you should capitalize the first letter on all. This sounds like common sense, but it is worth mentioning because I have seen many applications that do not use consistent capitalization. The best way to ensure that your applications use consistent fonts and capitalization is to decide on your scheme before you begin your development. Better yet, consider using consistent schemes across most of your applications. This saves time when selecting fonts, ensures consistent style, and improves ease of use.

3.2.2 Control Labels

This section provides conventions for control label text.

 Rule 3.21 Use succinct, intuitive control labels and embedded text

Let us extend Rule 3.17 to controls. Create succinct and intuitive labels for controls and indicators, as well as any text embedded within them. Challenge yourself to think of one or two simple words that best describe each control, indicator, or designated action. Figure 3-10 illustrates this principle applied to a menu of Boolean controls. The menus are functionally equivalent. Figure 3-10A contains a cluster labeled **COLLECT MENU** and three lines of text embedded in each control, including a two- or three-word name in all capital letters, a phrase that briefly describes the action, and a shortcut navigation key. Figure 3-10B contains maximally succinct one-word embedded labels for each control. The one-word labels more accurately describe the function of each button than the phrases do. Also, the cluster label has been moved from the panel to the window title, and the menu bar has been removed, further minimizing the overall quantity of text. As a result, Figure 3-10B appears more intuitive than Figure 3-10A.

What happens when the user needs more information about the control, such as a description of the action it invokes or the assigned shortcut key? This type of information is more appropriately stored in the control's description and tip strip fields. The description is displayed in the Context Help window when the user drags the mouse over the corresponding control. The description also enables you to enter a full paragraph or more of text that describes the control's function. This gives us the flexibility to provide much greater detail than the embedded text. Unlike a free label or embedded text, the Context Help window can be closed when it is not needed.

The tip strip is a label that appears for a few seconds when the user places the mouse over the control, and then disappears. It is ideal for storing a phrase or one-line sentence that summarizes the control's purpose. Figure 3-10C shows the Context Help window and tip strip when the user places the mouse over the **Account** button. Hence, the succinct Boolean menu has *greater* descriptive functionality with *minimal* static text, compared to the textual menu of Figure 3-10A.

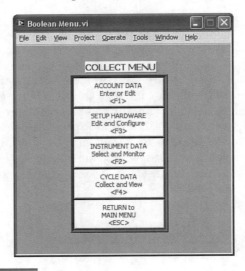

Figure 3-10A
This menu of Booleans contains excessive embedded text.

Chapter 3 • Front Panel Style

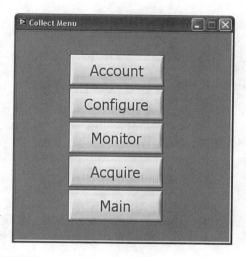

Figure 3-10B
This menu of Booleans contains succinct embedded text, hides the menu bar, and displays the menu name in the window's title.

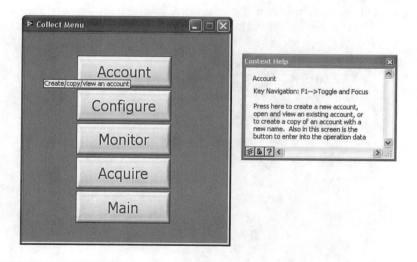

Figure 3-10C
This menu of Booleans has an informative description and tip strip visible.

One more comment regarding control labels. It is generally best to leave the background color transparent. This is the default setting of a front panel option called **Use transparent name labels** (accessed from **Tools»Options»Front Panel**). Older versions of LabVIEW applied raised labels to new controls by default. This is less aesthetically pleasing, and many of us used to manually color the background each of our labels transparent. This is no longer an issue, as long as **Use transparent name labels** is selected and we do not try to impose any unnecessary creativity with the labels.

3.2.3 SubVI Panel Text

This section presents rules for text applied to the panels of subVIs.

 Rule 3.22   Apply 13-point black Application font for most subVI panels

As noted previously, it is not necessary to get fancy with subVI panel text. Therefore, LabVIEW's 13-point black Application font is used in conjunction with default gray panels and default control appearances for most subVIs.

 Rule 3.23   Provide default values and units in parentheses at the end of owned labels

 Rule 3.24   Combine bold text labels with plain text parentheses, for control labels of commercial subVIs

Append the default value and units in parentheses to the end of the labels, if applicable. These helpful label extensions improve the readability of the subVI front panels and cause the data to also appear in the terminal labels displayed in the Context Help window. This helps expedite subVI learning and utilization. Commercial subVIs, such as instrument drivers and developer toolkits, require some additional effort. All control and indicator labels are bold, and default values and units are appended to the labels in parentheses using plain text. This convention is specified in NI's *Instrument Driver Guidelines*[2]. It makes the subVI panels appear more uniform. In Figure 3-11, the Interval Timer subVI from Figure 3-8 has been modified with bold text labels and plain text defaults.

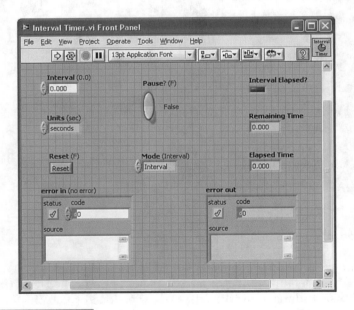

Figure 3-11
The label text style of Interval Timer VI has been modified for a commercial subVI. This consists of bold 13-point Application font, with default values in plain text enclosed in parentheses.

3.2.4 Industrial GUI VI Text

Visibility is very important for industrial GUI VIs. The following rules enhance the overall visibility of front panel text for multiuser industrial applications.

 Rule 3.25   Maximize the contrast between text color and background color

 Rule 3.26   Use large text size for command buttons and critical data

Important data and objects on the GUI VI panels of an industrial application should be readily visible from several feet away. Use large objects and text, and maximize the contrast between the text and background colors. Specifically, when using the default black text color, lighten the shade of the background and increase the size of the text. Increase the font size from the Font Style dialog, or select the text and press the **<CTRL>+<+>** keyboard shortcut. Each instance of the keyboard shortcut adds 1 point to the text size. Alternatively, use white, light, or bright-colored text on a dark background. Since many people have trouble distinguishing certain colors, the level of contrast is more important than the actual colors.

 Rule 3.27   Allow extra space between labels and objects for multiplatform applications

As discussed in Section 3.2.1, "General Rules," font appearance is platform and system dependent. The display resolution, video adapter, driver, and monitor settings, in addition to the operating system and user preferences, all affect the appearance of most fonts. If your application is intended for use on multiple platforms, allow extra space between your labels and other objects, to prevent them from overlapping.

3.3 Color

Most of us have experienced a GUI with a combination of colors that turned our stomachs. Even more common are GUIs with no color at all. Selecting colors is an art. Some great LabVIEW developers are very creative in a technical or scientific sense, but not in an artistic one. Furthermore, approximately 10% of the world's population is **color vision confused**, meaning they cannot differentiate two or more of the primary colors in the spectrum[3]. When you consider the color vision confused along with normal eyesight users operating software with poor color choices, color confusion might be an epidemic. In any event, color is one of the more subjective properties of a panel. Color combinations that are attractive to one person might be unattractive to another. However, some rules can really help.

 Rule 3.28   Apply color judiciously

- Select three complementary colors, including one highlight, one muted, and one grayscale

The GUI VIs that have turned my stomach all had one thing in common: too many bright colors. Try to judiciously select three or, at most, four colors that go well together, and stick with them for all the GUI VIs in your application. One technique that is relatively foolproof is to choose one highlight color, one muted color, and one shade of gray. This is very easy using LabVIEW's color picker, which contains separate rows for each category, as shown in Figure 3-12. The grayscale and muted colors are appropriate for larger objects and the panel background. You can select them from the top two rows of the color picker. The highlight colors are the bright colors, which you can select from the bottom row. Sparingly apply one or, at most, two highlight colors to help draw attention to smaller objects and animation.

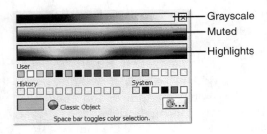

Figure 3-12
The LabVIEW color picker arranges colors into grayscale, muted, and highlight colors.

 Rule 3.29 Create a color theme

- Use common colors to associate related items
- Apply color consistently throughout all VI panels of an application

In Section 3.1, "Layout," we discussed grouping related items on a panel to indicate association, and spacing unrelated items to help disassociate them. Similarly, use color to reinforce and extend the association provided by the panel layout. Create a color theme that includes color coding of related items on a panel. Moreover, *extend* the color theme consistently throughout all the GUI VIs in an application. Consider providing a color legend to formally document the color theme.

 Rule 3.30 Follow universal conventions for green, yellow, and red

Green, yellow, and red have definitive meanings in most industrial environments. Green indicates normal operating conditions, yellow indicates caution or warning, and red indicates alarm or emergency. If your VI is an industrial GUI, heed these universal conventions.

 Rule 3.31 Leave the panel and objects of subVIs gray

Panels of subVIs need not be colored. As stated previously, it is a convention that subVI panels maintain their default appearance to help identify them as subVIs. If you are developing a commercial product such as a developer toolkit or instrument driver, you might want to incorporate a modest color scheme to help associate the panels of the toolkit. In this case, I recommend using a single color to add some limited color markings. However, attractive and meaningful icons are much more important with commercial subVIs because icons are visible from the caller's diagram. Therefore, I recommend investing time in a nifty icon convention and leaving the panels and objects with their defaults.

Rule 3.32 Keep the color schemes simple and time limited

Some developers might find themselves expending substantial time and effort experimenting with color schemes in search of the optimal GUI. Prioritize the application's functional requirements, including the GUI appearance, and budget your time accordingly. For commercial applications, a professional GUI is important, and effective color schemes are often a high priority. However, the primary purpose of most LabVIEW applications is the underlying tests, measurements, and control. Be sure your application functions properly before you expend excessive time and effort on color schemes. My preference is to identify, develop, and test the requirements in order of priority. Also, I often start with a combination of colors that I have used successfully in the past, and if time permits after the critical requirements have been satisfied, I add further enhancements.

Section 3.5, "Examples," presents specific examples of GUI VI and subVI panels. Because this book is printed in black and white, you cannot see the full effect of the color examples. Color versions of those example panels can be freely downloaded from the Publisher's website page for this book at `www.prenhallprofessional.com/title/0131458353`. An electronic (eBook) version of this book containing full-color illustrations is available for purchase from `www.prenhallprofessional.com/title/0132414813`.

3.4 GUI Navigation

Theorem 3.1: *Reliability is the developer's responsibility.*

GUI navigation pertains to how the user interacts with the controls and menus of a GUI VI to operate the software. If you are developing an application for use by anyone other than yourself, reliability and ease of use are extremely important. Moreover, it is important to recognize that reliability is the *developer's* responsibility. Never rely on the user to operate the GUI in the manner that you intend. Never present the user with options that could cause the software or hardware to malfunction. For each GUI VI, consider what combination of operator actions could cause a problem. For example, if the GUI VI contains a button for resetting a set of instruments, what would happen if the user clicked this button during a test or measurement sequence? Unless the control is disabled, your application or one or more of the instruments might hang up. I would like to make a few points, one philosophical and the others practical. On the philosophical side, it should never be the user's responsibility to avoid certain control values or operations. It is the developer's responsibility to ensure that the user cannot cause a malfunction inadvertently. LabVIEW provides the developer with all the tools needed to develop a reliable GUI. For the previous example, the button for resetting the instruments should not be active during the measurement sequence.

3.4.1 Control Scope

 Rule 3.33 Limit the number of controls that are visible and enabled at any one time

In general, limit the controls that are available to the user to controls that are relevant and will not trigger any undesirable actions. Consider the different operating modes of a typical application, such as configure, acquire, analyze, and present. Controls that are required during the configuration mode might not be relevant during the latter three modes. Consider using separate GUI VI panels or a tab control to group controls that the user needs access to on separate panels or tabs for each mode. The application programmatically selects the desired panel or tab and prevents the user from changing it, or limits which panels can be navigated during a given mode. Specifically, use the Visible and Disabled control properties to programmatically show, hide, enable, and disable tabs or controls based on the current context or state of the application. Hence, once the application begins acquiring, the user is not allowed to access the Configure panel. Section 3.5.2, "Dialog Utility VI," presents an analogous GUI with tab control.

Additionally, a GUI VI might have some parameters that are applicable for specific configurations. For example, consider an application that performs current, voltage, capacitance, or resistance measurements using a digital multimeter. The panel has a measurement configuration dialog VI that contains a **Resistance Type** control with selections for **2-wire** and **4-wire** measurements. Because this control applies only to resistance measurements, it is inactive when the measured parameter is not resistance.

At the same time, avoid excessive showing, hiding, or positioning of the controls. Your GUI VI panels should not resemble a magic show. An application is easy to use if the objects are present in consistent locations that are easy to find. Consistency is the key. My favorite strategy is to keep the inactive controls visible and to alternate the value of the Disabled property between 0-Enabled, and 2-Disabled and Grayed Out. This way, the operator always knows what parameters exist and where to find them. In the resistance measurement example, the **Resistance Type** control remains visible at the same place at all times, but it alternates between Enabled and Disabled and Grayed Out.

 Rule 3.34   Restrict the range of all controls to values that are relevant to the application

 Rule 3.35   Set the Data Range property of numeric controls

 Rule 3.36   Use ring or enumerated controls over string controls when feasible

Another consideration is what happens if the user enters *any* value into a numeric or string control. Are there some values that do not make sense, that will throw an error, or that will cause some other problem? This issue is usually easy to resolve for a numeric control. Simply specify the control's **Data Range** settings from the **Properties** dialog or programmatically using a Property Node. For a string control, you might need to create a subVI that searches for invalid characters or combinations. LabVIEW has many string functions, such as Match Regular Expression and Search and Replace String, that enable you to create a custom filter for parsing invalid characters. If the number of permissible string values is discrete, consider a text ring, list box, enum, or combo box control. Enter each valid string as a separate item, forcing the user to select from a discrete list of valid strings. The former three controls map the strings to integers; the combo box maps a string to the same or another string. Text ring controls are preferred for instrument drivers, per NI's *Instrument Driver Guidelines*[2]. Personally, I am partial to enums because they have a special property when wired to a Case structure's selector terminal: the enum strings become the Case structure's selector labels. This construct is called an **enumerated Case structure**. Chapter 4, "Block Diagram," and Chapter 9, "Documentation," discuss this in more detail.

When using a text ring, list box, or enum over a string control, the string of the selected item can be accessed on the diagram by indexing the control's Strings[] property. This is shown in Figure 3-13.

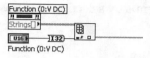

Figure 3-13
The string selected from a text ring is indexed from the Strings[] property using the value read from the terminal.

 Rule 3.37   Set the tabbing order and use the Key Focus property to aid navigation

If you would like to guide the user through the GUI, use a combination of tabbing order and the Key Focus property. The tabbing order specifies the order in which controls are traversed using the Tab key. Set the tabbing order by selecting **Edit»Set Tabbing Order**. Key Focus is a control property that highlights a specific control and makes it capable of receiving input from the keyboard without the user having to select it first. For example, a common technique for dialog VIs is to use Key Focus to programmatically select and highlight an initial control; then the user navigates to subsequent controls using the Tab key, in the order specified by the tabbing order. If you have any controls that are not intended to be operated by the user, such as an error cluster or some controls that have been scrolled off the visible area of the panel, remember to remove them from the panel's tab order. This is accomplished from the control's properties dialog, which is opened by selecting **Properties** from the control's shortcut menu. Select the **Key Navigation** tab and choose **Skip this control when tabbing**. Finally, specify key assignments for various actions of any controls that users must operate without a mouse. Control actions vary by control type. Key assignments are specified from the **Key Navigation** tab.

 Rule 3.38   Customize the runtime menus for top-level VIs

 *Rule 3.39   Always include a **Help** menu or button*

LabVIEW's default menu bar is marginally useful, at best, for applications deployed to users. The majority of the selections are intended to help developers edit source code within the LabVIEW environment. These selections are confusing for end users who are unfamiliar with LabVIEW. Instead, create custom runtime menus for the top-level VIs that are relevant to the application. Eliminate all the developer items and provide only the specific items required for the user to operate the application. At a minimum, always include a **Help** menu with functional selections for providing the user help. This might include **Show Context Help**, as well as one or more online documents. The **Show Context Help** menu item opens the Context Help window, which displays any descriptions available from the controls, indicators, and VI as the user scrolls the mouse over them. Compiled help files are documentation files compiled into CHM format, as discussed in Section 3.2.2, "Control Labels." Custom runtime menus are configured by selecting **Edit»Run-Time Menu**.

Menus are not appropriate with dialog VIs. Instead, provide Boolean command buttons for any required menu selections. This includes a Boolean for **Help** or **?** that can launch the Context Help window or an online document. Also, create shortcut menus for any actions related to a specific control. You can create shortcut menus by selecting **Advanced»Run-Time Shortcut Menu»Edit** from a control's shortcut menu.

3.4.2 Consistency

You might not apply or agree with every rule I have presented in this chapter, and that is fine. GUI style, in particular, is a subjective topic. Moreover, the most important rule regarding front panel style is to be consistent in whatever you do.

Rule 3.40   Be consistent!

- Maintain consistent appearance and location of similar controls
- Save controls with customized appearance as strict type definitions

If you choose a particular layout or set of fonts, colors, and control types, use them consistently throughout your application—and perhaps multiple applications. This will make it easy for the users to operate the software. By learning how to operate one panel, users should understand where to look for certain controls and indicators, and how to navigate the GUI. By using similar layouts, fonts, and colors, you reinforce to the user that each panel behaves similarly. Additionally, it is beneficial to be consistent throughout an organization. This can involve collaborating with your peers to set some GUI standards. Hopefully, this book provides a good starting point. If you create a series of VIs that have some common functionality, use identical controls to expose that functionality on each panel and keep those controls in identical locations for each VI. For example, if you have a succession of dialog VIs such as a wizard, which you can navigate or exit by clicking buttons labeled «**Back**«, »**Next**», and **Cancel**, apply the same buttons at the same locations of each VI. Consistency aids GUI navigation.

You can use LabVIEW's control editor to customize the appearance and properties of a control, specify the **Type Definition Status**, and save it to a CTL file. You access the control editor from the control's shortcut menu by selecting **Advanced»Customize**. A custom control or type definition enables you to reuse the customized control in multiple places throughout the GUI and maintain a consistent data type and appearance. Simply add the type definition to your LabVIEW project and drag and drop it from the project tree to your VI front panels. The control loads from the file that contains the customized properties specified in the file.

The **Type Definition Status** is specified using the ring control at the top of the control editor. The choices are **Control**, **Type Def**, and **Strict Type Def**. A control is simply a custom control with appearance and properties specified in the control file. The custom control can be edited on the panel after loading from file, without affecting the appearance, properties, and data type of the source file. Each instance of a type definition can be customized as well, with the exception of the data type. A strict type definition maintains the exact appearance, properties, and data type that are defined in the source file, for each instance of the strict type definition.

I normally create multiple instances of the controls that I customize and often make changes to their appearance. Within a project, I normally need the changes to be made to all instances. Therefore, I prefer to use the strict type definition. Instead of searching for every instance of the control, edit the strict type definition; all instances then are updated immediately. Also, because a strict type definition maintains the appearance of the control, the strict type definition is appropriate for maintaining consistent appearance with GUI VIs. Consistency aids GUI navigation, while inconsistency hinders GUI navigation. Chapter 6, "Data Structures," discusses type definitions in more detail.

3.5 Examples

This section presents specific examples of GUI VI and subVI front panels. Color versions of these panels can be freely downloaded from the Publisher's website page for this book at www.prenhallprofessional.com/title/0131458353. An electronic (eBook) version of this book containing the full-color images is available for purchase from www.prenhallprofessional.com/title/0132414813.

3.5.1 SubVI from Selection

A subVI from selection is created by selecting a section of the block diagram and choosing **Edit» Create SubVI**. Like magic, the selected code is replaced with a new subVI. This is a nifty tool but is also the world's most flagrant style offender. The resulting front panel, block diagram, icon, and connector pane require significant manual cleanup. For example, the front panel in Figure 3-14A violates several rules. The controls are arranged horizontally instead of resembling the connector assignments (Rule 3.15). Standard controls, such as task IDs and error clusters, are not in their standard locations at the top and bottom of the panel (Rule 3.40). The panel is not sized for a snug fit (Rule 3.16). Moreover, the control labels are counterintuitive (Rule 3.21). All three controls, which are the inputs to the subVI, contain **out** or **output** in their labels. A number is used to distinguish similarly named controls and indicators. Consequently, manual cleanup is essential. Figure 3-14B addresses these issues. The block diagram, icon, and connector pane of this example are discussed in Chapters 4, 5, and 9.

> **Application:** SubVI writes audio data to the computer's sound card
> **Conditions:** Created using the subVI from selection utility and contains multiple style rules violations.
> **Color scheme:** Default appearance of panel and all controls
> **Layout:** Controls are arranged horizontally instead of resembling the connector assignments
> **Heading text:** Not applicable
> **Boolean text:** Not applicable
> **Numeric and String Control Data:** Default 13-point normal black Application font
> **Secondary Fonts:** Not applicable
> **Navigation**: Not applicable for a subVI

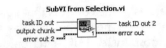
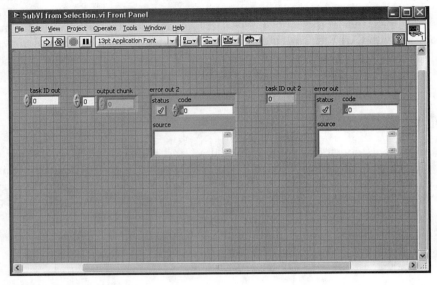

Figure 3-14A
SubVI from Selection VI violates rules 3.15, 3.16, 3.21, and 3.40.

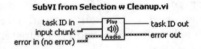
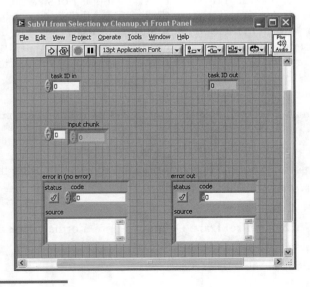

Figure 3-14B
SubVI from Selection w Cleanup VI conforms to the rules for subVI front panels.

3.5.2 Dialog Utility VI

The RS-232 Parse and Transmit Utility VI is a dialog VI that prompts the user to configure a serial messaging protocol. Figure 3-15A has characteristics similar to a dialog, such as the window appearance, layout of controls, and data types. However, the control styles, fonts, decoration, and colors do not conform to standard dialog conventions. Specifically, 13-point plain black Application font is used for all control data, and 16-point plain black Trebuchet MS is used for all labels, instead of using Dialog fonts for both. The panel is colored dark gray and sized small to resemble a dialog. The control labels and Boolean text are legible because of high contrast between black text and light gray and white backgrounds, but the black control data text is not sufficiently legible on dark gray background. The controls are navigated using the Tab key.

> **Application:** Utility for configuring periodic messages transmitted to an instrument
> **Conditions:** Not conforming to dialog conventions
> **Color scheme:** Light gray decoration on dark gray panel with dark gray controls and black text
> **Layout:** Related controls grouped together using a decoration
> **Heading text:** 16-point plain black Trebuchet MS
> **Boolean text:** 13-point normal black Application font on white background
> **Numeric and String Control Data:** 13-point normal black Application font on white background
> **Secondary Fonts:** Not applicable
> **Navigation:** Tab key navigation

Figure 3-15B contains a functionally equivalent dialog VI that conforms to dialog style conventions. The **Dialog** option for **Window Appearance** properties has been chosen, and all controls and decorations are selected from the **System** palette. A phrase describing the dialog's function replaces the VI name in the title bar. Note that the VISA Resource Name control has been replaced with a System Combo Box, to maximize the native dialog appearance. Also, the **Quit** and **OK** buttons have been replaced with **OK** and **Cancel** in the appropriate locations.

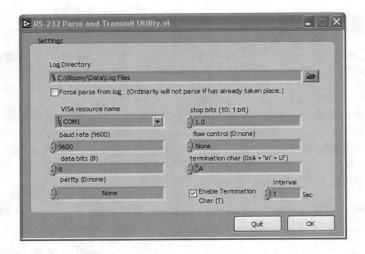

Figure 3-15A
RS-232 Parse and Transmit Utility VI does not have a native operating system appearance.

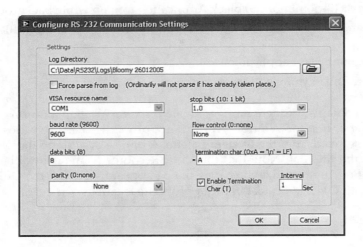

Figure 3-15B
This revision uses Dialog window title and appearance, Dialog fonts, and System controls conforming to the conventions for good dialog VI style.

3.5.3 Capacitor Test & Sort

Figure 3-16A shows the main GUI VI panel of a Capacitor Test & Sort application. This is an industrial test application operated by semiskilled users in a production environment. The layout is logical and intuitive, with heading data at the top, a graph prominently positioned in the center, statistical data grouped in clusters below the graph, and large Boolean controls for navigation at the bottom. There is excellent contrast between the foreground text and background color, such as the white text on the charcoal and blue backgrounds. However, it contains more fonts than necessary. For example, the four large Boolean controls contain four different font styles. **STOP** is a single succinct and intuitive word that uses 29-point bold Dialog font and all capital letters. **Change Work Order** uses 16-point bold font, **Print** uses 29-point bold Dialog font, and **Exit System** uses 20-point bold font. Hence, there are four different fonts and four different succinct/wordy/caps text conventions on four similar command-style Boolean controls. The following rules are violated by the Booleans alone: 3.17, 3.19, 3.21, and 3.40. Additionally, the panel contains the standard LabVIEW menu bar, which is not appropriate for a deployed application because most of the menu selections apply only to developers. Finally, a numeric indicator and Boolean LED overlap a graph indicator, violating Rule 3.11.

> **Application:** Capacitor Test & Sort top-level VI panel.
> **Conditions:** 1024×768 resolution, medium object density.
> **Color scheme:** Blue, charcoal, white.
> **Layout:** Graph is prominently sized and centered on the panel. Numeric indicator and Boolean LED overlap a graph indicator, in violation of Rule 3.11.
> **Boolean text:** Four different fonts and four different succinct/wordy/caps conventions. White text on charcoal background provides good contrast.
> **Heading text:** 16-point bold white Application font on charcoal background.
> **Numeric Control Data:** 16-point normal black Application font on white background.
> **Secondary fonts:** Control labels and heading data are 13-point black Application font. Labels are bold; data is normal.
> **Navigation:** Five large Boolean controls at the bottom of the panel and the standard menu bar facilitate navigation. The standard menu bar is not appropriate for the application's users.

Figure 3-16B contains the same panel, with improvements made to the Boolean text, menu bar, and overlapping indicators. All Boolean controls contain a single succinct command word with first-letter capitalization, using 29-point bold white Dialog font. A custom menu bar has been applied providing selections for **File**, **Work Orders**, **Print**, and **Help**. These selections further assist the application's users in navigating the software. Additionally, the numeric indicator and LED have been moved above the graph, eliminating the overlap. This improves efficiency by reducing the required drawing operations.

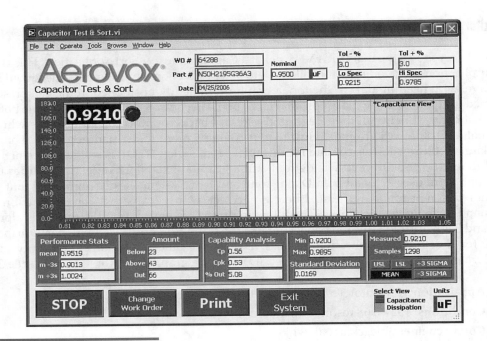

Figure 3-16A
The main GUI VI panel of a Capacitor Test & Sort application has a logical and intuitive layout, but it violates several rules regarding text, menu bars, and overlapping controls.

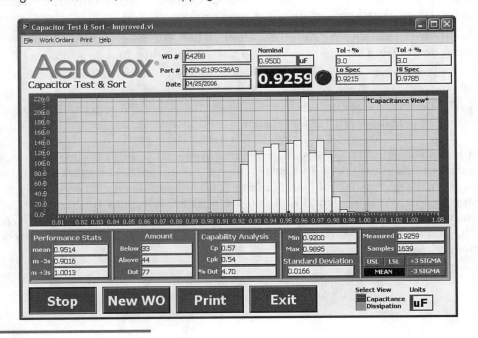

Figure 3-16B
The panel is improved, with succinct Boolean commands and consistent fonts, a custom menu bar for user navigation, and nonoverlapping indicators.

3.5.4 Centrifuge DAQ

Figure 3-17 shows the GUI VI panel for a centrifuge data acquisition (DAQ) and control system. This application is used to perform experiments on soil used for earthquake research. The operators are engineers and scientists. It has hundreds of configurable parameters. The corresponding controls are organized using a nested tab control. The outermost tab control is visible and directly operated by the user. It organizes controls into the categories **Configure**, **Monitor**, **Acquire**, and **Analyze**. A second tab control is embedded within the **Configure** tab and colored transparent. It organizes DAQ channel configuration parameters by transducer type and is programmatically navigated based on the value of the **Sensor Selection** control. Chapter 6 examines the embedded tab control and corresponding data. Aside from the various logos, this application uses only the Application font, with three different styles.

Application: Centrifuge DAQ top-level VI panel.

Conditions: 1024×768 resolution, high object density.

Color scheme: Maroon, gray, white.

Layout: Nested tab control with outermost tab control organizing controls into tabs for each operational mode. Additionally, hundreds of configurable parameters are organized using an invisible tab control within the **Configure** tab.

Heading text: Tab control headings are 18-point bold black Application font on a default gray background.

Boolean text: 13-point bold black Application font on default gray background.

Numeric and String Control Data: 13-point normal black Application font on a white background.

Secondary Fonts: 13-point normal white Application font on a maroon background.

Navigation: A combination of menus, tab control, listbox control, and Boolean command buttons are used for navigation.

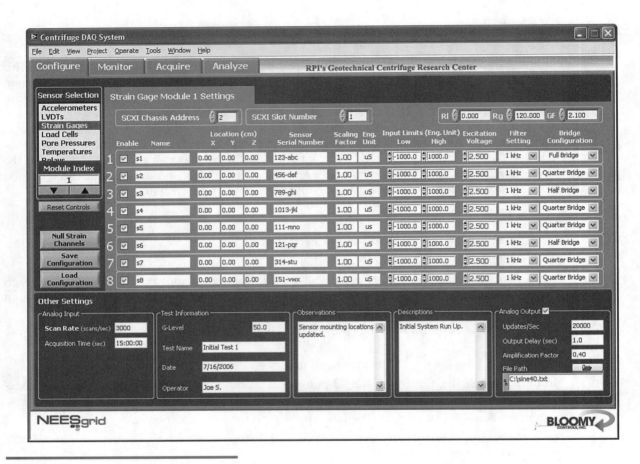

Figure 3-17
Centrifuge DAQ research application for seismic event simulation

3.5.5 Spectralyzer

The GUI VI panel shown in Figure 3-18 is a commercial application for diagnosing mechanical wear of industrial equipment. The operators are primarily maintenance technicians. A tab control groups relevant controls together. The FFT graph is prominently displayed in the middle of the screen. Note that the Boolean controls in the upper-left corner are always visible and are programmatically **Disabled and Grayed Out** when not active. The company logo appears in an open area on the right side of the graph. The GUI conforms to most of our rules and appears very intuitive.

Application: Virtual spectrum analyzer.
Conditions: Multiple display types and resolutions, medium object density.
Color scheme: Blue, white, default gray.

Layout: Graph is prominently sized and centered on the panel. Navigation is restricted via the Disabled property.
Heading text: 18-point normal black Application font on a light gray background.
Boolean text: 13-point normal black Application font on default gray background.
Numeric & String Control Data: 13-point normal black Application font on white background.
Secondary Fonts: 12-point normal black Application font on white or gray background.
Navigation: Tab control and Boolean command buttons.

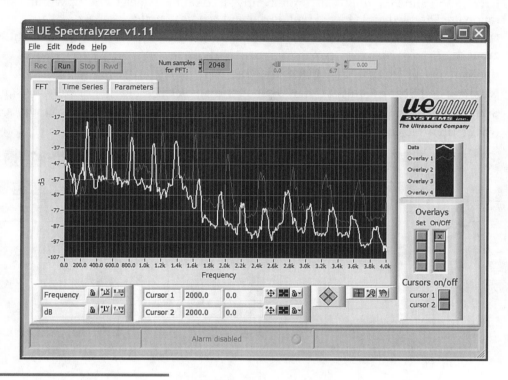

Figure 3-18
Spectralyzer is a commercial application for diagnosing mechanical wear of industrial equipment. The GUI conforms to most of the rules and appears very intuitive.

3.5.6 Parafoil Guidance Interface

The Parafoil Guidance Interface, shown in Figure 3-19, is a flight control interface that allows a pilot on the ground to interactively fly and collect parafoil performance data. The GUI resembles a virtual cockpit that contains four groups of indicators: Vertical and Horizontal Situation Indicators (**VSI**, **HSI**), **Status**, and **Control**. Each group of indicators is delineated spatially and using decorations. Simple numeric and string indicators, which contain minimally thin borders, are used to maximize the density of indicators without sacrificing visibility. Arial font is used throughout the application to maintain consistent alignment and appearance, regardless of the computer, operating system, and user preferences. Normal black text is used on a white or very light gray background to maximize the contrast and clarity. Icons are effectively used to maximize ease of use.

Application: Parafoil guidance interface.

Conditions: Virtual cockpit with maximum object density and visibility.

Color scheme: White or light gray background with black text.

Layout: VSI and HSI graphs are prominently sized on the panel. Controls within **Control** decoration are Disabled and Grayed Out during auto control mode.

Heading text: 15- and 18-point bold black MS Sans Serif font on a white or light gray background.

Boolean text: 13-point normal black Arial font on default gray background.

Numeric and String Control Data: 13-point normal black Arial font on white background.

Secondary Fonts: 14-point normal black Arial font on white or gray background.

Navigation: Tab controls and menu bar.

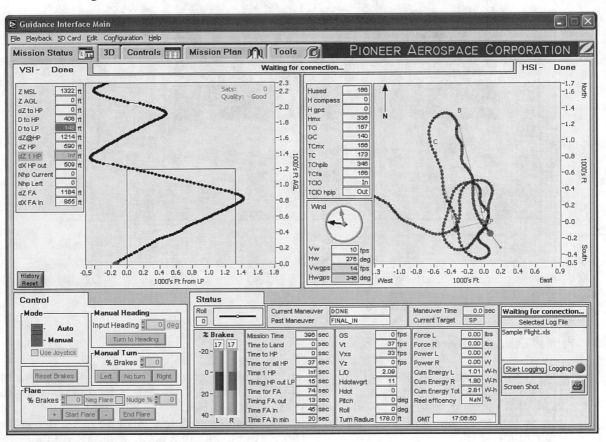

Figure 3-19
Parafoil Guidance Interface is a virtual cockpit. It contains a high density of data while maintaining good readability.

Endnotes

1. Ritter, David. *LabVIEW GUI Essential Techniques*. New York, NY: McGraw-Hill, 2002.
2. *Instrument Driver Guidelines* online tutorial, available from NI Developer Zone.
3. Rosenthal, Odeda, and Robert H. Phillips. *Coping with Color-Blindness*. New York, NY: Avery Publishing Group, 1997.

Block Diagram

4

The LabVIEW block diagram excels at conveying source code. A really good diagram is enlightening, even awe-inspiring, like a work of art. A careless diagram, however, can appear as jumbled as a bowl of spaghetti. Indeed, these two extremes are depicted by Meticulous VI and Spaghetti VI in Chapter 1, "The Significance of Style." Somewhere in the middle between artwork and spaghetti is where most applications reside. Some developers have neat wiring practices but large, flat diagrams. Others have overly modular diagrams that disguise the architecture. Still others prefer variables over data flow. Many, many developers skimp on documentation to save time. Moreover, most diagrams are characterized by tradeoffs between good style and shortcuts deemed necessary to get the job done. The overall outcome is a compromise among attractive appearance, personal preferences, and functional performance.

> ***Theorem 4.1:*** *Great LabVIEW diagrams can be expeditiously developed.*

Many developers wrongfully assume that attractive diagrams require a level of toil that is impractical for real-world applications that have tight deadlines. It seems faster and more productive to avoid getting caught up in diagram aesthetics. Indeed, it is possible to expend excessive time optimizing the appearance of a complex diagram, and most of us must plead guilty for doing this on occasion. However, it is always much more time consuming, in the long run, to debug and modify sloppy code. Per Theorem 1.1, applying good style significantly reduces time and effort throughout an application's life cycle. Additionally, neat development practices need *not* be overly time consuming. If you know the style rules and how to implement them, you eliminate the toil.

This chapter presents style rules that ensure neat and organized diagrams that are practical to implement in real applications with tight deadlines. Combined with the rules in other chapters, they ensure readable and maintainable LabVIEW source code. Moreover, mastery of these style rules may lead to *awe-inspiring* LabVIEW diagrams.

4.1 Layout

This section covers rules for block diagram layout, including layout basics, and subVI modularization.

4.1.1 Layout Basics

The following rules pertain to the general layout of the block diagram.

Rule 4.1   Use 1280 × 1024 display resolution

The display resolution affects the visible area the developer has to work with and how the diagram appears when opened on a given target computer. It is beneficial to standardize on one display resolution so that the diagram window maintains a consistent appearance when opened on PCs with similar display capabilities. The higher the resolution setting, the smaller the diagram objects shrink relative to the screen size, and the more code fits on one screen. A fairly high resolution is recommended to maximize the viewable diagram area without straining your eyes. The LabVIEW development environment is designed for a minimum 1024×768 resolution. A resolution of 1280×1024 provides additional real estate while maintaining compatibility with mainstream PC display technology. Avoid resolutions much higher than 1280×1024 because higher resolutions are less universally supported, and the larger work area promotes larger diagrams and potentially less modularity. Also, depending on the monitor size, very high resolutions may strain your eyes. Although I have 20/20 vision, several years ago, I went through a phase where I wore tinted prescription glasses during LabVIEW development. An adjustment to the resolution setting, along with general improvements in display technology, eliminated this problem for me.

Today many computers support multiple monitors. It is particularly useful to utilize two monitors for LabVIEW development. This allows you to dedicate one monitor to the front panel and the other monitor to the block diagram, and have both windows simultaneously visible without having to navigate between them.

Rule 4.2   Leave the background color white

Rule 4.3   Use a high object density

Rule 4.4   Limit the diagram size to one visible screen, or limit scrolling to one direction

Chapter 4 • Block Diagram

Do not color the diagrams. Leave the background of the diagram, and every subdiagram of every structure, default white. Data flow must be easy to visualize. A high density of objects is desired, without crowding objects too close and causing wires and objects to overlap. In general, try to limit the diagram size to one display screen. In some situations, it is difficult to work within this constraint, such as a complex diagram containing multiple parallel loops. In this case, organize the large diagram so that it may be viewed by scrolling in only one direction, or modularize the loops into subVIs to reduce space. Loop-subVIs are discussed in Chapter 8, "Design Patterns." Avoid large diagrams that require both horizontal and vertical scrolling because this is cumbersome to navigate.

Figure 4-1 contains two functionally equivalent implementations of a VI that evaluates a calibration interval. The diagram of Figure 4-1A is overly dense and sloppy. As you can see, several of the functions and wires overlap, and the diagram appears confusing. In Figure 4-1B, the same VI is revised for improved readability. The functions are neatly spaced, the wiring is clear, a few comments and enumerations are included, and the error cluster propagates throughout the diagram. The implementation of Figure 4-1B is much more readable than Figure 4-1A. We revisit this example very shortly.

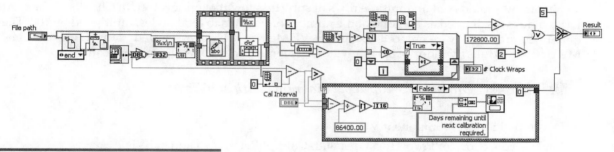

Figure 4-1A
The diagram for a VI that evaluates a calibration interval is overly dense and confusing.

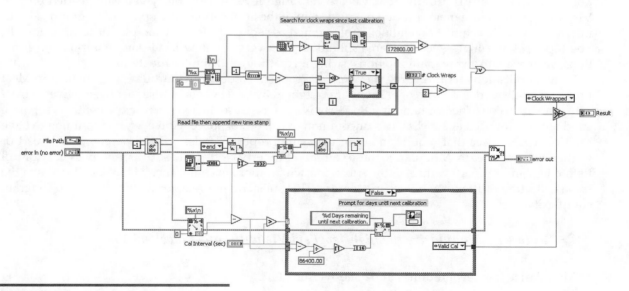

Figure 4-1B
A different implementation of the Calibration Interval VI contains appropriate spacing, clear wiring, and documentation.

4.1.2 SubVI Modularization

If your diagrams commonly extend beyond one window, maximized on a monitor with 1280×1024 resolution, your source code is not sufficiently modular. If the Navigation Window is integral to your development, you *definitely* need more subVIs!

 Rule 4.5   Create a multilayer hierarchy of subVIs

- Strive for a modularity index greater than 3.0

Develop your applications as a multilayer hierarchy of subVIs using a combination of top-down and bottom-up design and development techniques. The VI Hierarchy is viewed by selecting **View»VI Hierarchy**. Deselect **Include VI Lib**, **Include Globals**, and **Include typedef** from the toolbar to remove these items from the window, and view only the hierarchy of user VIs that you provided. Common geometries include pyramid, diamond, and oval. Except for very simple applications, the VI Hierarchy should contain multiple rows of subVIs below the top-level VI. In Chapter 1, the modularity index was defined as the ratio of the number of user VIs to total nodes, multiplied by 100. These quantities are quickly referenced using the VI Metrics window, selected from **Tools»Profile»VI Metrics**. A modularity index of 3.0 or greater is recommended for a typical application.

 Rule 4.6   Modularize top-level diagrams with subVIs

- Develop high-level component VIs
- Replace collections of Property Nodes with Control References and subVIs

Depending on the design pattern, most top-level diagrams should consist of structures, wires, component VIs, and subVIs. **Component VIs** are very high-level subVIs, or dynamically loaded plug-in VIs, that encapsulate a major portion or subsystem of the application. An application's graphical user interface and data acquisition engine, implemented as separate VIs, are examples of component VIs. The top-level and high-level diagrams should contain very few low-level data-manipulation functions, such as math, array manipulations, string formatting, and similar functions.

Some applications require large numbers of Property Nodes for controlling GUI behavior. Most Property Node read and write operations are triggered by GUI events. Consequently, the Event structure is an ideal construct for handling Property Nodes. Because the Event structure contains separate subdiagrams for each event case, it is uncommon to run out of space. However, it is common to have multiple event cases that require many of the same Property Nodes, with different values read or written to them in each. Modularize these common Property Nodes into subVIs, and pass the Control References and property values to the subVI in each location. Each instance of the subVI refers to the same collection of Property Nodes in memory. This substantially reduces memory use and diagram complexity.

 Rule 4.7   Modularize the high-level subVIs with lower level subVIs

- Modularize low-level routines into cohesive subVIs
- Use or develop instrument drivers and utility VIs

Likewise, modularize the diagrams of your high-level component VIs into lower-level subVIs. Using the top-down design and development approach, modularize any low-level routines into cohesive subVIs. Anywhere you have a collection of related functions that work together to perform a specific routine, replace them with a subVI. A subVI is cohesive if you can clearly describe its purpose in two or three sentences, such as when entering the subVI's description.

> **Cohesion Test:** A subVI is cohesive if its purpose is clearly described using two or three sentences.

Also, using the bottom-up approach, develop or reuse instrument drivers and utility VIs. Instrument drivers encapsulate the low-level device communications, including command string assembly, VISA functions, and response string parsing. Utility VIs complement or extend the capabilities of the built-in LabVIEW functions and VIs available on the **Functions** palette. Thousands of reusable instrument drivers and utility VIs are available for free download on the Web[1]. Code reuse is discussed in Chapter 2, "Prepare for Good Style."

In Figure 4-2, we revisit the calibration interval diagram from Figure 4-1. This is an example of a high-level subVI. Inspecting Figure 4-2A, we observe three low-level routines: reading and appending the time stamps to file, searching and counting clock wraps, and prompting the number of days until the next calibration. In Figure 4-2B, each routine is modularized into a cohesive subVI. The subVIs remain in the same respective locations as the code they encapsulate. In Figure 4-2C, the three subVIs are placed in series, the error cluster propagates between each subVI, and the Merge Errors VI is eliminated. The resulting diagram is now a simple set of three subVI calls. We can see that the diagram is much neater, simpler, and organized with subVIs.

   Rule 4.8   Do not create subVIs just to save space

   Rule 4.9   Avoid trivial subVIs containing few nodes

SubVIs are advantageous because it is easier to develop, test, debug, maintain, and reuse software modules as subVIs versus sections of a large diagram. As shown in Figure 4-2C, they also provide a considerable space-saving benefit. In general, if the collection of functions or Property Nodes or other routine is used in more than one place, it is an easy decision: Replace the repeated code with a subVI. Likewise, if several nonrepetitive nodes are related to one another and work together to perform a specific task, modularize them into cohesive subVIs, whether they are needed in multiple places or not. However, do not randomly select areas of the diagram and create subVIs just to save space. SubVIs created in this manner are not cohesive, intuitive, or reusable. Also, do not create trivial subVIs that contain few nodes. In this case, the subVI icon unnecessarily masks the underlying code on the diagram. For example, the subVI in Figure 4-3 contains only a single Index Array function. However, the subVI's icon, name, and description disguise the function. The LabVIEW functions and shipping subVIs within **vi.lib** are universally recognizable, so avoid masking them within trivial subVIs.

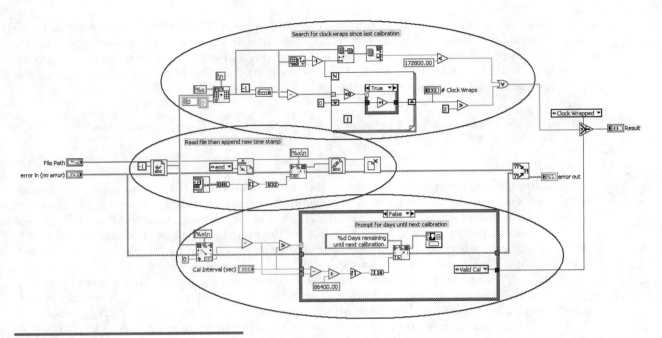

Figure 4-2A
The calibration interval VI contains three routines.

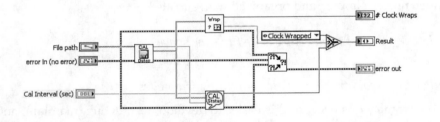

Figure 4-2B
Each routine is modularized into cohesive subVIs.

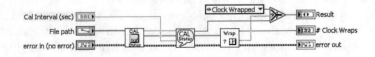

Figure 4-2C
The subVIs are rearranged into a dataflow sequence that propagates the error cluster and reduces wire clutter.

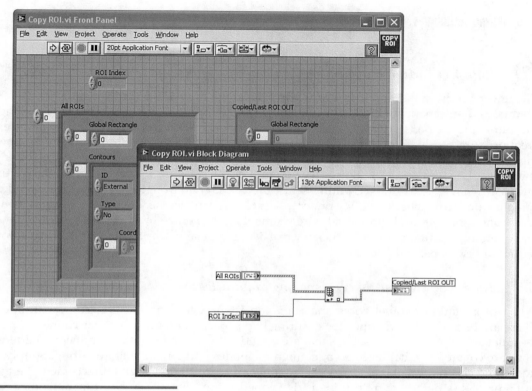

Figure 4-3
This subVI contains too few nodes. Its icon, name, and description disguise the Index Array function on the diagram of the calling VI.

 Rule 4.10   Create a meaningful icon and cohesive description for every subVI!

Always create a meaningful icon and cohesive description for every subVI. I cannot emphasize this enough. At Bloomy Controls, this is one of our most sacred precepts. The icon and description identify the subVI from the diagram of the calling VI through the Context Help window. The description enforces the cohesion test. If you cannot summarize its purpose in two or three sentences, it probably contains too much code for one subVI. Icons and descriptions are discussed in more detail in Chapter 5, "Icon and Connector," and Chapter 9, "Documentation," respectively.

4.2   Wiring

This section covers rules for wiring, including clear wiring techniques, and cluster modularization.

4.2.1 Clear Wiring Techniques

 Rule 4.11 Minimize wire bends; eliminate kinks and loops

The fewer bends, kinks, and loops your wires have, the more fluid your diagrams appear. There are manual and automatic methods of wire routing and cleaning. I prefer manual wire routing because the automatic methods prioritize horizontal terminal entries and avoiding overlaps, at the expense of creating extra bends. To minimize the bends, it is necessary to route wires manually. Auto wire routing can be disabled temporarily or for all new wires. To disable and enable auto wire routing on-the-fly, toggle the **<A>** key at any time after you initiate a new wire. To disable auto wire routing for all new wires, navigate to **Tools»Options**, select **Block Diagram** from the **Category** list, and deselect the **Enable auto wiring** check box. My preference is to first connect the wires manually and then consider repositioning nodes and individual wire segments to minimize overlaps. The result is fewer bends and overlaps. Additionally, right-click on an existing wire and select **Clean Up Wire** to automatically reduce kinks, loops, and bends.

 Rule 4.12 Maintain even spacing of parallel wires

When propagating parallel wires, maintain consistent, even spacing between each node or bend. There are two useful techniques for ensuring even spacing. First, use similar connector pattern and terminal assignments for all subVIs that are intended to be used together. Connector pattern conventions are covered in Chapter 5. Second, align the nodes horizontally before wiring. Simply select the nodes and choose any of the horizontal alignment tools from the **Align Objects** menu on the toolbar. In Figure 4-4, a data logging routine is developed using several File I/O functions that comprise a dataflow sequence. The file refnum and error cluster are propagated using parallel wires. The File I/O functions have similar connector patterns, which helps facilitate even spacing. However, even slight offsets in the horizontal alignment cause wire kinks. In Figure 4-4A, the diagram is initiated by dropping the functions from the palettes onto the diagram. In Figure 4-4B, the **Bottom Edges** alignment tool snaps the nodes into perfect horizontal alignment. You can also distribute the nodes with even horizontal gap using the horizontal distribution tool. In this example, an asymmetric horizontal gap is desired to provide extra spacing within the While Loop. In Figure 4-4C, clear wiring proceeds with even vertical spacing, without kinks or bends. In Figure 4-4D, the wiring is completed and the horizontal gaps between nodes are manually adjusted.

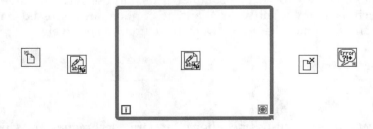

Figure 4-4A
A data logging routine consists of several functions containing similar connector patterns and terminal assignments.

Chapter 4 • Block Diagram

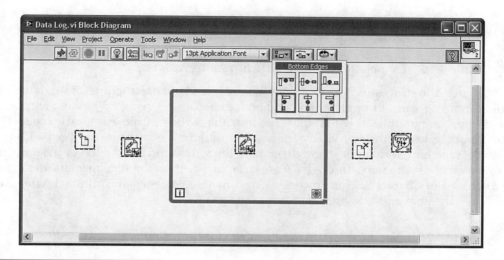

Figure 4-4B
The functions are aligned horizontally using the **Bottom Edges** alignment tool.

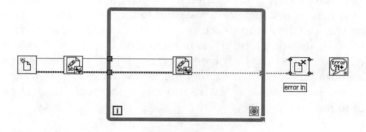

Figure 4-4C
Wires for file refnum and error cluster proceed with even spacing and without kinks or bends.

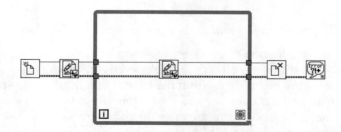

Figure 4-4D
The wiring is completed and the horizontal gaps between nodes are manually adjusted.

Rule 4.13   Tunnel wires through left and right borders of structures

 Rule 4.14   Do not wire through structures unnecessarily

Tunnel wires into structures through their left border, and out of structures through their right border. Avoid tunneling wires through the top and bottom borders. Also avoid passing wires through structures if they are not utilized within the structure, unless their purpose is clearly labeled. It is particularly annoying to flip through many frames of a multiframe structure, such as a Case or an Event structure, searching for places where the data in the wire is modified. However, sometimes it is useful to pass a few spare wires through a Case structure, such as the state machine design patterns that are discussed in Chapter 8. This practice reduces maintenance when additional wires are needed. Be sure to clearly label unused wires as `Unused`, or `Spare`.

 Rule 4.15   Never obstruct the view of wires and nodes

- Avoid overlapping diagram objects

Avoid obstructing the view of wires and nodes by overlapping them on the diagram. An occasional crossover of a wire routed horizontally with a wire routed vertically is unavoidable. For example, sometimes it is necessary to wire the iteration terminal of a looping structure, normally located at the bottom left of the structure, to a location above and to the right. Many of these same loops have wires routed horizontally across the entire structure, via either tunnels or shift registers. The error cluster is a prime example. If the iteration terminal is to remain on the bottom left, there may be no choice but to cross over the horizontal wires. The obstruction is minimized if the vertical wire is routed through a location of minimum wire density, overlapping as few wires as possible. Additionally, never overlap a wire with an object or a node. A wire running underneath a function or subVI resembles an input and output to the node.

 Rule 4.16   Limit wire lengths such that source and destination are visible on one screen

 Rule 4.17   Never use local and global variables for wiring convenience

 Rule 4.18   Label long wires and wires from hidden source terminals

Ideally, the source and destination of every wire should be readily visible without scrolling the diagram window. However, this is not always possible, even if the diagram is limited to one screen. For example, the terminals may be hidden within the frames of a multiframe structure. In this situation, be sure to label the wire in the frames or areas where the source terminal is not visible. While limiting wire lengths is desirable, long wires are preferred over no wires. Never use local or global variables as a method of reducing wire clutter. Variables increase processing overhead, memory use, and complexity. Moreover, variables undermine LabVIEW's dataflow principles by obscuring the actual source of the data. When variables are written and read from more than one location on the diagram, it becomes difficult to determine what is actually affecting the data. When wires are used, it is easy to trace the data to its unique source terminal. If the wires are long or the data source terminal is hidden, label them. Indicate the name of the wire's data source so that it is readily apparent when you view the diagram. Use the greater than sign (>) to reinforce the direction of data flow. Considerations with respect to local and global variables are discussed in Section 4.3, "Data Flow."

Rule 4.19   Place unwired front panel terminals in a consistent location

Place any unwired terminals of front panel objects in a consistent location on the diagram so that developers can find them easily. Note that any terminals not contained by a repeating structure are read only once. This is a problem for Boolean controls configured with latching action. In this case, the control is not able to reset itself. If the terminal's control is associated with an event that is registered by an Event structure, place the terminal within the corresponding event case. This ensures that the terminal's value will be read each time the event fires. For unwired terminals not associated with any events, simply place them to the left of the diagram's primary structure.

4.2.2 Cluster Modularization

Rule 4.20   Modularize wires of related data into clusters

It is much easier to implement clear wiring techniques and maintain organized diagrams if you reduce the overall number of wires you have to work with. Use clusters to group related data and reduce the quantity of wires. Wherever you have several wires of related data that are needed in the same areas of your diagrams, replace the individual wires with a cluster. This is analogous to modularizing low-level routines into cohesive subVIs. The data elements in the cluster should be related and serve a common purpose.

For example, consider the measurement routine from an optical filter test application, shown in Figure 4-5. The application prompts the user to define the laser scan parameters, configures a wavelength tunable laser source, measures the filter's transmission characteristics, graphs the data, and saves the data to file. Figure 4-5A contains the panel, nonvisible indicators, and Context Help window for Define Scan VI, the dialog used for selecting the laser scan parameters. This is the first subVI called in the measurement sequence shown in Figure 4-5B. Define Scan VI provides multiple parameters used by the subsequent VIs. As shown in the Context Help window, the connector pane is densely populated with individual terminal assignments for each parameter. In Figure 4-5B, there are many individual wires flowing through a relatively simple routine. The subVI connector terminal for Save Scan VI, the last subVI in the dataflow sequence, is also very densely populated. The wiring is kept reasonably neat because of even spacing and judicious terminal assignments. However, much toil is required to achieve this result, and even more toil is required to modify the VI. Specifically, any change to the required measurement parameters entails changes to the wires, subVIs, and connectors throughout the diagram.

Figure 4-5C contains an alternative implementation of Define Scan VI, with the laser scan parameters returned as a cluster. In Figure 4-5D, the measurement routine is revised using the cluster instead of individual wires. Wire clutter and development toil are substantially reduced. Additionally, parameters can be added and removed from the cluster without requiring any changes to the wires and subVI connector terminal assignments. Hence, clusters improve the diagram's appearance while reducing overall development and maintenance effort.

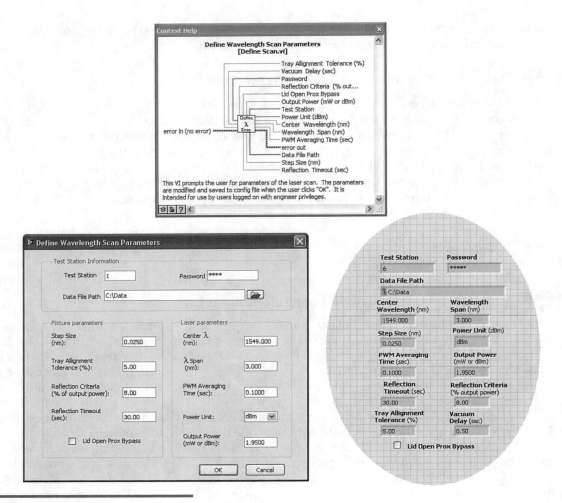

Figure 4-5A
Define Scan VI is a dialog that prompts the user to specify laser scan parameters. It returns 15 parameters through separate connector terminals for each.

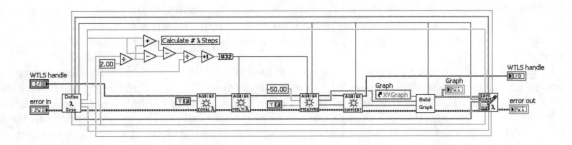

Figure 4-5B
An optical filter measurement routine calls Define Scan VI and propagates the laser scan parameters using individual wires. The subVI connector panes are densely populated, wiring is cluttered, and maintenance is tedious.

Chapter 4 • Block Diagram

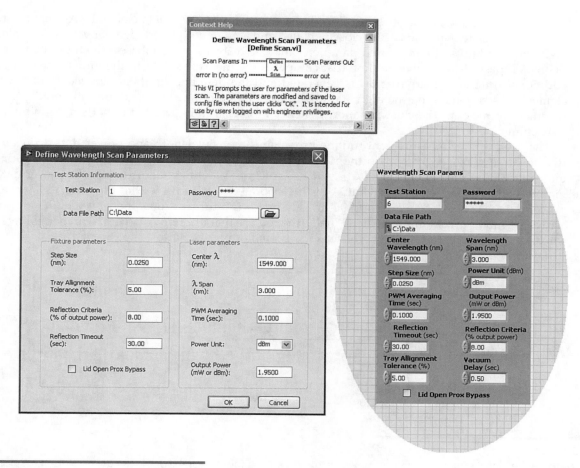

Figure 4-5C
Define Scan VI modularizes the laser scan parameters into a cluster. Terminal assignments are simplified.

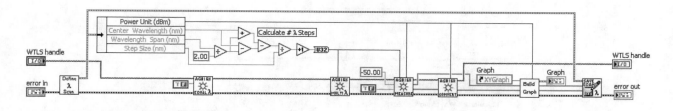

Figure 4-5D
The cluster of laser scan parameters propagates throughout the diagram. Wire clutter is reduced. Parameters can be added and removed from the cluster without affecting wiring and subVI terminals assignments.

In some situations, it makes sense to define a cluster for just two data elements. A small cluster is useful to associate data that is very closely related and is frequently used together. As an example, the high and low limits of a measured parameter may be read from a database, edited by the user in a dialog VI, passed to another routine that compares the measured data to the limits, and passed to additional routines where a report is generated and the data and limits are logged to file. A two-element cluster containing the high and low limits eliminates one wire and logically binds the data together. In this case, the cluster is more beneficial for associating related data than for eliminating wire clutter.

The optical filter measurement routine propagates the wavelength and power data arrays using two separate wires. As shown on the right half of Figure 4-6A, the two wires are used together in most places, with the exception of Convert Power Units VI, which requires only the power data. In Figure 4-6B, the Convert Power Units VI is incorporated as a subVI within Measure VI. Hence, the power data is converted to appropriate units before it is returned. Also, the wavelength and power arrays are modularized into a cluster. Finally, the routine that unbundles several scan parameters and calculates the number of wavelength steps has been modularized into a cohesive subVI.

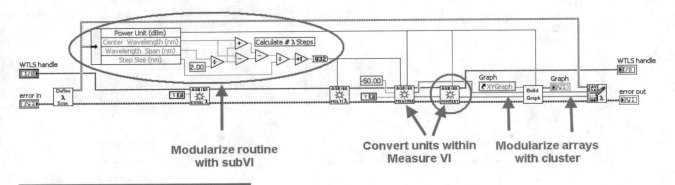

Figure 4-6A
Two separate wires for the wavelength and power arrays are used together in most places and may be modularized into a cluster. The routine for calculating the number of wavelength steps can be modularized into a subVI. The routine for converting power units can be performed within Measure VI.

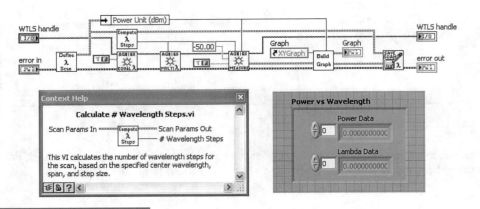

Figure 4-6B
The routine is further modularized with Calculate # Wavelength Steps VI and Power vs. Wavelength cluster.

Rule 4.21   Save clusters as type definitions

I cannot overemphasize this rule. (Consequently, it is further discussed in Chapter 6, "Data Structures.") Save every cluster as a type definition or strict type definition. A **type definition** (or **typedef**, for short) is a control that maintains its data type information in a CTL file. The typedef is copied onto any number of VI panels as a control or indicator, or diagrams as a constant, by either dragging and dropping from the project tree, or choosing either **Select a Control** from the **Controls** palette or **Select a VI** from the **Functions** palette. All instances of the typedef maintain the data type specified in the typedef's CTL file. Therefore, multiple instances of the control are maintained from one location via the Control Editor window. This is extremely beneficial for clusters because most clusters are used in multiple places, and the contents are subject to change. For example, in Figure 4-5C, new parameters are added to or removed from the **Wavelength Scan Params** cluster by simply adding or removing the corresponding controls to the typedef. In Figure 4-5D, all of the subVIs containing the typedef will automatically update to match the revised cluster.

A **strict type definition** (also known as **strict typedef**) maintains the control's properties, in addition to the data type, in the CTL file, so that all instances of the strict typedef maintain an identical appearance and behavior. I prefer the strict typedef because I often find that I need to maintain a common range, default value, and appearance among instances. To specify the type definition status, choose the corresponding item from the Typedef Status ring control on the Control Editor window's toolbar. Clusters and type definitions are discussed in greater detail in Chapter 6.

4.3 Data Flow

This section covers data flow, the fundamental principle of LabVIEW. It contains a brief review of basic principles and style rules, considerations regarding variables and Sequence structures, and techniques for optimizing data flow. The rules are illustrated with a combination of simple diagram snippets and a working application example.

4.3.1 Data Flow Basics

In LabVIEW, **data flows** along wires from source terminals to destination terminals. A block diagram node executes when data is received at all wired input terminals. Upon completion, data is supplied to its output terminals and propagates to the next node in the dataflow path. The dataflow principle distinguishes LabVIEW from traditional text-based software-development environments. The following are some basic rules regarding data flow.

 Rule 4.22   Always flow data from left to right

 Rule 4.23   Propagate the error cluster

With only one exception, always flow data through wires running from left to right. This is a sacred, age-old convention within the LabVIEW community. However, Feedback Nodes and Sequence Locals are built-in LabVIEW constructs that inherently violate this rule. In my opinion, the Feedback Node is a credible exception when the length of the right-to-left wire segment is very short, resulting in a reduction of wire clutter compared to a shift register. Sequence Locals are relevant only to Stacked Sequence structures, which should be avoided.

Zooming in on the diagram of Spaghetti VI, discussed in Chapter 1, we observe several violations of the left-to-right dataflow rule, including the wires highlighted in Figure 4-7. We also see unnecessary bends and kinks (4.11), overlapping wires and nodes (4.15), multiple local variables (4.17), and many other issues contributing to the spaghetti effect.

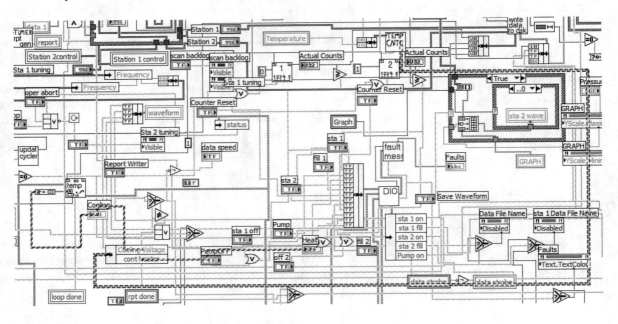

Figure 4-7
The diagram of Spaghetti VI contains several wires that violate the left-to-right dataflow rule.

Left-to-right data flow is accomplished by first positioning the order-dependent nodes from left to right so that interconnecting wires naturally flow left to right. In most cases, the order of execution is determined by data dependency. Simply propagate one or more common data elements, such as the error cluster, between the functions that require a specific execution order. This is similar to what is illustrated in Figure 4-4.

Rule 4.24   Avoid array and cluster coercions

Coercions are the unattractive dots that appear on terminals when there is a numeric representation mismatch. By default, they are colored gray prior to version 8.2, and red subsequent to version 8.2, but you can specify their color by selecting **Tools»Options**; choose **Colors** from the **Category** list, deselect **use default colors**, and click on the color next to **Coercion Dots**. Coercions indicate that LabVIEW is converting the data from one type to another. It is an additional operation that requires additional memory to store each representation. Eliminate coercions when possible, particularly with larger data structures such as arrays and clusters. Techniques for maintaining similar data types to prevent coercions are discussed in Chapter 6. Keep your eyes open for coercion dots, and try to eliminate them.

 Rule 4.25   Create controls and constants from a terminal's context menu

One simple way to avoid coercions and save development time as well is to create controls, indicators, and constants from a node's terminals on the diagram. This is accomplished by right-clicking on the desired node's connector terminal and selecting **Create»<Control/Indicator/Constant>** from the context menu. The corresponding control, indicator, or constant is automatically created, having the data type that matches the terminal it is wired to. Additionally, the wiring assignment is completed and the item inherits the label of the node's terminal. Hence, several editing steps are completed from just one menu selection.

 Rule 4.26   Disable dots at wire junctions

Dots at wire junctions have no functional purpose except for drawing attention to wire junctions. They are larger and usually more prevalent than coercion dots. During the development of a dense diagram, they help distinguish junctions from overlapping wires. However, they can mildly interfere with identifying coercion dots. Also, if you generally avoid overlapping wires, it is not necessary to highlight junctions. My personal opinion is that dots at wire junctions appear a tad obnoxious on a diagram with clear wiring and efficient data flow. Because wire junctions can be identified by triple-clicking on a wire of interest, I recommend disabling dots at wire junctions. This selection is available from **Tools»Options**; select **Block Diagram** from the **Category** list and deselect **Show dots at wire junctions**.

 Rule 4.27   Avoid Sequence structures unless required

Avoid using Sequence structures, unless they are required. In particular, never use a Sequence structure to force the execution order of functions that can execute in parallel. This is a common tendency of former text-based programmers that are accustomed to textual statements executing in the order in which they appear. Challenge yourself to avoid Sequence structures. Instead, try to keep independent functions and routines parallel to each other. For example, the Flat Sequence shown in Figure 4-8A is unnecessary. Additionally, the variables are unnecessary. The diagram shown in Figure 4-8B is functionally equivalent but cleaner and more efficient. Specific rules on the practical and impractical use of Sequence structures and variables are discussed in the sections that follow.

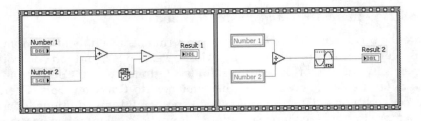

Figure 4-8A
This Flat Sequence structure is unnecessary because the contents need not execute sequentially.

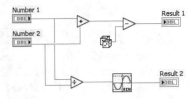

Figure 4-8B
The Sequence structure is replaced with parallel nodes, which LabVIEW executes efficiently.

 Rule 4.28   Avoid nesting beyond three layers

Nesting is the placement of structures within structures. Nesting beyond two or three layers begins to obscure the underlying logic and data flow, as shown in the Nested VI example from Chapter 1. It is difficult to visualize all of the possible logical branches and data paths in a highly nested diagram. Excessive nesting is normally caused by faulty logic, excessive use of Sequence structures, and lack of standard design patterns. Faulty logic is an important concern because the more nesting, the more difficult it is to comprehend the diagram to identify and debug the problem. Spend some time creating a flow chart, truth table, or Karnaugh map to understand the desired logic before implementing the source code. In Chapter 8, specific architectures that minimize nesting are presented. However, some level of nesting is required even for the most common and useful design patterns. Because the structures that comprise the design pattern are recognized as the design pattern, they need not count toward the excessive nesting layers. In this case, always save the VI with the most important frames of the nested structures selected so that the diagram opens this way by default. The most important frame is either the most frequently executed logical branch or the frame that reveals the most nodes. Finally, note that maximizing data flow and minimizing Sequence structures goes a long way toward preventing excessive nesting.

4.3.2   Practical Variables and Sequence Structures

Throughout this book, I generally recommend avoiding local and global variables and Sequence structures because they undermine dataflow principles. Now let us take a moment to consider their practical applications. In some circumstances, they serve important and useful purposes.

 Rule 4.29   Use write local variables for initializing control values

Rule 4.30   Use global variables for simple data sharing between parallel loops or VIs

Write local variables are the best method for writing to controls from the diagram. This is necessary to programmatically initialize control values, such as configuration parameters that are read from file. Also, local and global variables represent a fast and easy method of sharing data between parallel processes such as continuous loops or VIs. However, read local and global variables have no inputs, and write local and global variables have no outputs. Therefore, the execution order of a sequence containing local and global variables cannot be specified using data flow. Property Nodes and Shared Variable nodes have error terminals and are viable dataflow alternatives to local and global variables. A Property Node configured to read or write a control's Value property is similar to a local variable, but it causes the diagram to switch to the user interface thread to read or write the value directly from the front panel, in a synchronous manner. Local variables read or write from the control's terminal on the block diagram, which does not trigger an immediate thread change and user interface update. A single process shared variable functions similarly to a global variable that performs error checking. Local and global variables are generally simpler and more efficient than Property Nodes and Shared Variable nodes. If you must use local and global variables and you must specify the order in which they are written or read, then you must use a Sequence structure.

 Rule 4.31   Use a Sequence structure to order operations if no data dependency exists

 Rule 4.32   Use only Flat Sequence structures when required

Use a Sequence structure to specify execution order when a specific order is required and no data dependency exists. Specifically, Sequence structures can order initialization routines, including write variable operations, to ensure that they are performed at the very beginning of the application. Likewise, Sequence structures can force shutdown routines, such as Quit LabVIEW, to occur at the very end of the application. Figure 4-9 illustrates an example of a diagram that utilizes an initialization routine, two parallel loops, and a shutdown routine. The application's primary purpose is to run a test involving a lengthy data acquisition process upon user command via the **Run Test** Boolean control. Because the data acquisition process is lengthy, the application is divided into separate parallel master and slave loops, including an Event Loop for processing user interface events, and a DAQ Loop for performing the data acquisition task. The Event Loop is the master that triggers the data acquisition task within the slave DAQ Loop. The parallel loops enable the user interface to remain responsive while the data acquisition task runs in the background. LabVIEW spawns separate execution threads for each parallel loop.

A local variable is used to initialize the **Sensor Scaling** cluster with values read from file. Global variables are used to share Boolean data between parallel loops. The **Acquire** global variable triggers the data acquisition task from the Event Loop. The **Stop** global variable stops the DAQ Loop when the Event Loop stops, and vice versa. The application has a Boolean control that writes to the **Stop** global from within the Event structure's **Quit Value Change** event case, not shown. Specifically, when either loop stops, it passes a Boolean TRUE value through a While Loop tunnel to a **Stop** write global variable. Upon the next iteration, the other loop reads this value from its corresponding **Stop** read global variable, and stops. Global variables require less configuration and effort than alternative constructs such as shared variables, occurrences, notifiers, and queues. If one simple and unambiguous data item is being written, with only one or two instances of a write global variable and on only one occasion throughout the application, the global variable is a good choice. It is unnecessary to set up a shared variable, occurrence, notifier, or queue for such a limited scope.

In Figure 4-9A, two single-frame Sequence structures are placed on each side of the While Loops, and wires are used to create data dependency between each structure. The large Sequence structure on the left initializes the variables and passes two wires out of tunnels to the While Loops. Because data will not pass through the tunnels of the Sequence structure until all its operations have completed, via the principle of data flow, the write local variable operations must occur before the data propagates through the tunnels. Because the While Loops receive data from the initialization Sequence structure, the While Loops cannot begin executing until the initialization Sequence structure has fully completed. This condition is known as **data dependency**. Notice also that the Boolean data that is passed to the border of the Event Loop is not actually used within the loop. Instead, the wire terminates on the While Loop's border. Nonetheless, this wire ensures that the Event Loop begins executing only after the initial Sequence structure completes. This is known as **artificial data dependency**.

The Shutdown routine consists of a Sequence structure on the right containing the Quit LabVIEW function. It receives data from the two parallel loops, ensuring that the Shutdown routine executes last. Also note that the General Error Handler VI propagates the error cluster as an input from the DAQ Loop and an output to the Shutdown Sequence structure. This forces the General Error Handler VI to execute after the DAQ Loop stops, but before the Shutdown Sequence structure can begin. The diagrams in Figure 4-9B and Figure 4-9C are functionally equivalent to the diagram from Figure 4-9A, except that they utilize three-frame Stacked and Flat Sequence structures, respectively. Stacked Sequence structures are less desirable than Flat Sequence structures because only one frame is visible at a time, they require sequence local terminal for data flow between frames, and the multiple frames hide the data flow. Flat Sequence structures facilitate data flow between frames using tunnels, and all frames are simultaneously visible. Also, each frame of a Flat Sequence structure has a uniform appearance and does not require artificial data dependency. Therefore, the Flat Sequence implementation in Figure 4-9C is preferred in this example.

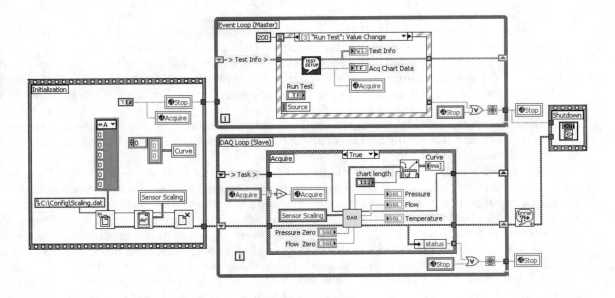

Figure 4-9A
Two single-frame Sequence structures are used for forcing the execution order of the Initialization routine, two While Loops, and Shutdown routine.

Chapter 4 • Block Diagram

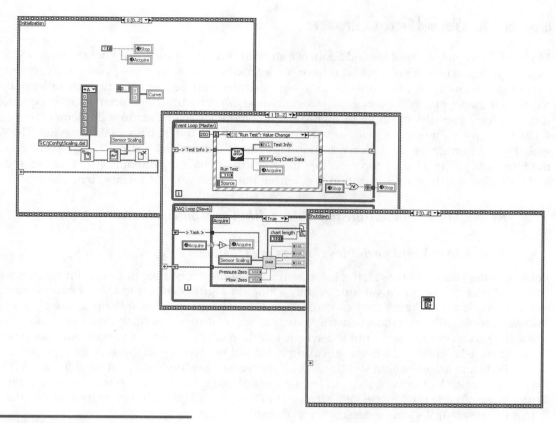

Figure 4-9B
A Stacked Sequence structure requires sequence local terminals to pass data between frames.

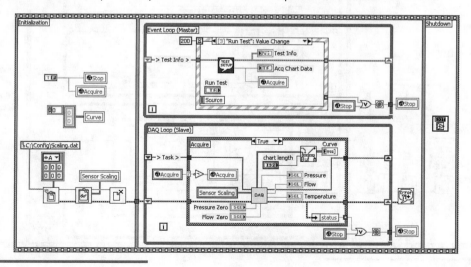

Figure 4-9C
A three-frame Flat Sequence structure facilitates data flow between frames via wires and tunnels, and all frames are visible. This is the preferred implementation.

4.3.3 Impractical Variables and Sequence Structures

Most local and global variables and Sequence structures used in practice are not necessary. Most often, they are overused by developers who have not learned efficient dataflow principles. The best way to master data flow is to force oneself to avoid variables and Sequence structures unless absolutely required. Indeed, mastering data flow is synonymous with minimizing variables and Sequence structures. In fact, even the example in Figure 4-9 has more efficient dataflow implementations. Figure 4-10A is a copy of Figure 4-9C with several unnecessary local and global variables circled. The **Acquire** global variable triggers the data acquisition task from the **Run Test Value Change** event case within the Event Loop. The DAQ Loop monitors the **Acquire** global variable until its value becomes TRUE and then executes the data acquisition task within the **True** frame of the Case structure.

 Rule 4.33   Avoid polling variables within continuous loops

 Rule 4.34   Avoid variables if wires are feasible

Polling is the condition in which a loop continuously monitors a resource until it reaches a specific value or state. Never poll a variable within a loop. In Figure 4-10A, the DAQ Loop executes at top speed, inefficiently utilizing the processor to detect a change in value of the **Acquire** global variable. Instead, use a synchronization construct, such as an occurrence, notifier, or queue. These functions allow the slave loop to sleep until the synchronization construct fires an event. Another observation from Figure 4-10A is that the **Sensor Scaling** write and read local variables can be replaced by a wire. The write local variable might still be useful if the **Sensor Scaling** data read from file needs to be displayed in a control. However, the read local variable inside the DAQ Loop is unnecessary. Finally, the **Curve** local variable is not necessary because it initializes a graph indicator to its default value, which is already the control's state when the VI loads into memory.

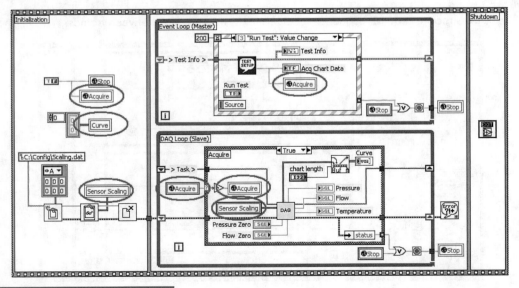

Figure 4-10A
Avoid polling variables in loops, such as the **Acquire** read global variable in the DAQ Loop. Additionally, the **Sensor Scaling** local variables can be replaced by a wire, and the **Curve** local variable is unnecessary.

In Figure 4-10B, the **Acquire** global variable is replaced with an occurrence, the **Curve** local variable is eliminated, and the **Sensor Scaling** local variables are replaced by a wire. The occurrence allows the DAQ Loop to sleep, not utilizing the processor, until the Set Occurrence function executes inside the Event Loop or a timeout occurs. The **timeout** input terminal of the Event Loop's Event Structure and the DAQ Loop's Wait on Occurrence function is set to 200ms, allowing each loop to periodically poll the **Stop** global variable and update the stop condition. Polling the **Stop** global every 200ms is more efficient than polling the **Acquire** global at full speed, but it is still a violation of Rule 4.33.

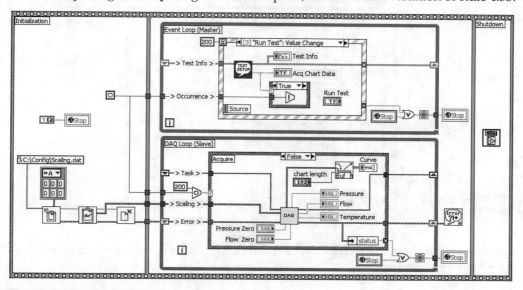

Figure 4-10B
An occurrence is used in place of the **Acquire** global variables, a wire replaces the **Sensor Scaling** local variables, and the **Curve** local variable is eliminated.

Applying a notifier instead of an occurrence further streamlines data flow and efficiency, as observed in Figure 4-11. A notifier is a synchronization construct that is similar to an occurrence. However, a notifier is programmatically released, causing the Wait on Notification to return from an indefinite wait. By comparison, the Wait on Occurrence function cannot terminate programmatically. Also, the notifier sends data from the sender to the receiver. In Figure 4-11A, the Event Loop's **Run Test** event case calls the Send Notification function along with a Boolean TRUE instructing the DAQ Loop to acquire data. The DAQ Loop's Wait on Notification function wakes up and returns the TRUE value from the **notification** output terminal to the case selector, and the Acquire routine runs. When any event fires that causes the Event Loop to stop, the Release Notifier function is called outside the Event Loop. For example, if the user clicks the **Quit** Boolean control, the **Quit Value Change** event case of the Event Loop fires, as shown in Figure 4-11B. This causes the Event Loop to stop, and the Release Notifier function runs. The DAQ Loop's Wait on Notification function immediately wakes up and returns a FALSE value from the **notification** output terminal, and an error from the **error out** terminal. The DAQ Loop executes the Case structure's **False** case, which clears the error and stops the loop. Hence, the Event Loop fully controls the DAQ Loop, without variables. Also, notice that an error in either loop stops both loops, without variables. Specifically, the error **status** is unbundled and wired to each loop's conditional terminal. A TRUE error status in the Event Loop stops the loop and calls the Release Notifier function, terminating the DAQ Loop as previously described. Additionally, a TRUE error status in the DAQ Loop terminates the loop, and the Property Node outside the DAQ Loop runs. This Property Node is associated with the **Quit** Boolean control and is configured to set the **Quit Value Signaling** property, which fires a **Quit Value Change** event within the Event Loop. The **Quit Value Change** event case runs and stops the Event Loop. Consequently, variable polling is no longer necessary, and all instances of the **Stop** global variable are eliminated. This implementation optimizes the master/slave synchronization efficiency, as shown in Figure 4-11B.

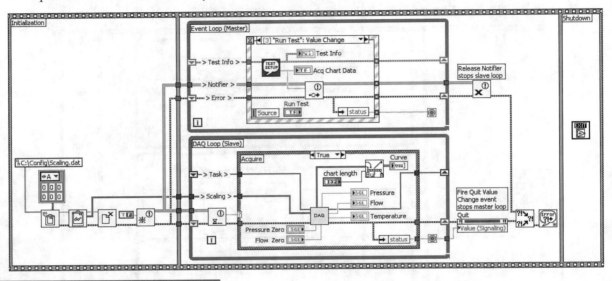

Figure 4-11A
A notifier replaces the occurrence, and the **Stop** global variables are eliminated. This implementation optimizes the synchronization efficienctly.

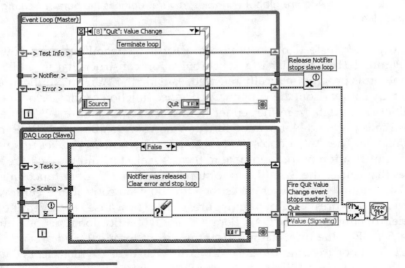

Figure 4-11B
The **Quit Value Change** event case stops the loop. The Release Notifier function is called outside the loop, causing the Wait on Notification function in the DAQ Loop to wake up and return an error. The error is cleared in the **False** case, and the DAQ Loop terminates.

4.3.4 Optimizing Data Flow

As discussed, mastering LabVIEW's dataflow principles is synonymous with eliminating local and global variables and Sequence structures, when feasible. The previous sections identify the practical and impractical uses of these programming constructs. This section presents alternatives that optimize data flow, including shift registers and looped Case structures.

 Rule 4.35   Use shift registers over local and global variables

 Rule 4.36   Group most shift registers near the top of the loop

 Rule 4.37   Label wires exiting the left shift register terminal

Shift registers are terminals on looping structure borders that shift data between loop iterations. They are functionally and conceptually similar to terminals that extend wires from the end of one loop iteration to the beginning of the next. Shift registers are viable alternatives to local and global variables when the required scope of data sharing is limited to a single While Loop, Timed Loop, or For Loop. Unlike variables, shift registers do not create copies of their data when read and are maximally efficient.

To avoid wire clutter and maintain organization, space most shift registers tightly, and group them near the top of the loop. This creates a data highway that is limited to an area near the top of the loop and minimizes wire crossovers. Leave just enough space between the shift registers to apply free labels on each wire near the left terminals. Exceptions to shift register grouping include error clusters and case selectors. Error clusters normally enter and exit near the bottom of loops. Also, case selectors are frequently positioned near the middle. Therefore, error clusters and case selectors are usually kept separate from the data highway at the top.

 Rule 4.38   Use looped Case structures over Sequence structures

A **looped Case structure** is a Case structure embedded within a loop. It functions similarly to a Stacked Sequence structure when the sequentially ordered code is placed in cases of the Case structure instead of the frames of the Sequence structure. However, the execution order of the cases is controlled programmatically via the Case structure's selector terminal. This is more flexible than a Sequence structure, for which the execution order is strictly determined by the frame numbers. Use shift registers on the loop to share data between cases of the looped Case structure. Shift registers promote better data flow than the Sequence Locals of a Sequence structure because the data enters and exits the cases in a consistent location and the data flows left-to-right.

Figure 4-12 provides four different implementations of a test sequence comprised of 12 sequentially ordered test VIs. Each test VI shares several common inputs and outputs, including a DAQmx task, a cluster of sensor scaling coefficients, a cluster of high and low test limits, and an error cluster. Additionally, each test appends a row to the **Test Results** table indicator for reporting the test results. These updates are performed immediately after each test is completed. Also, the limits for each test are calculated within the previous test VI, based on its test results.

Figure 4-12A utilizes a Stacked Sequence structure, in violation of Rules 4.27 and 4.32. The error cluster is passed between frames using sequence local terminals on the Sequence structure's inner border. Because sequence local terminals can be written to only once, 11 terminals are required to propagate the error cluster among 12 frames. This clutters the appearance and causes wiring and dataflow rules violations. Most frames contain one right-to-left data flow and a wire crossover caused by reading from a sequence local terminal on the right border or writing to a sequence local terminal on the left border. Additionally, the Stacked Sequence structure implementation contains 48 local variables. These include two local variables per frame for updating the **Test Results** table indicator, plus two local variables per frame for reading and writing the test limits.

Figure 4-12B utilizes a looped Case structure using a For Loop as the looping structure. The code within the cases of the Case structure is functionally equivalent to the code within the frames of the Stacked Sequence structure from the prior implementation. However, only five shift registers are required to replace 48 local variables and 11 sequence local terminals used by the Sequence structure implementation. The wiring is clear, with no right-to-left data flow or wire crossovers. Figure 4-12C contains a looped Case structure that utilizes a While Loop instead of a For Loop. This implementation is similar to the For Loop, with one important distinction: The While Loop is programmed to terminate if a test fails or an error occurs, without completing the full test sequence.

A looped Case structure need not be limited to sequentially ordered execution of numerically selected cases. A **flexible sequencer** utilizing the Classic State Machine design pattern is shown in Figure 4-12D. The **Classic State Machine**, discussed extensively in Chapter 8, consists of a looped Case structure with an enumerated data type for the case selector and a shift register for passing the next case selection between loop iterations. The cases are intuitively labeled based on the text items of the enumerated data type instead of integers. This eliminates the free labels within each case. Also, the Classic State Machine programmatically selects the next case, based on an operation performed in the previous case. In this manner, the test sequence is formed dynamically. This implementation maximizes flexibility compared to the previous implementations.

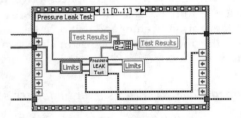

Figure 4-12A
A test sequence is implemented using a 12-frame Stacked Sequence structure. It contains 48 local variables plus 11 sequence local terminals, and right-to-left data flow.

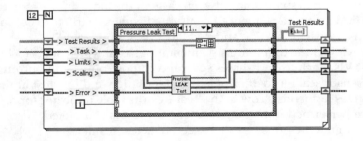

Figure 4-12B
The test sequence is implemented using a looped Case structure with a For Loop as the looping structure. Five shift registers replace the local variables and sequence local terminals.

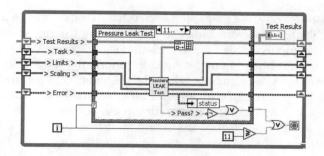

Figure 4-12C
An alternate looped Case structure utilizes a While Loop that terminates the test sequence if a test fails or an error occurs.

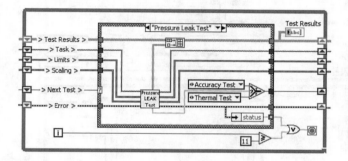

Figure 4-12D
A flexible sequencer is implemented utilizing the Classic State Machine design pattern. An enumerated data type and shift register are utilized for the case selection. The sequence is formed dynamically.

Let us now apply the flexible sequencer to a larger application. Consider the master/slave test VI from Figure 4-11. The slave loop is modified to perform the sequence of 12 tests from Figure 4-12 instead of the single data acquisition task. On the surface, it appears as though the test sequence should be placed inside the existing Case structure, resulting in four layers of nesting, a violation of Rule 4.28. However, the slave loop itself can be modified into a flexible sequencer. Specifically, in Figure 4-13A, the slave loop has been converted into a **Queued State Machine** design pattern, for which a queue replaces the notifier for synchronization and messaging between loops, an enumerated data type replaces the Boolean case selector, and the Case structure contains multiple cases corresponding to the tests. The queue is similar to a first-in, first-out buffer that stores multiple enumerations representing the slave loop's case selections or states. As shown in Figure 4-13A, the master loop's **Run Test** event case calls the Enqueue Element function in a For Loop, enqueuing the 12 cases that comprise the test sequence. Additionally, in Figure 4-13B, notice that the File I/O functions that read the **Sensor Scaling** data from file appear within the **Initialize** case of the slave loop's Case structure. The Enqueue Element function to the left of the slave loop enqueues the **Initialize** case via an enumerated constant, which ensures that the **Initialize** case executes first. Therefore, the functionality of the Queued State Machine is not limited to the test sequence, but also performs initialization and other routines as required. This implementation is neater and more flexible than the notifier implementation from Figure 4-11.

Figure 4-13C illustrates how the loops stop each other in the event of a user interface event or error. If the user presses the **Quit** Boolean control, the **Quit Value Change** event case adds the **Shutdown** state to the front of the queue via the Enqueue Element at Opposite End function. This causes the Test Sequence Loop to run the **Shutdown** state and terminate the loop. If an error occurs in the Event Loop, the **Error** case of the Case structure releases the Queue, causing the Dequeue Element function in the Test Sequence Loop to return with an error and execute the **Shutdown** state, since it is the Case structure's Default case. Finally, if an error occurs in the Test Sequence Loop, it calls the **Shutdown** state which releases the Queue, clears a DAQ task, and fires the **Quit Value Change** event, causing the Event Loop to terminate. The Queued State Machine design pattern is described in greater detail in Chapter 8.

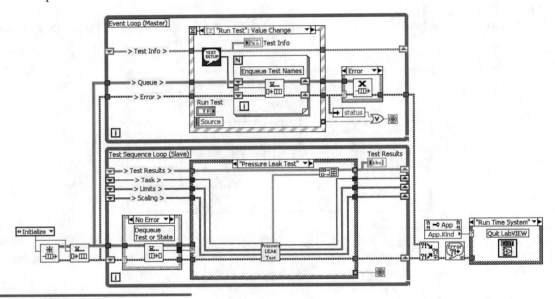

Figure 4-13A
The master/slave application from Figure 4-11 is modified to execute the 12-step test sequence from Figure 4-12, utilizing the Queued State Machine design pattern for the slave loop.

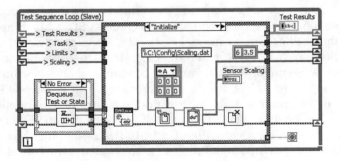

Figure 4-13B
The functionality of the Queued State Machine design pattern is expanded to accommodate an initialization routine in addition to the test sequence. The **Initialize** case contains the File I/O functions that read the scaling data.

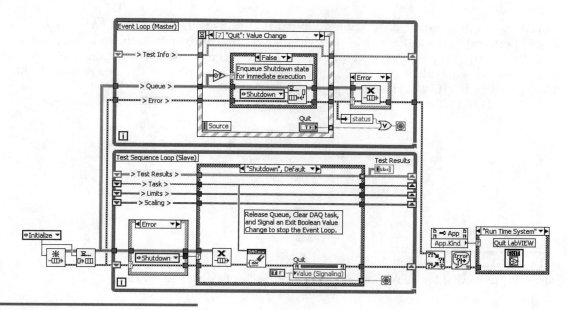

Figure 4-13C
The loops stop each other if the **Quit Value Change** event occurs in the Event Loop, or an error occurs in either loop.

4.4 Examples

In this section, a variety of block diagram examples are presented, both good and bad. Let us begin with the bad and gradually transition to the good. The bad examples are particularly effective at illustrating the reason for many of the block diagram style rules presented thus far.

4.4.1 SubVI from Selection

As observed in Chapter 3, "Front Panel Style," the SubVI from Selection utility is the world's most flagrant style violator. This is the process of selecting an area of the diagram and choosing **Edit»Create SubVI**. The worst possible programming practice is to create a subVI from selection and not clean up the aftermath. The terminal locations and labels, wiring, connector assignments, icon, and description all require corrective action. Sometimes the resulting subVI diagrams appear as if a bomb went off inside them. SubVIs created using this tool *never* conform to good style, and rework is mandatory.

Figure 4-14A corresponds to the diagram of the SubVI from Selection VI front panel example presented in Chapter 3. It contains several telltale signs of a subVI from selection. The wiring is kinked, the objects are tightly spaced, and the control labels are nonsensical. For example, the task ID and error control terminals are improperly labeled **task ID out** and **error out**. Additionally, one of the indicators has the generic label **Numeric**. What is curious is that the subVI has a custom icon and description, and the control terminals are within reasonable proximity to the structure. Perhaps it was partially repaired. The wiring and terminal labels have been cleaned up in Figure 4-14B. Also, Figure 4-14C contains equivalent code using the waveform data type and the newer sound output VIs released with LabVIEW 8. This is the preferred implementation.

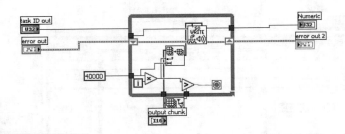

Figure 4-14A
This diagram is the product of the SubVI from Selection utility, without the mandatory repairs.

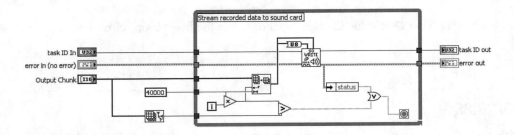

Figure 4-14B
The VI is revised to incorporate appropriate control labels, object spacing, and clear wiring.

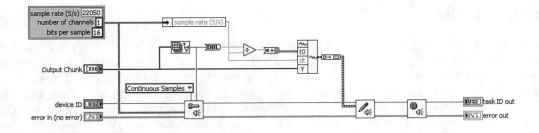

Figure 4-14C
The VI is rewritten to utilize the waveform data type and LabVIEW 8 sound VIs.

Layout: The diagram (4-14A) is small and the objects are poorly spaced.

Modularity: Not applicable because there are only a few nodes. However, the VI itself is the outcome of the developer's attempt to modularize the calling application.

Wiring Scheme: The wires are excessively kinked, a common byproduct of the subVI from selection utility. Additionally, one wire is not lined up with the tunnel in which it enters the loop.

Data flow: Data flows vertically as well as horizontally, including through two tunnels on the bottom border of the While Loop. Also, there is one coercion dot as sound data is converted from I16 to U8.

4.4.2 Excessively Nested VI

The diagram of Excessively Nested VI, shown in Figure 4-15, is oversized and severely nested. The top illustration is the Navigation window, which, unfortunately, is the only way to view the whole diagram. The bottom illustration is a highly nested section of the diagram. Less than half of the overall diagram is visible on the screen with 1280×1024 resolution. Navigation entails much scrolling and even more flipping through the frames and cases of the nested structures. The quadrant shown contains 11 layers of nesting. It is not possible for even the most advanced developers (and blackjack players) to comprehend all the logical branches represented. Avoid large diagrams and avoid excessive nesting. Chapter 8 presents standard diagram architectures that help prevent large and unwieldy diagrams like this one.

Layout: The diagram is oversized, overly nested, and unwieldy.

Modularity: There are only 40 user VIs out of 2,040 nodes, for a modularity index of just 1.9. Hence, it is not modular.

Wiring Scheme: There are many long wires due to a large diagram, along with several unnecessary bends.

Data flow: Nesting obscures data flow. There are several instances of right-to-left data flow, as well as data entering structures through vertical tunnels.

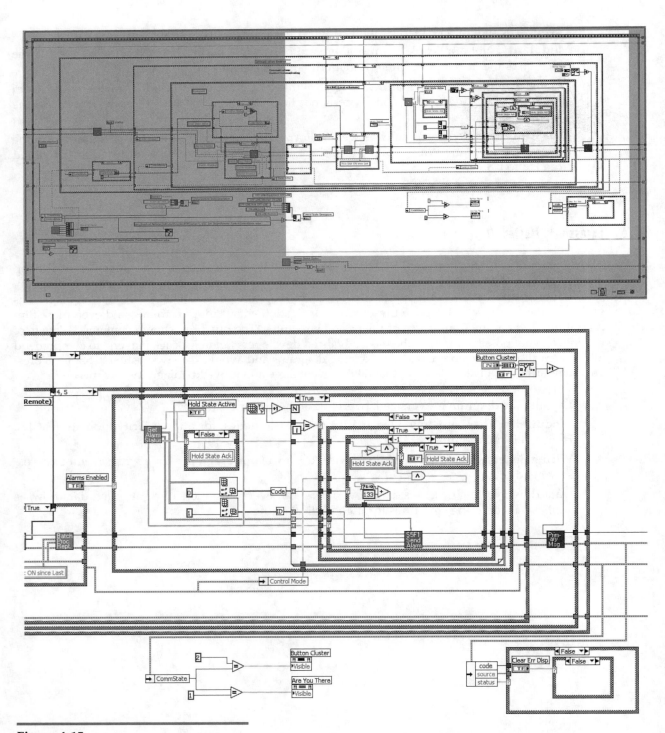

Figure 4-15
The Navigation window shown at the top is required to navigate the oversized diagram. The subsection at the bottom magnifies 9 of the diagram's 11 layers of nesting.

4.4.3 Haphazard VI

The diagram of Haphazard VI, shown in Figure 4-16, contains haphazard wiring and data flow. More wires enter and exit structures vertically than horizontally. There are many wire bends and kinks. Data is flowing left to right, right to left, up, down, and all around. Additionally, several local variables are improperly initialized outside the While Loop in parallel with code that is reading the values from the corresponding control terminals inside the loop. Because there is no data dependency between the initialization operations and the While Loop, the order of execution is not specified. Therefore, these controls may not be correctly initialized before they are accessed within the looping structure. This is a situation in which a single-frame Sequence structure with artificial data dependency—or a multiframe Flat Sequence structure—is required, similar to Figures 4-9A and 4-9C, respectively. The unusual icon convention is another matter that is discussed in Chapter 5.

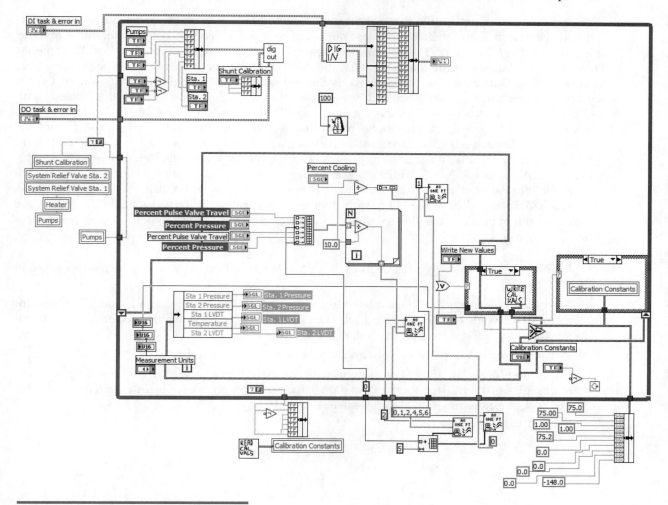

Figure 4-16
This diagram has excessive wire bends, haphazard data flow, and unusual icon convention. The order of execution between the local variable write operations and the While Loop is not specified.

Layout: The diagram is sized appropriately, with medium to low density.

Modularity: There are 13 user VIs out of 396 nodes, for a modularity index of 3.3. Clutter outside of the loop should be combined into one initialization subVI.

Wiring Scheme: Wires contain excessive bends and kinks. Many wires enter and exit structures vertically. Several wires are too long.

Data flow: Data flows in *all* directions. Initialization routines are not explicitly ordered to execute before the main While Loop. There are excessive coercions.

4.4.4 Right to Left VI

Right to Left VI is an application that acquires waveform data from pressure and volume transducers until an event occurs, and then logs the last sample of each waveform to a comma-delimited text file. It is illustrated in Figure 4-17. The diagram utilizes efficient data flow, and contains liberal comments. However, it contains one wire that is routed right to left, as shown near the center of the primary Case structure's **Log Data** case. The wire is cleverly labeled to indicate the right-to-left data flow. On the surface, it appears to represent good right-to-left data flow style. In a pinch, one neatly labeled right-to-left wire never hurt anyone. However, it is still right-to-left nonetheless, a rule violation.

Close inspection reveals that the right-to-left wire is a file path that is formed by some low-level functions within an inner Case structure. Whenever a collection of low-level functions that work together to perform a cohesive routine appears on the diagram of a high-level VI, it is a good opportunity for a subVI, as per Rule 4.7. Additionally, the Case structure's **False** case contains a read local variable for the **File Path** indicator. This represents an opportunity for a shift register to replace a variable. Finally, a wire is routed from the **Lot #** control terminal vertically through a tunnel in the inner Case structure's bottom border. This violates Rule 4.13. Hence, one right-to-left-flowing wire leads us to *four* style rule violations.

Layout: This is a sparsely populated Classic State Machine design pattern, which is discussed in Chapter 8.

Modularity: This application is relatively simple, containing only 6 user VIs and 173 nodes, for a modularity index of 3.5. As seen in the **Log Data** case, low-level functions for file path forming should be modularized into a subVI, as per Rule 4.7.

Wiring Scheme: Wires are straight and some overlapping is required.

Data flow: Most data flows left to right, with the exception of one wire that flows right to left. Shift registers are utilized to maintain data between loop iterations. The **File Path** local variable can be replaced with another shift register.

Chapter 4 • Block Diagram

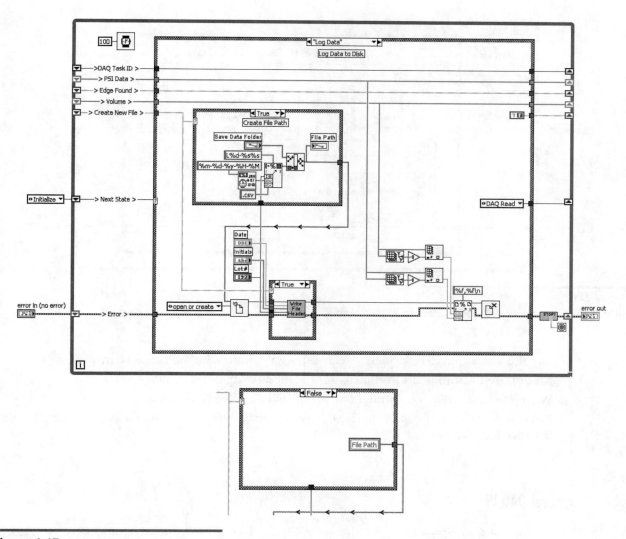

Figure 4-17
Right to Left VI is a top-level VI containing low-level functions for forming the file path, a local variable, a vertical tunnel, and right-to-left data flow.

4.4.5 Left to Right VI

Figure 4-18 shows the same VI from Figure 4-17, formerly known as Right to Left VI, modified with several enhancements. It contains a subVI for forming the file path, and a shift register for maintaining the file path between loop iterations. The revised diagram no longer requires the local variable, tunnel on the bottom border, or right-to-left data flow. Additionally, the object spacing and structure sizes have been reduced, providing a higher overall object density. An important lesson learned in this example is that the style rules are interrelated. The more you follow on a consistent basis, the easier it becomes to maintain good style throughout an application.

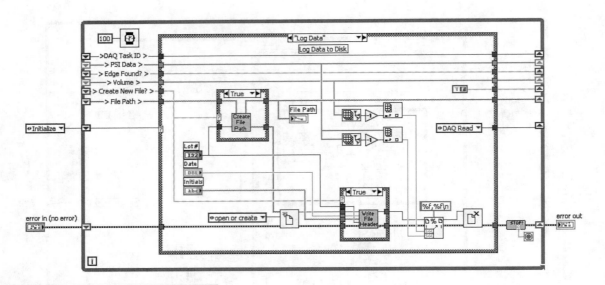

Figure 4-18
Left to Right VI is functionally equivalent to Right to Left VI but contains multiple style enhancements.

Layout: This is the Classic State Machine design pattern with high object density.

Modularity: Additional subVI increases the modularity index to 4.0.

Wiring Scheme: Wires are straight and some overlapping is required.

Data flow: All data flows left to right. An additional shift register maintains the file path value between loop iterations.

4.4.6 Centrifuge DAQ VI

The diagram of Centrifuge DAQ VI shown in Figure 4-19 is neat, intuitive, and very well documented. The primary data elements are read from shift registers and wired through tunnels into the primary Case structure, facilitating good data flow. Each wire is labeled outside the Case structure and then bent upward to create additional space within the Case structure. Near the bottom of the diagram, an array of control references is passed to a subVI for manipulating properties. The Queued State Machine design pattern is readily expandable via adding cases to the primary Case structure and items to the enumerated data type that is wired to its selector. More features of this application are discussed in Chapters 6 and 8.

Chapter 4 • Block Diagram

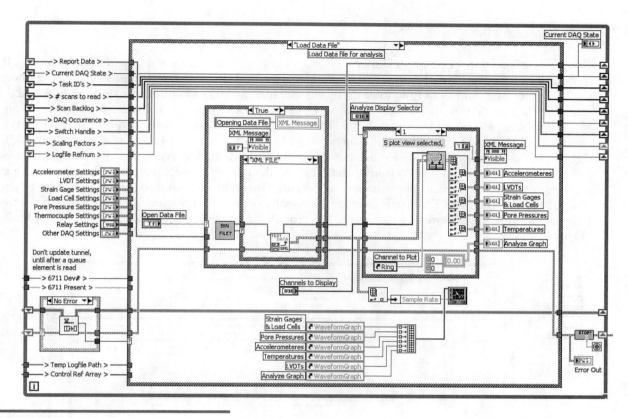

Figure 4-19
Centrifuge DAQ VI contains many advanced features, including control references and the Queued State Machine design pattern.

Layout: This diagram actually has two parallel While Loops, only one of which is visible on the monitor and included in this illustration. It utilizes the Queued State Machine design pattern.

Modularity: There are 113 user VIs out of 2,887 nodes, for a modularity index of 3.9. SubVIs are used for manipulating control properties, reducing real estate occupied by Property Nodes.

Wiring Scheme: Clusters are utilized to avoid wire clutter. Wire bends create additional space within the main Case structure.

Data flow: Data flows strictly from left to right. Queues are used to pass data between parallel loops without variables. Shift registers are used over variables wherever feasible. The application contains 50 variables, primarily used to initialize control values.

4.4.7 Screw Inspection VI

The Screw Inspection VI diagram shown in Figure 4-20A is clear, with a staircase dataflow appeal. It is modular with a particularly attractive icon convention that is featured in Chapter 5. At first glance, the layout, wiring, and data flow look pretty good. Shift registers are utilized over variables, maximizing data flow. Wires and objects are evenly spaced. However, close inspection reveals multiple rules violations, as shown in Figure 4-20B. These include unnecessary wire bends and kinks (4.11), overlapping objects (4.15), unlabeled long wires (4.18) including wires exiting shift registers (4.37), right-to-left data flow (4.22), discontinuous error clusters (4.23), and coercion dots (4.24). There is unnecessary clutter, including constants and labels to the left of the shift registers. The constants are unnecessary because they are the same as the default values for each shift register's data type when first loaded into memory. Error handling is functionally nonexistent. Figure 4-20C contains the same VI with multiple improvements. However, the error cluster propagation is deferred until Chapter 7, "Error Handling."

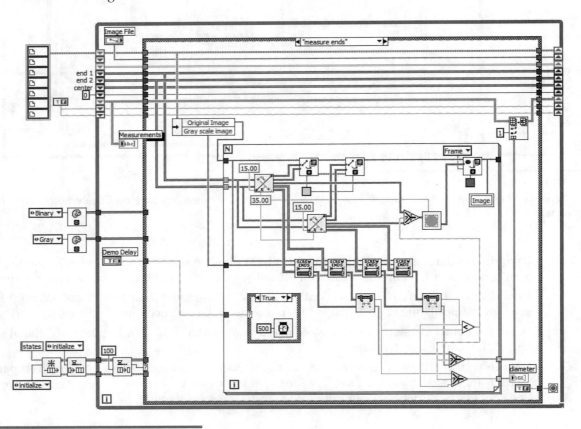

Figure 4-20A
Screw Inspection VI is neat, with cute icons and a staircase wiring appeal. At first glance, it appears to have satisfactory layout, wiring, and data flow.

Chapter 4 • Block Diagram

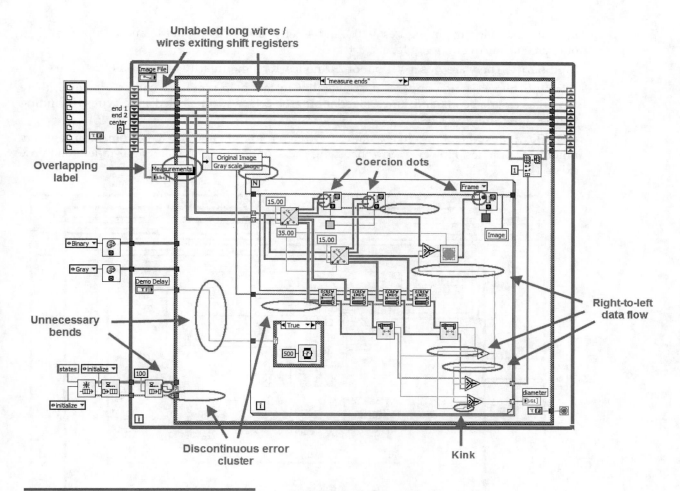

Figure 4-20B
Close inspection reveals multiple violations of the block diagram style rules.

Layout: The diagrams in Figures 4-20A and 4-20B are sufficiently dense and occupy less than one screen. One overlapping label exists.

Modularity: There are 7 user VIs out of 292 nodes, for a modularity index of 2.4.

Wiring Scheme: Wiring appears neat overall. Clusters help minimize clutter. However, there are unnecessary bends and kinks, overlapping objects, and unlabeled long wires, including wires exiting shift registers.

Data flow: Most data flows like a staircase, from left to right and top to bottom. Shift registers minimize variables. However, error handling is discontinuous and some right-to-left data flow exists.

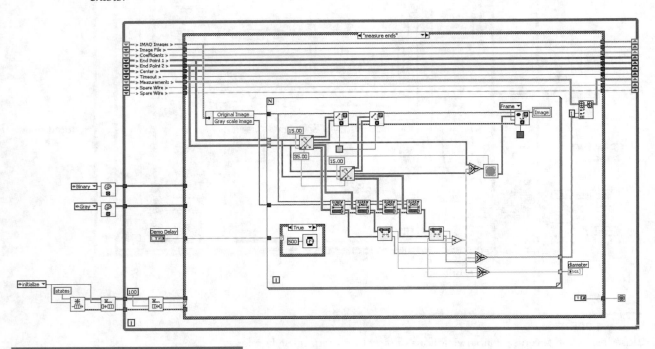

Figure 4-20C
Multiple improvements are made to the VI. Error cluster propagation is discussed in Chapter 7.

4.4.8 Optical Filter Test VI

The optical filter test application introduced in Section 4.2.2 is continued in Figure 4-21. This illustration represents the top-level VI that calls the filter measurement routine as a state within the Queued State Machine design pattern. Specifically, the filter measurement routine from Figure 4-6A is incorporated in the **Normal Sweep** case of the state machine in Figure 4-21A. The application's execution order is controlled programmatically using the queue. Shift registers facilitate data flow. However, the shift registers are randomly spaced (4.36) and not labeled (4.37). This results in wires flowing through the middle of each state, dividing the diagram into multiple sections. Additionally, there are multiple data flow and wiring rules violations, including wire crossovers (4.15) and right-to-left data flow (4.22). Specifically, the error cluster crosses over several shift register wires to reach the Dequeue Element function on the left of the Case structure. Also, the low-level routine that calculates the number of wavelength steps propagates the result to two instrument driver VIs below it. Finally, readability is hindered by several low-level routines on a high-level VI (4.7), in addition to the data flow and wiring issues.

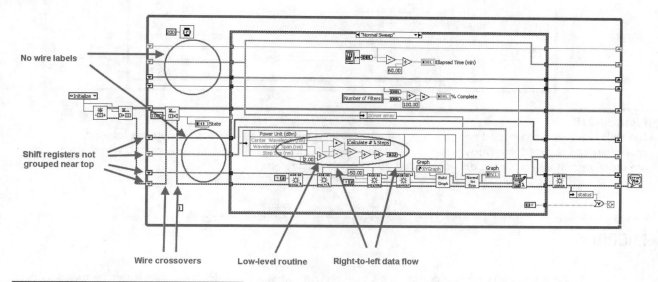

Figure 4-21A
The optical filter test application contains a Queued State Machine design pattern. The shift registers are not grouped, the wires exiting the shift registers are not labeled, there is right-to-left data flow, and there is clutter from low-level routines.

In Figure 4-21B, the shift registers are labeled and grouped near the top, forming a nonintrusive data highway and increasing the space available for nodes in each of the Case structure's cases. The queue functions are moved to the bottom of the diagram, eliminating crossovers of the error cluster wire. A subVI replaces the low-level routine for computing the number of wavelength steps, reducing clutter and eliminating the right-to-left data flow. Additionally, the wavelength and power data are modularized into clusters. Consequently, the diagram of Figure 4-21B is much more readable than the one in Figure 4-21A.

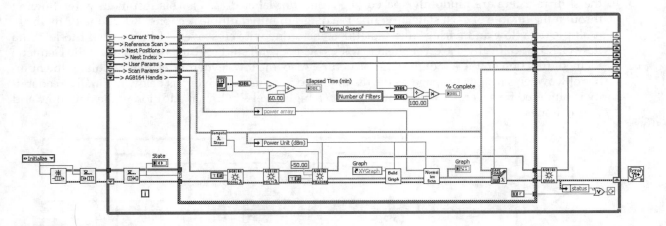

Figure 4-21B
The shift registers are grouped near the top, the wires exiting the left shift register terminals are labeled, wire crossovers are reduced, the array size computation is modularized into a subVI, the power and wavelength arrays are modularized into a cluster, and right-to-left data flow is eliminated.

Endnotes

1. NI's Developer Zone contains a Code Sharing area and an Instrument Driver Network containing free downloads, including drivers for more than 5,000 instruments. The URL is `www.ni.com/devzone/`.

Icon and Connector

5

Imagine for a moment the appearance and development techniques of traditional text-based source code. Thousands of lines of plain black text with some colored keywords and comments interspersed. Myriad cryptic symbols, function names, modifiers, and hex codes. Tracing data flow entails repeated searches for text-based variable names. Source code consists of text, colored text, and cryptic text, while editing involves manipulating text. Let us just say that the overall zest and appeal is somewhat lacking. Enter LabVIEW. Colorful and intuitive diagrams comprise the source code. Functions and routines are represented by icons. Data flow is observed by visual inspection. *Welcome LabVIEW, to the formerly mundane world of software engineering!*

Icons are a distinguishing feature of graphical source code. They are fun to create and are much more intuitive and visually appealing than lines of text. If a picture is worth a thousand words, then an icon is worth a thousand lines of code in a conventional text-based language. It is not often that engineers, programmers, and scientists have the opportunity to draw cute little pictures. LabVIEW is truly a unique environment.

The icon and connector are often the finishing touches in VI development, enabling the VI to be called as a subVI. The icon is a graphical representation of a subVI call. The connector contains terminals that allow data to flow from the wires of the calling VI's diagram, through the controls on the subVI's panel, to the terminals, wires, and nodes of the diagram; upon subVI completion, data returns from the diagram, through the indicator and connector terminals, and back to the wires on the caller's diagram. This fundamental process is illustrated in Figures 5-1A–5-1D. If the subVI's front panel window is not open and the diagram does not utilize Property Nodes or local variables, the front panel is not actually loaded in memory. Instead, the data passes directly from the subVI's connector terminals to its diagram.

The LabVIEW Style Book

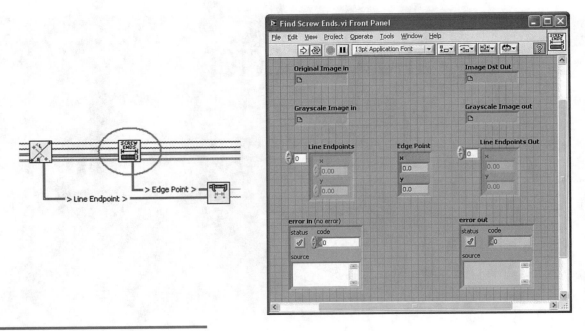

Figure 5-1A
A diagram subset contains a subVI call to Find Screw Ends VI. The subVI's front panel is open to monitor data flow. The controls and indicators are initially empty.

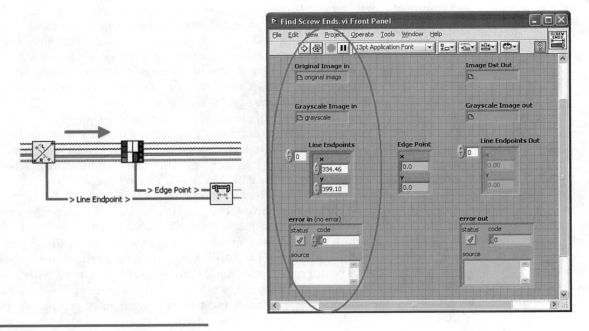

Figure 5-1B
The subVI call with terminals visible. Data flows from the input connector terminals through the controls on the front panel to the corresponding terminals on the diagram.

Chapter 5 • Icon and Connector

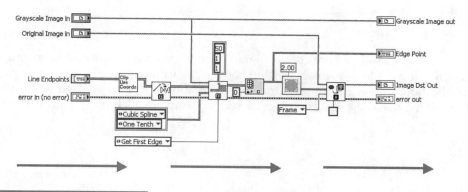

Figure 5-1C
Data flows through the terminals, wires, and nodes of the subVI's diagram.

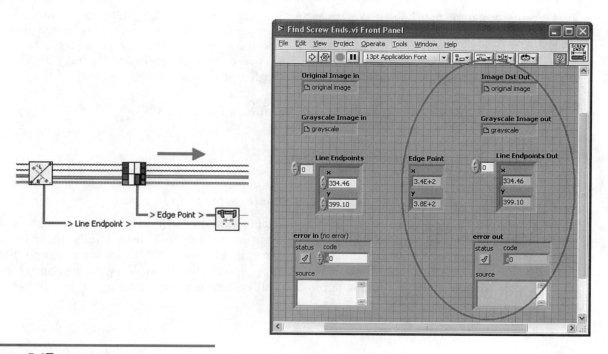

Figure 5-1D
The subVI passes data through the indicators to the output connector terminals and wires on the calling VI.

Hence, the icon and connector facilitate subVI calls. SubVIs promote modularity, organization, and code reuse. Moreover, icons and connectors make software development fun. This chapter provides style rules for the icons and connectors.

5.1 Icon

The icon is a 32×32 pixel symbol containing any combination of graphics and text. Icons are created from scratch using the tools available within LabVIEW's icon editor, or copied from an existing image file or application. Consequently, any drawing application may also be used to create or manipulate icons. Most of the rules described in this section can be implemented using LabVIEW's icon editor.

5.1.1 Icon Basics

We begin with some basic style rules for icon development.

Rule 5.1   Have fun creating icons

It is readily apparent that LabVIEW's inventors intended for us to enjoy developing software. As much as text-based programmers may try to steal our joy by teasing us for drawing pretty pictures, we know that, deep down inside, they are jealous because they are not having as much fun as we are (nor are they nearly as productive). Creating icons is an enjoyable experience for most developers. The more tips you learn, the more efficient and rewarding it becomes.

Rule 5.2   Create a unique and meaningful icon for every VI

Rule 5.3   Never use LabVIEW's default icons

Rule 5.4   Save VIs with subVI icons visible instead of terminals

Icons graphically identify the subVI, promote readability, and contribute to documenting the diagrams of the caller VIs. Consequently, it is important to create icons that meaningfully depict the VI's purpose for each and every VI. The default icons generated by LabVIEW are not unique and meaningful. Always replace them with something more appropriate. Save your VIs with their subVI icons visible on the diagram instead of using terminals. This ensures the icons remain visible each time the diagram window opens.

Rule 5.5   Use a black border

A black border drawn around the perimeter of the icon helps to clearly delineate the icon from the calling VI's diagram. Always use a black border unless you have a compelling reason not to, such as an imported graphic that requires every pixel.

Rule 5.6   Combine a glyph with color and text for best style

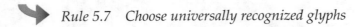

Rule 5.7   Choose universally recognized glyphs

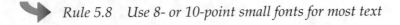

Rule 5.8   Use 8- or 10-point small fonts for most text

Because icons are only 32×32 pixels, their contents must be small and punchy. Create graphics and fonts that are legible and meaningful within these space constraints. A **glyph** is a recognizable graphical symbol, such as a public sign or traffic signal. Signs are excellent examples of graphics and text that are combined to provide distinct meaning in a small area. Libraries of glyphs, graphics, and art are available with most drawing applications and online as well. For example, NI Developer Zone hosts an Icon Art Glossary at www.ni.com/devzone/idnet/library/icon_art_glossary.htm.

Applying the default 13-point application font, we are limited to about five characters per line of text, filling one-third of the icon's height per line. This font is effective for icons that contain just a few large characters, such as **F -> C**, representing a Fahrenheit to Celsius unit conversion. Use 8- or 10-point small fonts for whole words and acronyms because this maximizes the number of characters that will fit on the icon while maintaining legibility. Using uppercase 8-point small fonts, we can squeeze seven characters of text in one line, occupying just one-quarter of the icon. Note that I am not advocating that we fill up our icons with four lines of text. Graphics and some empty space are desirable, as per Rule 5.6. However, 8- or 10-point small fonts provide more room for graphics and white space for the same amount of text as any larger font. Figure 5-2 provides a comparison of several icon fonts and capitalization.

Figure 5-2
This code snippet provides a comparison of several icon fonts and capitalization. Eight-point small font with all characters capitalized is the smallest legible font.

The most intuitive and attractive icons combine a glyph, succinct text, and two or three colors. The glyph comprises about half or two-thirds of the icon, with one to three succinct words or meaningful acronyms. Any color can be used for any portion of the icon, but a high level of contrast between foreground and background is desired. This is the case for Find Screw Ends VI, shown in Figure 5-3. A glyph of a screw dimension comprises the bottom two-thirds of the icon, and two words appear across the top third. Black and gray are used for the screw and text, and a bright yellow background is used for contrast.[1] Not the preferred color scheme for the panel of a GUI VI, but great for an icon. The combination of text and graphics is important because they reinforce each other. A glyph can be incomplete or subject to interpretation, and succinct text is an excellent enhancement.

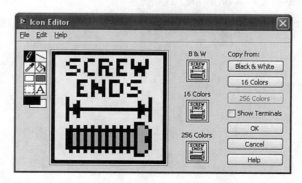

Figure 5-3
The icon of Find Screw Ends VI combines a glyph of a screw with two words created using 8-point small fonts.

Rule 5.9   Choose a unified style for related VIs

If you have a collection of VIs that work together to perform a specific task, such as data logging or instrument control, choose an icon convention that associates those VIs. One simple approach is to choose a common background color, glyph, and font, and vary the text to signify the unique meaning of each VI. For example, Figure 5-4 contains several sets of VIs, each with its own icon convention. The pressure and flow instrument icons contain a glyph of the manufacturer's logo for each instrument and a unique word across the bottom describing each VI. The configuration file icons contain black text on a light blue background, each containing the word "GET" across the top, followed by the name of the configuration parameter across the bottom. The report generation icons resemble a page on a brown border containing some text. The unit conversion icons contain a two-colored bidirectional arrow, with the measured parameter in capital letters across the top, and the word "UNIT" in the middle. Hence, graphics, color, and fonts are used to associate each set of related VIs, and text is used to distinguish the specific purpose of each VI.

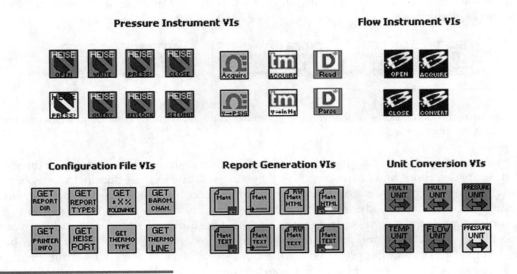

Figure 5-4
An icon convention with unified styles for several sets of related VIs

Rule 5.10 Budget your time in proportion to intended VI reuse

Budget the time and effort for icon development in proportion to the intended scope of VI reuse. If you are creating a developer toolkit or instrument driver for distribution as a product or shareware, it is worthwhile to invest significant time and effort to create a well-conceived icon convention. If you are developing a very large application with several hundred subVIs, a tight deadline, and no specific plans for reuse, you might need to apply the most expedient method possible for creating unique and meaningful icons. Icon shortcuts are discussed next.

5.1.2 Icon Shortcuts

Creating really cute icons from scratch is an art. However, some developers are not artistically inclined. Also, an artist's creativity can be compromised by deadlines and time constraints. Realistically, it is impractical to create an imaginative new icon for every VI within a large application. Shortcuts are essential. The following rules help to minimize icon development time, while ensuring meaningful and effective icons.

 Rule 5.11 Use textual foreground, colored background, and a black border for fastest results

 Rule 5.12 Contrast the text and background colors

 Rule 5.13 Choose a color convention for the type of VI

From my experience, the fastest way to create a unique and meaningful icon for every VI is to create icons that consist of a text-based foreground, a colored background, and a black border. In other words, eliminate the glyph. Choose text and background colors that provide sufficient contrast, such as black or dark text on a white or light background, or vice versa. Choose the background color to designate the type of VI, via color code convention. In the Icon Editor, start with a new VI containing a default icon, select the desired background color, and double-click the **Filled Rectangle** tool, as shown in Figure 5-5. Note that black should already be selected as the default foreground color, unless you have changed it. This operation effectively replaces the default icon contents with an empty icon filled with the background color and bounded with a black border. This is how I begin most of my icons. Type the VI's name or a succinct description using 8- or 10-point small fonts. I find that this technique minimizes the required imagination, time, and effort. If you can name the VI, then you can create an icon containing its name or a meaningful subset of characters abbreviating the name. Later, when your application is nearing completion, you can return to them and enhance the icons, if desired. If you run out of time and cannot return to it, as we often do, then you can rest assured that you have satisfied the requirements that unique and meaningful icons were created via text, and related VIs are unified via the background color.

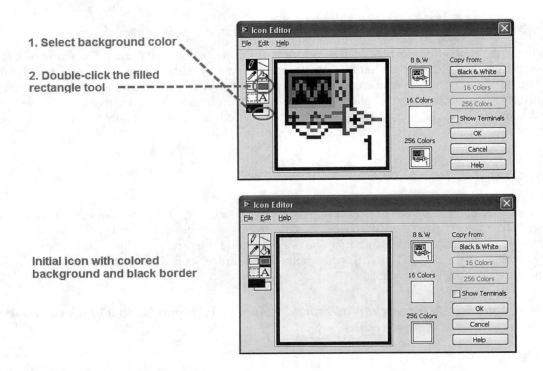

Figure 5-5
The top illustration is a new VI containing a default icon. Select the desired background color and double-click the **Filled Rectangle** tool to initiate an icon with a colored background and black border.

 Rule 5.14   Create an icon template for related VIs

 Rule 5.15   Reuse one glyph, color scheme, and font for related VIs

Commercial developer tools such as instrument drivers and toolkits demand a professional-grade icon convention. Create a project library or VI template containing a standard icon for use throughout the toolkit. For a project library, simply edit the library's default icon, and the icon is automatically applied to all new VIs in the library. For a template, create a VI containing the desired default icon, connector, controls, and indicators, and save it with the VIT extension. Create one glyph that depicts the type of operations that the VIs are used for, and leave room for a line of text across the top. Create each new VI from the library or template, and edit only the text portion of each icon. This way, you invest your time on only one glyph, and you maximize the return by applying it to all related VIs. Additionally, note that instrument drivers have some conventions already established for icons and connectors. Refer to the NI Instrument Driver Development Guidelines[2] document for more information.

Chapter 5 • Icon and Connector

Figure 5-6A contains the icon of the VI template used for a commercial developer toolkit that controls a semiconductor wafer probe system. It consists of a glyph of a probe tip contacting a semiconductor device, and some text containing an acronym for the product manufacturer and model across the top. This icon required a significant amount of time, effort, and creativity to create. As shown in the VI Tree in Figure 5-6B, all VIs in the toolkit contain the same icon convention, based on the template. Hence, only one creative icon design was developed, and the resulting template is reused throughout the toolkit. This example is revisited in Section 5.3, "Examples."

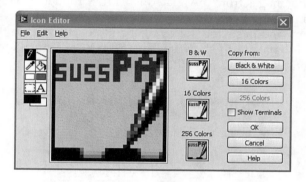

Figure 5-6A
The icon template for the Suss Interface Toolkit contains a glyph of a probe tip contacting a semiconductor device, and an acronym describing the instrument manufacturer and model.

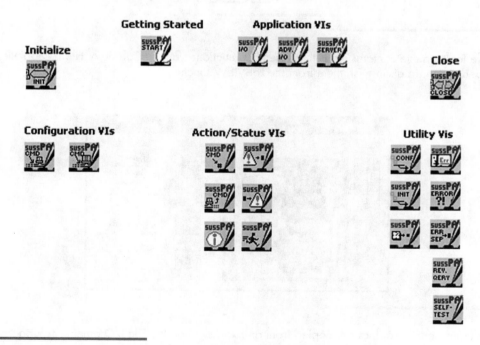

Figure 5-6B
All VIs contained in the toolkit are based on the template, as shown in the VI tree.

Rule 5.16 *Copy graphics*

Within legal limits, copy your graphics instead of drawing them from scratch. Copy and paste portions of other LabVIEW icons, or import images from a variety of applications and image files. When you copy portions of other icons, make your icons distinctively different so that your subVIs are not confused with the standard VIs in LabVIEW's `vi.lib` folder. For example, the OpenG File Tools contain icons that resemble LabVIEW's File I/O palette, as shown in Figure 5-7. These icons are formed by simply copying the images from the related LabVIEW functions. However, the OpenG File Tools are shaded green, for a distinctively different appearance. As another example, when I create file logging VIs, I like to copy and paste the floppy disk and pencil symbols that appear on other VIs and functions in the **File I/O** palette, as shown in Save Scan VI in Figure 5-8. The text, background color, and border help distinguish my VI from the LabVIEW shipping VIs.

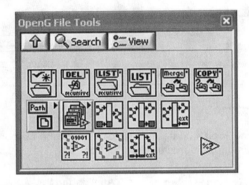

Figure 5-7
The **OpenG File Tools** palette contains icons that resemble functions from the LabVIEW **File I/O** palette. The OpenG icons are shaded green, to distinguish them from the LabVIEW functions.

Figure 5-8
The floppy disk and pencil graphics are copied from an Express VI on the **File I/O** palette. The background color and text are customized, for distinction.

Many free icon galleries are available on the Web. Most web browsers allow you to copy or save the graphics right out of a web page. Using Microsoft Internet Explorer, simply right-click on a graphic and choose **Copy**. Then open the icon editor in LabVIEW and paste the image from the clipboard onto the icon. Alternatively, drag and drop the graphic from the web browser to the icon in the upper-right corner of the target VI's front panel window. If you need to rescale the image to fit into the 32×32 pixel area, first save the image to file using a drawing application such as Microsoft Paint. Scale or edit the image within the drawing application, save the modified image file, and then import the image onto the clipboard for use in LabVIEW's icon editor using **Edit»Import Picture to Clipboard**. LabVIEW supports several standard graphic formats, including bitmaps (BMP), JPEG (JPG), Portable Networks Graphics (PNG), Multi-Image Network Graphics (MNG), and Graphics Interchange Format (GIF). Additionally, you can drag and drop the image file from Microsoft Explorer into the icon on the front panel window. Using the latter method, LabVIEW automatically resizes the image as necessary to fit the 32×32 pixel area. Finally, note that the NI Developer Zone provides an online icon art glossary[3]. The corresponding glyphs are already sized appropriately for a LabVIEW icon.

5.1.3 International Icons

LabVIEW is used internationally, with localized versions available for multiple languages. Because icons are images, they appear the same, regardless of the native language of the LabVIEW development system they are loaded into. This is an important consideration if you happen to be developing VIs intended for use by developers on an international scale. This may include the software reuse library for a multinational organization or a commercial toolkit intended for international distribution. In these instances, observe the following rule:

 Rule 5.17   Avoid text and graphics not universally understood on international icons

Although text alone makes for fast and easy icons, text is language specific. A glyph combined with a line of text is preferred, as long as the text and glyph are chosen to be as universal as possible. For example, all LabVIEW Plug and Play–compatible instrument drivers contain an acronym representing the instrument manufacturer and model. This is known as the instrument prefix, a universally recognized convention with roots from the VXI*plug&play* Consortium.

Many LabVIEW developers like to create cute glyphs describing a word or an action indirectly referencing the desired word or action. For example, an axe-wielding lumberjack or a pile of timber may be used to describe data logging, a gun may represent arming a measurement trigger, or a toilet may be used to describe the flushing of a buffer or file. Additionally, Figure 5-9 contains a fisherman that is used to represent the word **cast**, which indirectly references the VI's file casting operation. All of these examples are cute and fun, but not universal. Feel free to develop such icons for your own applications, but avoid them for developer tools intended for international distribution.

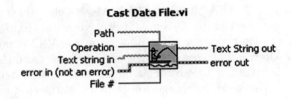

Figure 5-9
A graphic of a fisherman is used to describe a file casting operation. The association between a fisherman and the term **cast** exists only in the English language.

5.2 Connector Pane

The connector pane is the subVI's wiring terminal interface. Conventions for connector panes are presented in this section.

 Rule 5.18   Choose connectors and terminal assignments that promote clear wiring and proper data flow

Judicious terminal assignments promote clear wiring and proper data flow on the diagrams of the subVI's callers. Style rules related to wiring and data flow are presented in Chapter 4. A few of them are repeated here, for convenience.

    Rule 4.11   Minimize wire bends, eliminate kinks and loops
    Rule 4.12   Maintain even spacing of parallel wires
    Rule 4.22   Always flow data left to right
    Rule 4.23   Propagate the error cluster

We begin with some connector rules that help prevent wire clutter.

 Rule 5.19   Select a pattern with extra terminals

It is common to add and remove inputs and outputs from subVIs throughout an application's development and maintenance cycles, which necessitate changes to their terminal assignments. A change in the connector pattern is required when the number of inputs and outputs exceeds the number of available terminals. Changes to the connector pattern cause the connected wires in the callers' diagrams to move, become misaligned with their terminals, or break altogether. Additionally, every instance of the subVI call must be manually relinked by right-clicking and selecting **Relink to SubVI** from the shortcut menu on each instance. These undesirable side effects can ripple throughout an application, requiring close inspection and fine adjustments to the subVI calls. However, connector pattern changes can be avoided by always selecting a pattern that contains more terminals than you will likely need, starting from the outset.

Chapter 5 • Icon and Connector

 Rule 5.20   Choose a unified pattern for related VIs

Related VIs that are commonly used together typically share some common inputs and outputs that are normally wired together. For example, data acquisition VIs must share their task and error cluster by propagating wires between the corresponding input and output terminals of each VI. Likewise, instrument control VIs share a common VISA resource name and error cluster. Wire bends are avoided if the related VIs have the same connector pattern, with the common inputs and outputs assigned to the same terminals. As we can see from Figure 5-10, the DAQmx VIs all use the 5×3×3×5 connector pattern, with the task assigned to the top left and right terminals, and the error cluster assigned to the bottom left and right terminals. In Figure 5-11, the niDMM VIs all use the 4×2×2×4 connector pattern.

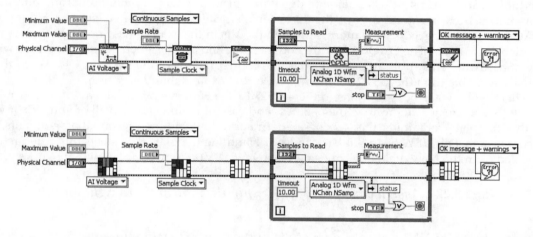

Figure 5-10
Cont Acq&Graph Voltage-Int Clk VI is a LabVIEW shipping example containing several DAQmx VIs. The standard icon view is shown at the top. These VIs utilize the 5×3×3×5 connector pattern, as shown at the bottom.

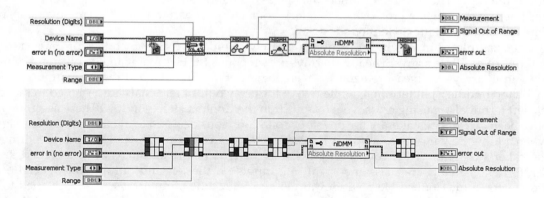

Figure 5-11
DMM Measurement VI is an example VI from the niDMM instrument driver. These VIs all utilize the 4×2×2×4 connector pattern, as shown at the bottom.

Rule 5.21 Use the 4×2×2×4 connector pattern for most VIs

We can take the unified pattern rule one step further and standardize on the same pattern for most VIs we create, regardless of their purpose and association. If all of our VIs have similar connector patterns and terminal assignments to the maximum extent possible, the benefit to our diagrams is maximized. As of LabVIEW version 8.2, 4×2×2×4 is the default pattern applied to new VIs. I recommend this pattern because it promotes better modularity and neater diagrams than patterns with more terminals, while providing ample room for expansion compared to patterns with fewer terminals. If you need more than 12 connector terminals, you are probably performing too many tasks within one subVI, or you may need to modularize some of your wires with a cluster. As an example, Find Screw Ends VI from Figure 5-1 uses the 4×2×2×4 pattern. It propagates an array of cluster named **Line Endpoints** and returns a cluster named **Edge Point**, in addition to six other inputs and outputs. These parameters occupy a total of 9 of the 12 terminals, leaving 3 spares. As seen in Figure 5-10, the DAQmx VIs use the 5×3×3×5 pattern, although the VIs in the example use no more than 11 of the 16 available terminals. The DAQmx VIs are polymorphic and are designed to support a very wide variety of hardware and applications. Indeed, some configurations of the DAQmx VIs exceed the 12 terminals available from the 4×2×2×4 pattern, and future expandability is critical.

There will always be exceptions to the 4×2×2×4 rule. However, you can save *at least* 1 minute on every VI you create *and* improve the neatness and ease of maintenance of all of your VIs if you start with a template containing the 4×2×2×4 pattern and a few common terminal assignments, such as error clusters. LabVIEW has a SubVI with Error Handling template that meets this criteria. It may be selected from the New dialog by choosing **File»New…»VI»From Template»Frameworks**.

 Rule 5.22 Assign controls to left terminals, indicators to right terminals

 Rule 5.23 Never cross wire stubs in the Context Help window

The terminal assignments of the connector pane must promote left-to-right data flow, without wire bends and crossovers. Therefore, controls must be assigned to terminals on the left and indicators to terminals on the right. Otherwise, input wires flowing in from the left will cross output wires flowing out to the right. The middle terminals may be assigned to controls or indicators, as long as they are not intermixed, causing crossovers. To check your terminal assignments for crossovers, inspect the wire stubs in the Context Help window.

The example VI in Figure 5-12A uses a densely populated 5×3×3×5 connector pattern to receive and return multiple arrays. The input arrays are set points for FieldPoint analog output modules, and the output arrays are measurements from analog input modules. One of the array inputs is assigned to the top middle right terminal, while two of the array output terminals are assigned to the top middle left terminals, causing crossovers to appear in the Context Help window. These assignments may have been careless mistakes by the developer or a conscious decision to avoid terminal reassignments when adding new inputs or outputs. Indeed, each instance of this VI would require rewiring within the caller's diagram if the connector pattern is changed or the terminals are reassigned. In Figure 5-12B, the terminals are reassigned with the inputs on the left and outputs on the right, and no crossovers in the Context Help window. In Figure 5-12C, the input arrays are all assigned to the left side and bottom middle terminals, and the output arrays are all assigned to the top middle and right side terminals. Figures 5-12B and 5-12C are equivalent methods of eliminating crossovers. However, the dense population of the 5×3×3×5 connector pattern limits expandability and maintainability while promoting wire clutter. In Figure 5-12D, the input and output arrays have each been combined into clusters, reducing the required number of terminals while increasing expandability and maintainability. Additional arrays can simply be added to the clusters without adding or rearranging terminals. Also, the connector pattern is reduced to the preferred 4×2×2×4 pattern.

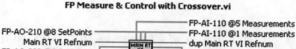

Figure 5-12A
This VI uses the 5×3×3×5 connector pattern. The Context Help window reveals wire stub crossovers.

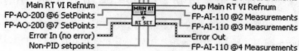

Figure 5-12B
Connector assignments with middle terminals divided, with controls on the left and indicators on the right and no crossovers

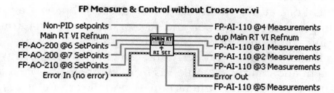

Figure 5-12C
Connector assignment with controls assigned to bottom middle terminals and indicators assigned to top middle terminals, and no crossovers

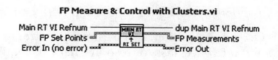

Figure 5-12D
Input arrays of set points and output arrays of measurements have been combined into clusters, and the 4×2×2×4 pattern has been applied. The number of assigned terminals is greatly reduced, decreasing wire clutter on the calling VI diagrams.

Rule 5.24 Specify terminal assignments resembling the panel layout

 Rule 5.25 Assign error clusters to bottom left and right terminals

 Rule 5.26 Assign references and I/O names to top left and right terminals

It is a long-standing convention that error clusters are always assigned to the bottom left and right terminals, and references and I/O names are always assigned to top left and right terminals. References and I/O names include file refnum, DAQmx task and channel names, VISA resource name, VI Server application and VI references, control references, and similar items. The remaining controls and indicators may be assigned to the remaining terminals, in the same respective locations as the object positions on the front panel. Therefore, the terminal assignments should resemble the front panel layout. Recall that panel layout rules were provided in Section 3.2, "Text."

Consistent placement of controls and indicators on the subVI panels, and consistent connector terminal assignments improve the recognition and ease of use of subVIs throughout your applications. Additionally, assigning common controls to consistent terminals of a standard connector pattern ensures that the terminals will line up properly and can be wired together without wire bends. In the FieldPoint measurement and control example VIs from Figure 5-12, note that the refnums are appropriately assigned to the upper left and right terminals, and the error clusters are assigned to the lower left and right terminals.

 Rule 5.27 Choose left and right vertical edge connector terminals for high priority inputs and outputs

 Rule 5.28 Choose top and bottom horizontal edge connector terminals for lower priority inputs and outputs

Be sure to prioritize your terminals appropriately. Priority can be specified explicitly using the **required**, **recommended**, and **optional** priority settings from the context menu, and implicitly using the locations of the terminal assignments. The most commonly used inputs and outputs are normally assigned to terminals on the left and right vertical edges of the connector pane, and the less commonly used inputs and outputs are on the top and bottom horizontal sides. The front panel layout should reflect these priorities, as was described in Chapter 3, "Front Panel Style."

 Rule 5.29 Specify required priority for critical inputs and outputs

 Rule 5.30 Specify optional priority for inputs and outputs that are normally not used

If a subVI performs file I/O, data acquisition, or instrument control with a resource that was opened prior to the subVI call, a valid reference, task, or resource name is required or the subVI cannot operate properly. Specify **required** priority for any critical terminals that the subVI cannot run without, such as reference inputs. Additionally, specify **optional** priority for any inputs or outputs that are normally not used because the default values are generally acceptable. This way, the user can quickly identify the important terminals by the bold appearance of the labels in the Context Help window.

However, do not set priorities unless they are really warranted. Leave the default **recommended** priority for most terminals.

Figure 5-13 shows the Context Help window containing Find Screw Ends VI. **Original Image in** and **Grayscale Image in** are references to images that are required for the VI to run properly. The corresponding terminal labels are boldface, indicating that the priority of these two terminals is required. All other terminals are recommended. All input terminals on the left side are propagated through the subVI's diagram to corresponding output terminals on the right side. This provides consistency as well as symmetry. The **Edge Point** is an important output but is assigned to a bottom middle terminal, to allow consistent assignment of common terminals among related VIs.

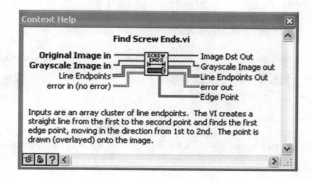

Figure 5-13
Context Help window for Find Screw Ends VI indicates required priorities for **Original Image in** and **Grayscale Image in**, and recommended priority for all other terminals.

Note that many subVIs have an input for the index of looping structures. These subVIs may initialize or reset a resource or variable upon the first iteration of the loop (iteration = 0). Because the index terminal is normally at the bottom left corner of a looping structure, assigning the control to a bottom middle terminal helps minimize wire bends and overlaps.

5.3 Examples

This section contains examples of icons and connectors, including some that highlight, reinforce, and extend the syle rules from the prior sections, as well as a few exceptions to the rules. Some are quite clever, while a few are downright obnoxious. We begin with the latter.

5.3.1 Obnoxious Examples

Figure 5-14A contains the icon, terminals, and connector assignments for SubVI from Selection VI, previously discussed in Chapters 3 and 4. A subVI from selection is created automatically by selecting an area of a diagram and choosing **Edit»Create SubVI**. They are characterized by LabVIEW's default icon, unconventional front panel layout, counterintuitive control and indicator labels, and sloppy wiring on the diagram. As discussed in the previous chapters, these VIs never conform to good programming style without significant cleanup. Specifically, the terminal labels in Figure 5-14A are inappropriate, including **task ID out**, **task ID out 2**, **error out 2**, and **output chunk**. Input terminals should not contain the word **out**, and duplicate terminal labels should not be distinguished using numbers. Additionally, it contains the default icon assigned by LabVIEW instead of a unique and meaningful icon. Finally, it utilizes the 3×2 connector pattern, with no spare terminals. These are violations of Rules 3.21, 5.2, 5.3, 5.19, and 5.21.

SubVI from Selection w Cleanup VI, shown in Figure 5-14B, is an improved revision of the VI. The duplicate terminal labels are distinguished using the suffix **in** for the inputs and **out** for the outputs. The input array label has the prefix **input** instead of **output**. The VI has a unique and meaningful icon that depicts what the VI actually does. It utilizes the standard 4×2×2×4 connector pattern, with plenty of spare terminals available for future expansion.

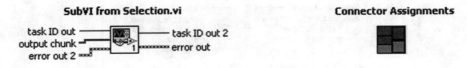

Figure 5-14A
SubVI from Selection VI violates multiple style rules, including default icon and counterintuitive terminal labels.

Figure 5-14B
SubVI from Selection w Cleanup VI contains a meaningful icon, intuitive terminal labels, and the standard 4×2×2×4 connector pattern.

On more than one occasion, I have seen source code in which the developer creates icons by drawing the text freehand instead of using the Icon Editor's text tool. Figures 5-15 and 5-16 are two such examples. Figure 5-15A is a subVI extracted from the Haphazard VI from Chapter 4. We can see that it also contains an output assigned to a terminal on the left, and the data acquisition task reference has been bundled with the error cluster into one input. These elements are unrelated and should not be bundled together. Figure 5-15B is a revision containing an improved icon, connector pattern, and terminal assignments. The icon utilizes 8-point small fonts for the complete words **DIGITAL INPUTS** and a glyph of a pair of glasses copied from the DAQmx Read VI. The standard 4×2×2×4 connector pattern has been applied, with the **DAQmx Task** passed through the top left and right terminals.

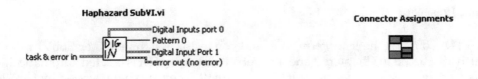

Figure 5-15A
Icon contains text drawn freehand, and connector assignments are unconventional.

Chapter 5 • Icon and Connector

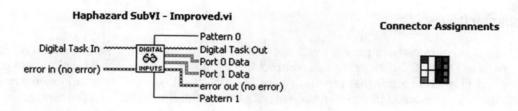

Figure 5-15B
The subVI icon has been improved using 8-point small fonts and a glyph of a pair of glasses. The DAQmx task passes through the top left and right terminals of the 4×2×2×4 connector pattern.

Figure 5-16A is a VI icon containing a custom font that the engineering staff at Bloomy Controls affectionately refers to as **Bob's Bold**, after its creator, Bob Hamburger. Bob is a highly regarded Business Development Manager at Bloomy Controls who also developed the Find Screw Ends VI icon in Figure 5-4, which exemplifies good style. I would like to thank Bob for contributing his greatest as well as his most obnoxious icons for public display. The characters of Bob's Bold are so big and bold that the icons in an application appear to be shouting at you, as seen in Figure 5-16B. Bob's Bold is a great way to get your point across, but it should be used sparingly, to keep your diagram's decibel level within legal limits.

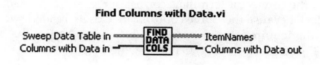

Figure 5-16A
This icon illustrates Bob's bold font.

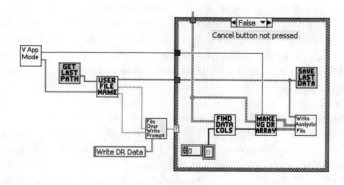

Figure 5-16B
Bob's bold font on several icons in a diagram section appears to shout.

5.3.2 Instrument Drivers

This section contains examples of icons and connectors corresponding to instrument drivers. The NI Instrument Driver Guidelines[2] covers comprehensive requirements for the development of LabVIEW Plug and Play–compatible instrument drivers. Most of the Plug and Play requirements are consistent with the rules presented in this chapter, as well as throughout this book, with a few minor exceptions that are noted where they appear.

Figure 5-17A shows the Initialization VI for a fictitious medical instrument, the PM-1000. The icon consists of three lines of text and a glyph of a bedridden patient on a plain white background. The glyph is cute, but the icon is a bit too text heavy, giving it a complicated appearance. The connector uses the 5×3×3×5 connector pattern, and 12 of the 16 terminals are assigned to controls and indicators with recommended priority. If all 12 of these terminals are wired on the caller's diagram, it will appear cluttered.

Closer inspection of the icon's text reveals that all 14 characters on the first two lines are dedicated to identifying the instrument's manufacturer and model. This can be improved. There exists a long-established convention, originating from the VXI*plug&play* consortium, for which a very succinct **instrument prefix** is used to represent the instrument's manufacturer and model number. The first two or three characters of the prefix identify the manufacturer, and the following four or five characters identify the instrument model or family. The resulting six- to eight-character acronym is used in the VI or library name and appears as a **banner** across the top of the icons throughout the driver. Additionally, five of the connector terminals correspond to serial port settings, in addition to a terminal for the **Serial Port**. Because these five controls are closely related, we can modularize them into a cluster and reduce the required number of terminals, and also use the preferred 4×2×2×4 connector pattern. The Serial Port terminal is an I/O name, which is rightfully assigned to the top left terminal, per Rule 5.26.

In Figure 5-17B, the icon has been improved by replacing the top two lines of text with a banner containing **PHPM1k** as the instrument prefix. Additionally, the glyph is centered, for better visibility. The icon appears clear and simple as a result of these changes. Also, five serial port settings have been modularized into a cluster, and the 4×2×2×4 connector pattern is applied. The **ID Query** and **Reset** controls have been assigned to the middle terminals of the icon, which implies a slightly lower priority than the **Serial Port Settings** and other input terminals on the left border.

Figure 5-17C contains the icon and connector assignments for the Initialize VI from the instrument driver template. This VI is generated automatically by the Create New Instrument Driver Project utility that is selected from **Tools»Instrumentation»Create Instrument Driver Project**. The **Reset** and **ID Query** controls are assigned to terminals on the left border of the connector, and the **Serial Port Settings** cluster is assigned to the top middle terminal. These assignments are equivalent but more consistent with instrument driver standards than Figure 5-17B. The icon contains a standard glyph that the wizard applies to all Initialize VIs. The standard glyph helps to improve the recognition of common instrument driver VIs using graphics in addition to text. Also, the standard glyph reduces icon editing effort because it is created automatically by the utility. However, the standard glyph reduces the association of the VI with the specific instrument, as well as the association of the VIs that comprise the instrument driver. Instead, the banner is the only means by which the instrument is identified. Hence, a tradeoff exists between recognizing common function types using the function-specific glyphs and recognizing the instrument using a common instrument-specific glyph. The NI guidelines recommend the former, whereas I generally prefer the latter.

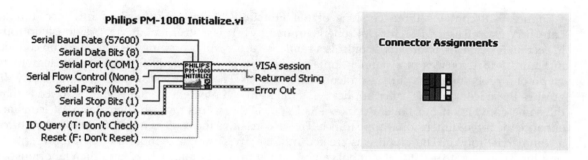

Figure 5-17A
Instrument driver VI for a fictitious medical instrument has a text-heavy icon and too many terminals.

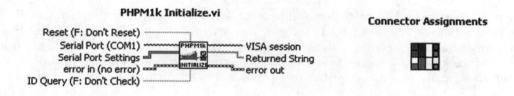

Figure 5-17B
The VI has been improved with a banner containing the instrument prefix, a centered glyph, cluster input, and fewer terminals.

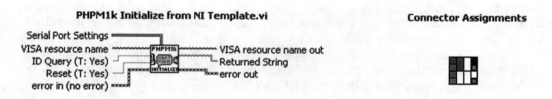

Figure 5-17C
The icon and connector of the Initialize VI generated automatically using the Create New Instrument Driver Project utility. The utility uses a template to create a standard glyph, connector assignments, and terminal labels.

Before embracing the PHPM1k Initialize VI revisions in Figures 5-17B and 5-17C, there are a few considerations. Depending on the application's data structures, the callers may now have to perform a bundle operation to program the serial parameters. A tradeoff exists between reducing wire clutter and adding cluster operations to the caller's diagram. According to the NI Instrument Driver Development Guidelines, passing cluster information between VIs increases the complexity of the calling application. However, my recommendation is to use consistent style for the instrument driver VIs, as for their calling applications. Specifically, I recommend using clusters if the data elements are closely related and are likely to be used together in more than one place. Also, judicious use of clusters in an instrument driver promotes good data structures throughout the calling applications. For example, the user of the revised VI may now be inclined to use the same **Serial Port Settings** cluster to pass this data among other VIs in the call chain. Data types and structures are discussed in greater detail in Chapter 6, "Data Structures."

Figure 5-18 contains three versions of the **Functions** palette for the Suss Interface Toolkit for LabVIEW, as well as the icon for DoProberCommand VI. This instrument driver, previously introduced in Section 5.1.2, "Icon Shortcuts," controls a semiconductor wafer probing system. The icon convention in Figure 5-18A consists of a banner containing the instrument prefix **SussPA** and a demonstrative glyph of a probe tip contacting a semiconductor device. The bright yellow background clearly distinguishes these icons on the caller application's diagrams. The palette contains submenu icons that use the same convention but have a darker shade of yellow, to distinguish them from VI icons and to enhance the three-dimensional appearance. These icons have the professional look and feel required for a commercial toolkit. The glyph was created by first envisioning a symbol that best represents wafer probing, searching to see if that symbol can be copied from a public archive, and then laboriously editing that graphic for best appearance on a 32×32–pixel icon. It is important to note that the same glyph appears on the icons of all VIs of the toolkit, and the toolkit is a commercial product. Therefore, a high-quality icon convention is required, and it is worthwhile to invest the necessary time and effort.

DoProberCmd VI is one of the VIs from the **Action/Status** palette. The icon in Figure 5-18A contains the palette's icon convention, with the addition of an artful silhouette of a runner. The VI name and the icon could more closely resemble each other. Why not name it SussPA *Run*ProberCmd VI? Alternatively, the runner could be replaced with the word "Do." Moreover, the toolkit is intended for international use, and the glyph of a runner is not a universal symbol for the VI's primary purpose of executing a command. Finally, note that **Registration In** and **Command** are input terminals with required priority.

Figure 5-18B contains the palette of a similar instrument driver developed from the LabVIEW instrument driver template, using the Create New Instrument Driver Project utility. It contains standard glyphs for common functions and subpalettes, in place of the instrument-specific glyph. The banner is the only feature identifying the instrument on the icon. Consequently, the palette appears rather generic, with no direct association to wafer probing. DoProberCmd VI contains the green sideways triangle glyph commonly used to indicate Run, Start, or Initiate. This glyph is more appropriate than the runner in Figure 5-18A for describing the execute command function, but the association with wafer probing remains weak.

Figure 5-18C merges the standard glyphs from the instrument driver template with the custom icon convention for the prober into a new icon convention that provides the best of both worlds. The wafer prober glyph provides direct association with the instrument type. The template glyphs reinforce the consistency with LabVIEW Plug and Play standards and reduce the dependence on text. The resulting driver is intuitive and fully conforms to LabVIEW Plug and Play standards.

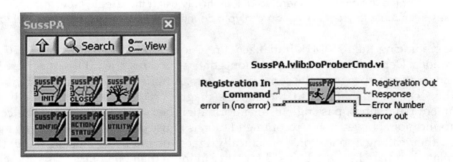

Figure 5-18A
The **Functions** palette for the Suss Interface Toolkit for LabVIEW, a commercial product that controls a semiconductor wafer probe system. All icons share a common glyph describing the instrument. DoProberCmd VI is a subVI from the **Action/Status** subpalette containing an artful silhouette of a runner and common glyph.

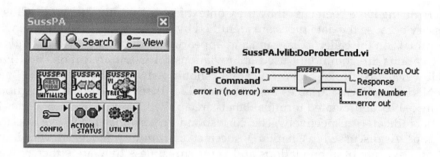

Figure 5-18B
An alternate icon convention consists of the standard glyphs from the instrument driver template. The icon for DoProberCmd VI contains a sideways triangle, a standard glyph for Run, Start, and Initiate. This convention weakens the association with the wafer prober.

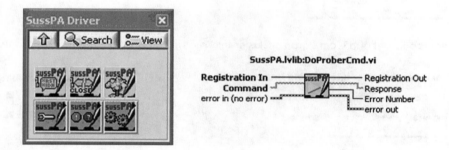

Figure 5-18C
The standard glyphs from the instrument driver template are merged with the instrument-specific glyphs to maintain strong association to both the wafer prober and instrument driver icon conventions.

Unfortunately, it is not always possible to merge multiple glyphs and a banner on a single icon that is only 32×32 pixels. In fact, it can be very tedious to attempt to do so. If your icons contain a banner with the instrument prefix, a color convention, and a reasonable balance between glyphs and text, and if all of the required VIs and palettes are organized according to the NI guidelines, the driver should satisfy the icon requirements for a LabVIEW Plug and Play instrument driver. It is more important for your VIs to be intuitive and strongly associated with the instrument than it is to share the same glyphs as the template. The users of your instrument driver can also identify the VIs based on the VI names and locations on the palette.

5.3.3 Miscellaneous Examples

This section presents some random examples that also may be improved upon. The icon and connector from Figure 5-19A belongs to a VI that converts a data structure into a formatted string. The icon consists of a clever glyph of a telephone handset and text. I commend the developer's creativity, or her ability to find and copy graphics, whichever the case may be. The connector, however, can use a few improvements. First, the labels of the control and indicator assigned to the input and output terminals are too long. In fact, the full labels **Communication Parameters Cluster** and **Communication**

Parameters String have been cut off by the Context Help window. The words **Cluster** and **String** are not necessary because the data types are identified by the design and color of the wire stubs provided by the terminals on the caller's diagram, and appearing in the Context Help window. Also, the terms **Com** and **Params** are commonly used abbreviations for Communications and Parameters, respectively. However, controls and indicators should always have unique names; otherwise, Property Nodes and local variables that are created for them may appear ambiguous.

In Figure 5-19B, abbreviated terminal labels are applied, along with the suffixes **in** and **out** for the control and indicator, respectively. The corresponding terminal labels are succinct, intuitive, and unambiguous. As discussed in Chapter 3, succinct and intuitive labels are very important. However, I am not an advocate of abbreviations and acronyms unless they are either universally familiar or extremely intuitive, as they are in this case. Additionally, error terminals have been added to promote proper data flow.

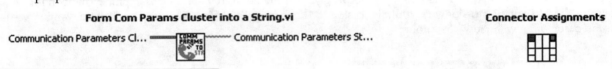

Figure 5-19A
The lengthy terminal labels of Form Com Params Cluster Into a String VI are truncated by the Context Help window.

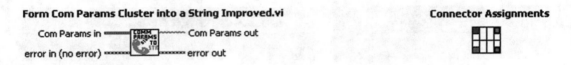

Figure 5-19B
The label lengths have been reduced, and error terminals have been applied to promote proper data flow.

The example shown in Figure 5-20 is the DAQ Assistant Express VI icon and connector assignments, with **View as Icon** selected from the shortcut menu. This particular configuration generates inputs assigned to terminals on the right instead of the left. This causes wire crossovers in the Context Help window, and promotes wire crossovers in the diagrams of the calling VIs.

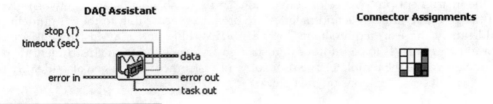

Figure 5-20
DAQ Assistant Express VI has inputs assigned to terminals on the right, causing wire crossovers on the calling VI diagrams.

Figure 5-21A shows the icon and connector of Confirm Quit VI. The icon contains a familiar Windows system symbol that accompanies various confirmation dialogs. The glyph appears big, bold, and unmistakable. However, it does not indicate the specific type of confirmation. It contains only one output assigned to the middle terminal of the 3×3 connector pattern. In Figure 5-21B, the Confirm Quit dialog is revised with error clusters, a 4×2×2×4 connector pattern, and an icon combining text with the graphic. The type of confirmation is specified on the icon by reducing the glyph size and adding the word **QUIT?** using 9-point small fonts. Error cluster terminals have been added to facilitate execution ordering via data flow. The 4×2×2×4 connector pattern allows the input and output terminals to line up with VIs utilizing this standard connector pattern. In the future, the VI could be enhanced to handle more than one type of confirmation, by adding an enumerated control to specify the confirmation type and changing the icon text as appropriate.

Figure 5-21A
Confirm Quit VI has a simple graphic that has been copied from a Windows system prompt. The type of confirmation is not identified by the icon.

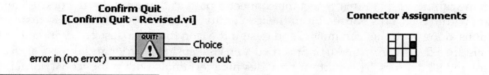

Figure 5-21B
The icon has been modified to indicate the type of confirmation. The standard 4×2×2×4 connector pattern has been applied, as well as the assignment of error terminals to facilitate execution ordering.

5.3.4 Clever Examples

Print VI in Landscape Mode VI, shown in Figure 5-22, is composed of three-quarters graphic and one-quarter text. The graphic consists of a rectangular control panel that represents a GUI panel. A single line of text in 10-point small font simply reads **Print VI**. The VI uses the default 4×2×2×4 pattern with the appropriate assignments. It is neat and effective, and looks great in color.

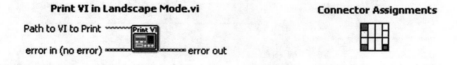

Figure 5-22
Print VI in Landscape Mode VI is composed of three-quarters demonstrative graphic and one-quarter text.

Dynamic VI Path Builder VI, shown in Figure 5-23, contains an attractive icon that is two-thirds graphic and one-third text. The graphic depicts a fork in a yellow road, with both roads leading through green grass into an aqua blue horizon. Again, you probably need to see the full-color version to appreciate it. The VI is used to build a file path. Beware that the relationship between a graphic of a foot path and a file path is language sensitive. The 3×3 connector pattern is used to center the single input and single output terminals. However, a standard 4×2×2×4 connector pattern with error terminals should be used for consistency and proper subVI error handling. SubVI error handling is discussed in Chapter 7, "Error Handling."

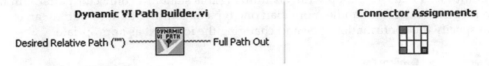

Figure 5-23
Dynamic VI Path Builder VI is composed of two-thirds demonstrative graphic and one-third text.

The next two examples utilize some special effects to achieve unique sizes and shapes. This is accomplished by erasing the default border and drawing a new border in the desired shape of the icon. LabVIEW uses the outermost nonwhite pixels to define the icon's border. It is important to note that all controls and indicators designed to pass data into and out of the subVI must be assigned to connector terminals that reside within the icon's border. All terminals outside the border are not visible on the calling VI diagrams. Select the connector pattern that provides the desired terminal configuration within the icon's border. From the icon editor, select **Show Terminals** to display the terminal partitions along with the icon image, and design the icon to fit the desired terminal locations.

In Figure 5-24, Clear Error All or Specified VI uses a custom border and the 5×3×3×5 connector pattern rotated 90 degrees. This pattern provides nine terminals within the icon's border, five of which are assigned. The connector pattern is rotated by selecting **Rotate 90 Degrees** from the connector's shortcut menu. This VI's function is discussed in Chapter 7.

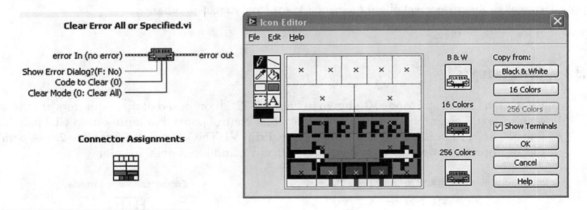

Figure 5-24
Clear Error All or Specified VI has a custom border providing a unique shape, and a 5×3×3×5 connector pattern rotated 90 degrees.

The VI in Figure 5-25 also has a unique shape formed by a custom border. Its purpose is to incorporate an in-line delay using data flow instead of a Sequence structure. Recall that good block diagram style involves using data flow over Sequence structures and variables. Wait n mSec VI promotes good style. Its small size allows it to be seamlessly inserted between the VIs of a dataflow sequence, while allowing other wires to be connected among the input and output terminals without bends. The connector pattern is 5×3×3×5, with three terminals assigned.

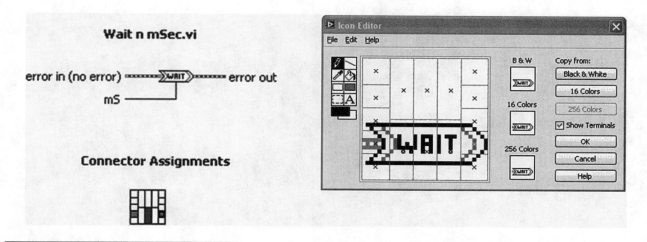

Figure 5-25
Wait n mSec VI has a custom border and promotes good data flow style. It has a 5×3×3×5 connector pattern with three terminals assigned.

Endnotes

1. Color versions of the example illustrations in Chapter 5 can be freely downloaded from the Publisher's website page for this book at www.prenhallprofessional.com/title/0131458353. For full-color illustrations, the electronic (eBook) version of *The LabVIEW Style Book* is available for purchase from www.prenhallprofessional.com/title/0132414813.
2. NI Instrument Driver Guidelines, online tutorial available from the NI Developer Zone. The URL is `www.ni.com/devzone/idnet/library/instrument_driver_guidelines.htm`.
3. An icon art glossary is available from the NI Developer Zone at `www.ni.com/devzone/idnet/library/icon_art_glossary.htm`.

Data Structures

6

The data structures of an application consist of LabVIEW's built-in data types and developer-defined data constructs. **Data types** are the fundamental data elements depicted by unique terminal or wire styles on the diagram. Data types define the memory size and functionality of the data. **Data constructs** are data sets that use the fundamental data types as elements, such as arrays and clusters. **Data structures** are the data types and data constructs used by an application. Data structures are defined by the developer's choice of controls, arrays, and clusters on the VI front panels, as well as the operations performed on the diagrams.

Control types determine how users understand and operate the front panels. This includes both GUI VIs that users interact with and subVIs that only the developers see. As presented in Chapter 3, "Front Panel Style," judicious selections for controls make the panels intuitive, user-friendly, and reliable. Additionally, each control type supports one or a limited set of data types. Consequently, the developer's choice of controls helps determine the application's data structures.

Data constructs are used to organize data and reduce wire clutter on the diagram. Consistent use of data constructs throughout an application help make it easy to understand and maintain, as well as memory efficient. Nested data structures have special considerations with respect to performance and memory use, as discussed later in this chapter.

> **Theorem 6-1:** *Because memory and data storage access rates are the principal performance barriers of modern computing devices, software execution speed is inversely proportional to memory use.*

It is a fact common to all modern software that execution speed is inversely proportional to memory use. Specifically, memory and data storage access rates are the principal performance barriers of modern computing devices. A LabVIEW application's memory use is defined by its data structures and the operations performed on the data. LabVIEW does not impose restrictions on the data structures that an application may use. Therefore, an application's performance is directly related to the developer's choice of data structures and corresponding operations. This chapter presents style rules for choosing data structures that provide intuitive, reliable, and efficient operation.

6.1 Data Structure Design Methodology

Evaluate the data structures during the design phase of application development. Describe the data required for each major subsystem, including the GUI, data acquisition and instrument I/O, analysis, report generation, file I/O, database queries, and network and interapplication communications. Examine all of the application's primary data sources and destinations. Prototyping front panel development is a very useful technique for specifying the GUI, as discussed in Chapter 2, "Prepare for Good Style." Additionally, use prototype front panels to specify the data required for each component of the application.

When coding begins, the data structures are implicitly declared by the developer's choice of controls and indicators on the front panels, as well as the inputs and outputs required by the nodes on the diagrams. Implementing the data structures for a VI is generally a three-step process. First, select the controls and data types. Second, configure the properties of those controls. Third, group the controls into data constructs, if appropriate. In this section, we start with some general rules to guide us through this process. The sections that follow discuss rules that apply to specific types of LabVIEW data structures.

6.1.1 Choose the Controls and Data Types

 Rule 6.1   Choose controls that simplify the operation of the panel

The term *simple* has several connotations with respect to LabVIEW controls and data structures. **Simple data structures** store data in contiguous memory addresses. They include all scalar data types, such as Boolean, numeric, and string; arrays of Boolean and numeric, and clusters containing only the aforementioned simple data types. **Simple controls** are controls that are intuitive and easy to operate, and that represent simple data structures. Simple controls possess properties that can be configured to help validate user and programmatic input. When properly configured, simple controls are reliable and memory efficient.

To some degree, control types can be classified in order of operational and data simplicity. For example, numeric controls are very intuitive, can be configured to restrict values to a specified range, and are stored efficiently in memory according to the representation. String controls can contain any alphanumeric data, without restriction, and are stored as blocks of contiguous bytes, 1 byte per character. Boolean controls are maximally restrictive, allowing only a value of TRUE or FALSE, and are stored as byte integers. Therefore, of these three control types, Boolean is the simplest, followed by numeric and then string. Table 6-1 contains the complete list of LabVIEW control types, listed in approximate order of operational and data simplicity. The controls at the top are the simplest, beginning with Boolean and numerics, and the controls at the bottom are potentially the most complex. Note that the list is both overlapping and subjective. For example, numeric simplicity depends on the representation. Arrays and clusters are data constructs, and their simplicity depends on their contents. Also, a string's memory use is the same as a one-dimensional array of byte integers of equivalent length. Therefore, the control types are approximately organized by data simplicity. Because arrays have the potential to contain larger and more complex data than strings, the string type is ordered simpler than array.

Table 6-1 *LabVIEW Controls and Descriptions, Listed in Top-Down Order of Operational and Data Simplicity*

	Boolean—Buttons, switches, and lights that are used to enter and display Boolean (TRUE/FALSE) values.
	Ring—Provides a user-selectable list of strings (text or menu ring) or pictures (picture ring) that are associated with numeric values.
	Enumeration—Provides a user-selectable list of strings that are associated with numeric values, similar to a text ring, except that the numeric values and string labels are part of the data type. Also called an enum.
	Tab—Consists of multiple pages for controls and GUI objects, and a selector for choosing the foreground page. Each tab is associated with a text label and numeric value, similar to an enumeration.
	Numeric—Simple method of entering a scalar numeric value. Can be configured for a variety of whole and fractional numeric representations.
	Listbox—Provides a user-selectable list of strings associated with numeric scalar (single selection mode) or array (multiple selections mode) values.
	Combo box—Provides a user-selectable list of strings associated with the same or an alternate list of strings.
	String—Used to enter or display alphanumeric data. Multiple display modes include normal \ codes and hexadecimal.
	Tree—User-selectable list of strings organized in a hierarchy with expandable and collapsible capability.
	Array—Groups multiple data elements of the same type. Can be one or multidimensional.
	Matrix—Two-dimensional array of real or complex numbers used for matrix operations such as linear algebra.
	Table—Two-dimensional array of strings arranged into a table with rows and columns of cells containing the text elements.
	Cluster—Groups multiple data elements of the same or mixed data types.
	Variant—Used to pass or display variant data consisting of a name, data type, data, and attributes.
	ActiveX container—Provides an ActiveX control or document on the front panel.
	.NET container—Provides a .NET control on the front panel.

Rule 6.2 Choose memory efficient data types

Because many of the control types can be configured to represent or contain a variety of data types, memory efficiency depends on the combination of the control type and data type. Therefore, it is useful to consider data type efficiency in addition to control type simplicity. Specifically, the data types define the size and simplicity of the data in memory. For example, we can organize the numeric data types in order of representation, for which the smallest integer representation is the most memory-efficient data type and the largest floating-point representation is the least efficient. Table 6-2 contains the complete list of LabVIEW data types, presented in approximate order of memory efficiency. As shown, the smallest numeric representation is an 8-bit (1-byte) integer, and the largest is an 8- to 16-byte extended-precision floating-point numeric.

Similar to the list of control types in Table 6-1, the order of data types in Table 6-2 is approximate and overlapping. For example, an enumeration consists of two components, an unsigned integer and a set of string labels. The efficiency of the unsigned integer component depends on its representation, either 8-bit, 16-bit, or 32-bit. The efficiency of the string component depends on the quantity and lengths of the labels. These properties are configured by the developer. Likewise, the more complicated data types listed near the bottom of the table, such as Dynamic and Variant, depend on the data and attributes they contain.

Table 6-2 *LabVIEW Data Types and Descriptions, Listed in Top-Down Order of Memory Efficiency*

TF	**Boolean**—Stores a TRUE or FALSE value as a byte integer.
U8	**8-bit unsigned integer**—Represents only non-negative whole numbers in the range of 0 to 255.
I8	**8-bit signed integer**—Represents positive and negative whole numbers in the range of –128 to +127.
⟨⟩	**Enumeration**—Provides a pick list of text items that map to 8-bit, 16-bit, or 32-bit unsigned integers.
U16	**16-bit unsigned integer**—Represents non-negative whole numbers in the range of 0 to 65,535.
I16	**16-bit signed integer**—Represents positive and negative whole numbers in the range of –32,768 to +32,767.
U32	**32-bit unsigned integer**—Represents non-negative whole numbers in the range of 0 to 4,294,967,295.
I32	**32-bit signed integer**—Represents positive and negative whole numbers in the range of –147,483,648 to +2,147,483,647.
SGL	**Single-precision floating-point**—Uses 4 bytes to represent fractional numbers conforming to the ANSI/IEEE Standard 754-1985.
▭	**Refnum**—Unique identifier for an object, such as a VI, file, .NET control, or network connection.
I/O	**I/O name**—Unique identifier for I/O resources, such as DAQmx and VISA, to communicate with an instrument or a device.
U64	**64-bit unsigned integer**—Represents non-negative whole numbers in the range of 0 to 18,446,744,073,709,551,615.
I64	**64-bit signed integer**—Represents positive and negative whole numbers in the range of –9,223,372,036,854,775,808 to +9,223,372,036,854,775,807.
DBL	**Double-precision floating-point**—Uses 8 bytes to represent fractional numbers conforming to the ANSI/IEEE Standard 754-1985.

EXT	**Extended-precision floating-point**—Uses 8 to 16 bytes, depending on the platform, to represent fractional numbers with maximum precision.	
CSG	**Complex single-precision floating-point**—Fractional number containing real and imaginary parts, each represented with 4-bytes of precision (SGL).	
CDB	**Complex double-precision floating-point**—Fractional number containing real and imaginary parts, each represented with 8 bytes of precision (DBL).	
CXT	**Complex extended-precision floating-point**—Fractional number containing real and imaginary parts, each represented with 10 to 16 bytes of precision (EXT).	
⌛	**<64.64>-bit time stamp**—Stores absolute time with very high precision. Actually stores the number of seconds that have elapsed since 12:00 a.m. Friday, January 1, 1904, Universal Time, using 16 bytes.	
[DBL]	**Real matrix**—Two-dimensional array of double-precision floating-point (DBL) numeric data saved as a type definition. Used with mathematics functions that require the real matrix type definition, such as linear algebra.	
[DBL]	**Complex matrix**—Two-dimensional array of complex double-precision floating-point (CDB) numeric data saved as a type definition. Used with mathematics functions that require the complex matrix type definition, such as linear algebra.	
abc	**String**—A sequence of ASCII characters stored as a 1D array of bytes. Can contain numbers and letters, both displayable and nondisplayable.	
	Path—Stores the location of a file or directory using the platform's native syntax.	
	Picture—Contains a set of instructions and data for displaying pictures created with LabVIEW's graphics VIs.	
0101	**Digital data**—Contains digital signal data.	
⎍⎍⎍	**Digital waveform**—Data structure consisting of an initial time (t0) represented as a time stamp, the time interval between samples (dt) represented as a double-precision floating-point numeric, and a digital waveform (Y) represented as a digital data type.	
∿	**Waveform**—Data structure consisting of an initial time (t0) represented as a time stamp, the time interval between samples (dt) represented as a double-precision floating-point numeric, a waveform (Y) represented as a 1D array of any numeric data type, and attributes.	
	Dynamic—Provides signal data and attributes represented as an array of analog waveforms, used only with Express VIs.	
	Variant—Encodes a name, data type, data, and attributes into a generic data type. Used with ActiveX, DataSocket, and subVIs requiring a fixed interface to handle a variety of data.	

Choose the control types and data types for intuitive operation and memory efficiency. Table 6-3 presents a matrix containing the control types and supported data types. The control types are organized along the top horizontal heading in left-to-right order of operational and data simplicity, similar to Table 6-1 turned sideways. The data types are listed along the left vertical heading in top-down order of memory efficiency, similar to Table 6-2. The cells of the table indicate the compatibility of each control type with each data type. Use this table to optimize the control and data type selections in order of operational simplicity and memory efficiency. Consider the type of data that is required, and choose the control type that is best suited to represent that data. Begin on the top left, and evaluate each control type from left to right. If your data element has two selections that are opposites, choose a Boolean control. If you need to provide a range of numeric selections, first consider a ring or an enumeration, followed by a numeric. Select the first control type that meets the operational requirements. Next, traverse the corresponding column of data type compatibility, from top to bottom, until you identify the first data type that provides the desired functionality. The result is the control and data type combination that provide the simplest operation and greatest memory efficiency.

Table 6-3 *The Simplicity and Compatibility of Each Control Type Listed in the Top Horizontal Heading with Data Types Listed in the Left Vertical Heading*

		Boolean	Ring	Enumeration	Tab	Numeric	Listbox	Combo box	String	Tree	Array	Matrix	Table	Cluster	Variant	ActiveX container	.NET container
Boolean	TF	✔									✔			✔	✔		
8-bit unsigned integer	U8		✔	✔		✔					✔			✔	✔		
8-bit signed integer	I8		✔			✔					✔			✔	✔		
Enumeration				✔	✔						✔			✔	✔		
16-bit unsigned integer	U16		✔	✔		✔					✔			✔	✔		
16-bit signed integer	I16		✔			✔					✔			✔	✔		
32-bit unsigned integer	U32		✔	✔	✔	✔					✔			✔	✔		
32-bit signed integer	I32		✔			✔	✔				✔			✔	✔		
Single-precision floating-point	SGL		✔			✔					✔			✔	✔		

Chapter 6 • Data Structures

		Boolean	Ring	Enumeration	Tab	Numeric	Listbox	Combo box	String	Tree	Array	Matrix	Table	Cluster	Variant	ActiveX container	.NET container
Refnum	[▭]										✔			✔	✔	✔	✔
I/O name	[I/O]										✔			✔	✔		
64-bit unsigned integer	[U64]		✔	✔		✔					✔			✔	✔		
64-bit signed integer	[I64]		✔			✔					✔			✔	✔		
Double-precision floating-point	[DBL]		✔			✔					✔	✔		✔	✔		
Extended-precision floating-point	[EXT]		✔			✔					✔			✔	✔		
Complex single-precision floating-point	[CSG]					✔					✔			✔	✔		
Complex double-precision floating-point	[CDB]					✔					✔	✔		✔	✔		
Complex extended-precision floating-point	[CXT]					✔					✔			✔	✔		

Table 6-3   *Continued*

		Boolean	Ring	Enumeration	Tab	Numeric	Listbox	Combo box	String	Tree	Array	Matrix	Table	Cluster	Variant	ActiveX container	.NET container
64-bit time stamp											✓			✓	✓		
Real matrix												✓			✓		
Complex matrix												✓			✓		
String								✓	✓	✓	✓		✓	✓	✓		
Path											✓			✓	✓		
Picture											✓			✓	✓		
Digital data											✓			✓	✓		
Digital waveform											✓			✓	✓		
Waveform											✓			✓	✓		
Dynamic											✓			✓	✓		
Variant															✓		

Note that the use of Rules 6.1 and 6.2, and Table 6-3 help prevent the application from generating invalid data. For example, you would never choose a string control for data that is strictly numeric. A numeric control is much easier and more reliable for a user to operate, and the data is stored efficiently. Also, if you have a discrete number of alphanumeric selections, use an enumeration or a ring control before using a string control. As we can see from Table 6-3, enumeration and ring controls are much simpler than string controls because they restrict user input to a limited number of discrete selections and are represented as integers.

Finally, note that enumeration and variant appear in each table as both controls and data types. This is because they are both. An enumeration is a special data type consisting of a numeric with associated text strings. A variant is a self-describing data type that encodes any LabVIEW data into a generic format. Similarly, arrays and clusters can be considered data types as well as controls. However, the data type of arrays and clusters is undefined until they are populated with specific data structures. Unlike enumerations and variants, wiring assignments cannot be made from the terminals on the diagram until the data type definition is completed by depositing controls into the array and cluster shells. Otherwise, any wires are broken. Therefore, arrays and clusters do not become valid data types until populated.

Rule 6.3   Choose controls and data types that facilitate consistent data structures throughout an application

Referencing Theorem 6-1, the most important consideration in choosing controls and data types is consistency. Dissimilar data types require conversions, either explicitly via the formatting and conversion functions or automatically via coercions that appear as dots on terminals. LabVIEW allocates separate memory buffers to store each representation of data. Therefore, dissimilar data types lead to extra programming, processing, buffer allocations, and memory consumption. Good LabVIEW developers strive to optimize the performance and memory use in their applications.[1] Consistent data types reduce the level of programming effort required, while ensuring efficient and reliable performance.

Rules 6.1, 6.2, and 6.3 must be considered together when developing an application's data structures. Controls that are simple to operate, data types that are memory efficient, and overall consistent data structures are desired. How do we ensure that each consideration is satisfied? Carefully examine the data requirements of each component or subVI, and choose data types that are intuitive, efficient, and consistent. When compromises between memory efficiency (6.2) and data structure consistency (6.3) are required, always choose consistency (6.3). Per Theorem 6-1, memory access time is the universal latency. The more memory operations are performed by our applications, such as buffer allocations, the greater the negative impact is on performance. All data type conversions entail buffer allocations that are susceptible to this issue. However, larger data types, such as DBL instead of I16, do not necessarily affect the number of memory buffer allocations. Indeed, allocating a 4,096 element array of DBL may require the same memory access time as allocating a similar array of I16. Rather, allocating an array of I16 and later converting to an array of DBL entails multiple buffer allocations.

Consider the example in Figure 6-1. An instrument driver for an oscilloscope acquires and processes a waveform through a hierarchy of subVIs, as shown in Figure 6-1A. The lowest-level VI is Get Raw I16 Waveform VI, shown in Figure 6-1B. This VI reads the raw data as a binary string, converts the data to an array of unsigned byte integers, combines every two successive integers into a single 16-bit signed integer, and returns the raw data as an array of 16-bit signed integers. Because there are normally 2 bytes per sample, 16-bit signed integer is the simplest and most memory efficient numeric data type for representing this data, per Table 6-3.

In Figure 6-1C, Fetch Waveform VI converts the raw data returned from Get Raw Waveform VI into a scaled array via several arithmetic functions. Because the offsets and scaling factors are double-precision floating-point numeric, the array of raw data must be coerced to double-precision floating-point numeric prior to the mathematical operations. This is indicated by the coercion dot on the subtract function. The coercion creates a new buffer of data, expanded to 8 bytes per array element, in addition to the previous buffer. Hence, the coercion increases the memory use by a factor of 4, in addition to allocating a new memory buffer.

Next, the scaled array passes through Read Waveform VI without modification, as shown in Figure 6-1D. Finally, the top-level VI, shown in Figure 6-1E, bundles the scaled array together with the **Initial X** and **X Increment** and updates the Waveform Graph. The bundle function causes the scaled array to be copied into the cluster that it builds, thus forming an additional memory buffer.

Each VI in this call chain has different data requirements, and the control and data types have been chosen for maximum simplicity and memory efficiency within each specific VI. However, the dissimilar data types in this call chain increase the memory consumption by a factor of 9 versus the original array of 16-bit signed integers. More important, the waveform's data type is converted five times throughout the call chain, adding unnecessary buffer allocations that reduce execution speed. Therefore, using the simplest and most memory efficient data types within each individual VI *reduces* the overall performance of the application because of the inconsistent data structures.

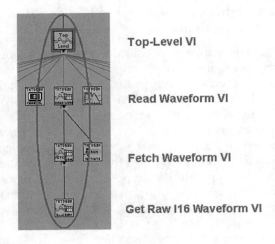

Figure 6-1A
A section of the VI Hierarchy window for a top-level VI that calls an instrument driver to acquire a waveform from an oscilloscope. The instrument driver uses a call chain of three subVIs, including read, fetch, and get.

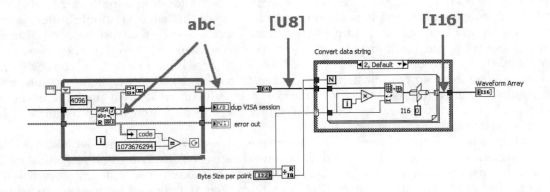

Figure 6-1B
Get Raw I16 Waveform VI returns an array of 16-bit signed integers. This is the most memory efficient data type for storing the waveform in this VI.

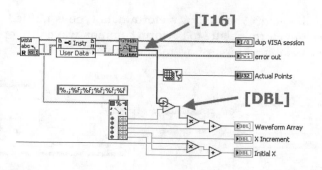

Figure 6-1C
Fetch Waveform VI scales the raw array to a scaled array of double-precision floating-point numbers. Arithmetic operations between dissimilar data types cause a coercion, which allocates a new memory buffer.

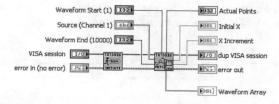

Figure 6-1D
Read Waveform VI passes the scaled array without modification.

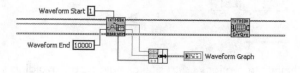

Figure 6-1E
The top-level VI bundles the scaled array and other waveform components into a cluster and updates a waveform graph.

Figure 6-2 contains the same example, except consistent data types are used throughout the call chain. In Figure 6-2A, every 2 successive bytes of the binary string are converted directly to an array of double-precision floating-point numbers. Note that this selection requires four times more memory than the 16-bit signed integer used in Figure 6-1A, but it improves consistency throughout the call chain. Also notice that the intermediate conversion of the binary string to 8-bit unsigned integers has been eliminated, partially offsetting some of the extra memory and processing required for the conversion to double-precision floating-point numbers. In Figure 6-2B, the raw array is scaled using arithmetic operations applied to consistent data types, eliminating the coercion and extra memory buffer versus Figure 6-1B. Also, a waveform data type is assembled. In Figure 6-2C, the waveform passes through Read Waveform VI without modification. Finally, in Figure 6-2D, the waveform passes up to the top-level VI and updates the waveform graph without modification.

In this example, consistent data types modestly reduce memory consumption, while substantially reducing the number of buffer allocations. Hence, overall performance is optimized. Additionally, the waveform data type is a convenient data structure that requires fewer terminals and wires for the multiple VIs of the call chain. When the waveform data type is formed, the developer can use the high-level VIs on the Waveform palette to perform common waveform operations. Hence, Rules 6.1, 6.2, and 6.3 have been combined for best overall performance.

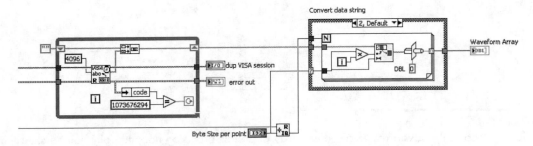

Figure 6-2A
Get Raw DBL Waveform VI returns an array of double-precision floating-point numbers, which compromises memory efficiency within this VI for data structure consistency throughout the application. Also, an intermediate conversion is eliminated.

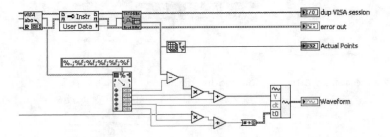

Figure 6-2B
Fetch Waveform VI performs arithmetic operations using similar data types, avoiding the coercion, and constructs a waveform data type.

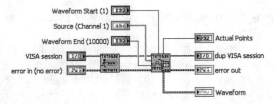

Figure 6-2C
Read Waveform VI passes the waveform without modification.

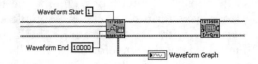

Figure 6-2D
The top-level VI simply updates the waveform graph. No modifications to the data are required.

It should be noted that, in some applications, the raw waveform is preferred in the smallest format instead of either waveform data type or double-precision floating-point numbers. For example, if the objective is to acquire and stream the raw data to file, the smallest file size and fastest disk streaming rate are desired. In this instance, streaming the raw binary data to file in the format that it is read directly from the instrument provides the optimum result. Be sure to log the waveform scaling information to convert the data to the desired engineering unit offline.

6.1.2 Configure the Properties

After the controls have been selected, the next step is to configure their properties. For Boolean command buttons, it is common to configure the Mechanical Action and Boolean Text properties. For numeric controls, configure the format, precision, and range. For strings, specify the display style, single line, and wrap behavior. Customize the appearance and default value properties of each control, or define your own custom controls using the Control Editor.

 Rule 6.4   Configure an appropriate default value for each control

To help promote proper VI use, configure appropriate default values for each control. Consider the common use cases, determine what typical value each control may utilize, and configure it as default if it exists. At the very least, the VI should be able to load from memory and run properly without modifying the control values. Required inputs such as refnums and instrument sessions are an exception. For subVIs, indicate the default value in parentheses at the end of the owned label so that it is visible from the panel as well as the Context Help window of the calling VI.

 Rule 6.5   Enter control descriptions

As discussed in Chapter 3, intuitive and succinct owned labels help document the controls and indicators. Additionally, enter one or two sentences for each control description that further describes the control's purpose, default value, and range, unless it is completely intuitive based on the label. Control descriptions provide documentation that is visible from the Context Help window. In Chapter 9, "Documentation," we discuss how a documentation set is generated containing the control labels and descriptions. It behooves us to enter the description, even if it seems unnecessary, or we might be disappointed when a formal document is required.

 Rule 6.6   Save custom controls as strict type definitions

A strict type definition is a control that is customized and saved as **Strict Type Def.** in the Control Editor window. The Control Editor forms a CTL file that maintains the data type and properties of the strict type definition. This allows the multiple instances to maintain the same customized properties, including appearance and behavior. The properties are then modified via the Control Editor, and changes are automatically applied to all instances. Strict type definitions are discussed in detail in Section 3.4.2, "Consistency." In general, if the control has one or more properties that have been configured and will have more than one instance requiring the same properties, save it as a strict type definition.

6.1.3 Create the Data Constructs

After the controls and properties are specified from the previous two steps, the final step is to implement the data constructs. This is primarily accomplished by grouping related controls into arrays and clusters, and saving as type definitions. Arrays are multivalued data sets of the same type. Clusters combine multiple controls of any type into a new data structure. Arrays and clusters are represented by a single terminal and wire on the diagram. They are very useful for organizing data and reducing subVI terminals and wire clutter.

  Rule 6.7   Create arrays and clusters that associate related data

Consider the Torque Hysteresis VI shown in Figure 6-3. First, the application prompts the user to enter some information about the unit under test (UUT). The application queries a database, and if the UUT is recognized, it prompts the user to enter a set of motion control parameters. Next, the VI runs a test consisting of a motion profile that acquires torque versus angle data. Upon test completion, the VI performs a statistical analysis to the data and generates a report. Finally, the application saves the data to file. Following the guidelines in Chapter 4, "Block Diagram," these requirements are implemented using a modular diagram containing separate subVIs for each task, as follows: UUT Information Dialog (1), Motion Parameters Dialog (2), Run Test VI (3), Compute Statistics VI (4), Generate Report VI (5), and Save Data VI (6). A simplified version of the top-level diagram containing these six subVIs is shown in Figure 6-3A, prior to the assignment of data structures. The data that is generated by the first four subVIs and processed by the latter two subVIs is shown in Figure 6-3B.

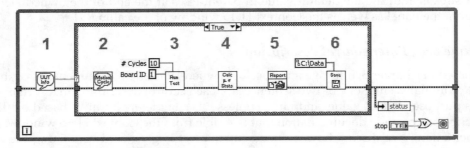

Figure 6-3A
Top-level diagram of Torque Hysteresis VI, prior to assignment of data constructs

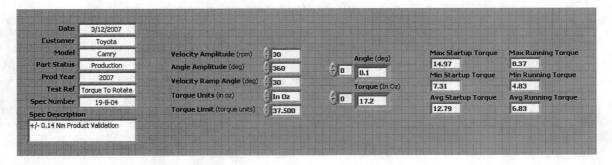

Figure 6-3B
Data generated by the first four subVIs and processed by Generate Report VI and Save Data VI

Figure 6-4A shows the diagram of Torque Hysteresis VI implemented without clusters. Many wires and terminals are required to pass the data from the first three subVIs to the successive subVIs. The subsequent subVI, Compute Statistics VI, generates six additional calculated values that are used by Generate Report VI and Save Data VI. However, the latter two subVIs do not have enough terminals to receive this data using wires, despite the use of the connector pattern with the maximum 28 terminals, as shown in Figure 6-4B. Instead, global variables are used to pass the excess data between subVI diagrams. Additionally, maintenance of this VI is a chore. If an additional parameter is added to the UUT Information Dialog, corresponding controls, terminal assignments, and wires must be added to three subVIs, or new global variables are required.

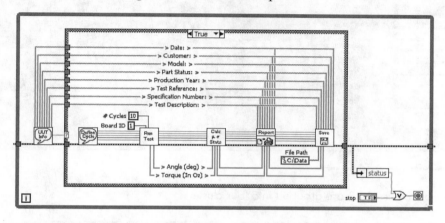

Figure 6-4A
Top-level diagram of Torque Hysteresis VI implemented without clusters. The wire and subVI terminal density complicates maintenance.

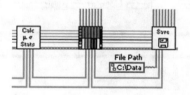

Figure 6-4B
Although Generate Report VI has the connector pattern containing the maximum 28 terminals, it is not sufficient for the six additional values calculated in Compute Statistics VI.

Applying Rule 6.7, the data from Figure 6-3B is grouped into the clusters shown in Figure 6-5A. The diagram of Torque Hysteresis VI is implemented using clusters, as shown in Figure 6-5B. Specifically, the eight parameters of UUT information are bundled into a cluster named **UUT Information**. Five motion control parameters are stored in a cluster named **Motion Parameters**. The **Motion Parameters** cluster is passed to Run Test VI. Upon test completion, the VI stores the raw data acquired during the test into a cluster named **Torque vs Angle**. Next, the **Motion Parameters** and **Torque vs Angle** clusters are both passed to Compute Statistics VI. The computed results are stored into a cluster named **Statistics**. Also, Compute Statistics VI passes the **Motion Parameters** and **Torque vs Angle** clusters unmodified to Generate Report VI. This VI receives the **UUT Information** cluster from the

UUT Information Dialog and creates the report. Finally, all clusters are passed to Save Data VI, where the data is saved to disk.

In this example, 21 parameters have been logically grouped into four clusters. All data is passed among subVIs using terminals and wires. All subVIs use the standard 4×2×2×4 connector pattern. The diagram appears neat and orderly due to the use of clusters. Each cluster is saved as a type definition, for ease of maintenance. It is simple to add and remove parameters by adding and removing controls to the type definitions.

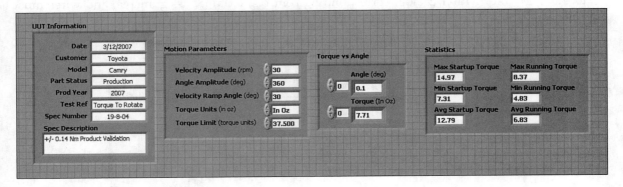

Figure 6-5A
The 21 parameters from Figure 6-3B are grouped into four clusters.

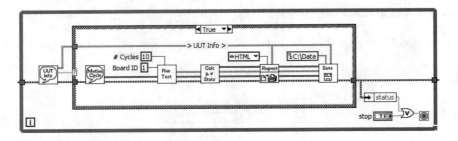

Figure 6-5B
Top-level diagram of Torque Hysteresis VI is implemented using clusters. The diagram appears neat and orderly, and maintenance is simplified.

The sections that follow contain more guidelines for simple data types and simple and complicated data constructs, as well as many more examples.

6.2 Simple Data Types

Simple data types are a subset of the fundamental LabVIEW data types that are stored in contiguous memory locations. They include Boolean, numeric, string, path, picture, and several special numeric types. In this section, we consider style rules pertaining to the simple data types.

6.2.1 Boolean

Boolean is the simplest control type in terms of both operational and data simplicity and has two possible values, TRUE or FALSE. Boolean data is actually stored as a byte integer. Hence, the Boolean data type is really no more efficient in terms of memory size than the smallest integer representation. However, Boolean controls are extremely intuitive. They have two clearly defined states and resemble objects commonly seen in everyday life. They include push buttons, toggle switches, command buttons, slide switches, LEDs, and radio buttons.

 Rule 6.8   Use Booleans if two states are logical opposites

The most important rule with Booleans is to use them to represent parameters that have exactly two states that are logical opposites—for example, On or Off, Yes or No, Open or Closed, and Stop or Go. When the two states are not opposites, use a text ring or enumeration to provide text selections that more accurately describe the choices.

 Rule 6.9   Assign names that identify the TRUE and FALSE value behavior

When a parameter's states are indeed opposites, it becomes easy to assign the control a name that clearly identifies its behavior in relation to the TRUE and FALSE Boolean states. For example, if we have a control named **Valve State**, the TRUE/FALSE behavior is unclear. Instead, consider a name such as **Close Valve**. In this case, we can determine from the name that the valve is closed when TRUE and open when FALSE. Per Rule 3.23, always enter the default value in parentheses for subVIs. This eliminates any chance of ambiguity. In the latter example, we have **Close Valve** (F = Open).

 Rule 6.10   Use command buttons for action, slide switches for parameter settings

Use command buttons to represent Booleans that invoke immediate action on GUI VI panels, such as **Trigger**, **Run**, **Cancel**, **Quit**, and **Close Valve**. Simply start with the **OK**, **Cancel**, or **Stop** button from the **Boolean Controls** palette, and customize it for your needs. This may involve simply changing the Boolean text or editing the size, fonts, and colors as well. Do not use command buttons for settings that do not generate immediate action, such as configuration properties. Instead, use the control that appears most intuitive for the users. NI's Instrument Driver Guidelines[2] recommends the vertical slide switch for all Boolean controls within an instrument driver. Most applications can be completed using just two types of Boolean controls: command buttons for action on GUI VI panels, and vertical slide switches for configuration parameters.

 Rule 6.11   Label the TRUE and FALSE states of slide and toggle switches

Always label the TRUE and FALSE states of toggle and slide switches with names corresponding to the behavior of each state, and position the labels next to the physical switch positions. Specifically, from the control's shortcut menu, choose **Advanced»Customize** to open the Control Editor window. Drop free labels adjacent to each of the control's switch positions and enter the text describing the TRUE and FALSE states. Edit the font, positions, and orientations of the labels so that they line up with the TRUE and FALSE switch positions. Close the Control Editor window and apply the custom changes. This method makes the label a permanent part of the control that you cannot accidentally move or delete while editing the VI. A faster alternative is to display the control's Boolean text. Boolean text is made visible by selecting **Visible Items»Boolean Text** from the control's shortcut menu, and it can be repositioned and edited. However, the Boolean text shows only one state at a time and is not as intuitive as the TRUE and FALSE state labels.

Figure 6-6 provides a vertical toggle switch and command button for a valve control configured with three different labeling schemes. The command button contains a glyph of a valve imported as a decal. In Figure 6-6A, the **Valve State** control labels are ambiguous in terms of the TRUE/FALSE behavior. In Figure 6-6B, the **Close Valve** control labels clearly depict the TRUE/FALSE behavior. Additionally, the vertical toggle switch contains visible Boolean text with default **ON/OFF** strings. In Figure 6-6C, the Boolean text of the vertical toggle switch has been replaced with more intuitive state position labels identifying the TRUE/FALSE behavior as **Closed** and **Open.** Additionally, the default value for each control has been appended to the control labels in parentheses. This additional text is distracting on a GUI VI but is helpful for reinforcing a subVI terminal's behavior via the Context Help window. Per Rule 6.10, the command button is preferred for controls that invoke immediate action, such as opening and closing an important valve. However, proper labeling makes the two control types approximately equivalent.

Figure 6-6A
Two alternatives for valve control include the vertical toggle switch and the command button with valve image decal. The label **Valve State** is ambiguous in this example.

Figure 6-6B
The owned label **Close Valve** clearly identifies the TRUE and FALSE behavior. The vertical switch has Boolean text visible with default TRUE/FALSE state labels. The command button is preferred for a GUI VI.

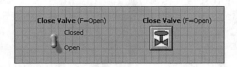

Figure 6-6C
The TRUE and FALSE states of the vertical toggle switch are appropriately labeled Closed and Open. The default value of each control is identified within the labels using parentheses.

 Rule 6.12 Avoid using buttons or switches as indicators, and LEDs as controls

LabVIEW's front panel objects have a useful property whereby any control can be converted to an indicator, and vice versa. In some instances, this does not make sense. For example, buttons and switches should not be used as indicators, and LEDs should not be used as controls. These reversed associations are counterintuitive and might confuse people, thereby violating Rule 6.1.

Boolean controls that are part of a cluster maintain the same appearance when the cluster is used as both control and indicator. In this case, it is difficult to avoid violating Rule 6.12 because consistent data structures are desired, per Rule 6.3. There are two alternatives. First, a command button can be customized to neutralize its appearance, similar to the status control of the error cluster. This is the preferred approach. Alternatively, separate clusters can be created for controls, containing the buttons or switches, and indicators, containing the LEDs. In this case, coercions will result if the clusters are saved as type definitions and then wired together. Because clusters are often saved as type definitions, this approach violates Rule 6.3 and should be avoided.

6.2.2 Numeric

Two categories of numeric data types exist: integer and floating-point number. Integers are used to represent whole numbers, while floating-point numbers are required for fractional data.

Rule 6.13 Use I32 representation for integers and DBL for floating-point numbers

Section 6.1.1, "Choose the Controls and Data Types," discusses the significance of simplicity, memory efficiency, and data type consistency. Per Rule 6.3, it is important to maintain consistent data structures throughout an application. In most situations, use 4-byte signed integer (I32) representation for integer data types, and 8-byte double precision (DBL) representation for floating-point numbers. Anything with physical units, such as voltage, power, frequency, length, angle, and similar quantities, should be DBL.

LabVIEW's numeric functions can accept and operate on any numeric data types. **Polymorphism** is a term that describes a function or VI with one or more terminals that can accept more than one data type. Polymorphic functions and VIs adapt to the input data type instead of breaking the wire or forcing a coercion to occur on the input terminal. More important, the function completes its operation successfully. Most of LabVIEW's built-in functions, VIs, and constructs that are not polymorphic consistently use I32 for integers and DBL for floating-point data. For example, the iteration and count terminals of looping structures are I32, as are the index terminals of array and string manipulation functions. DAQmx and most instrument driver VIs return DBL, or waveforms containing arrays of DBL. Likewise, the analysis VIs process arrays of DBL. Therefore, using I32 for integer and DBL for floating-point numbers helps maintain consistent data types throughout an application and avoids unnecessary coercions and conversions.

For completeness, two exceptions to this standard include digital pattern I/O and traditional data acquisition VIs. Reading or writing digital pattern data from a data acquisition device generally requires 1- or 4-byte unsigned integers (U8, U32). Also, the analog traditional DAQ VIs use 4-byte single-precision (SGL) numeric, if configured to read or write scaled data instead of waveforms. However, the DAQmx VIs and driver are both functionally and stylistically superior to the traditional DAQ VIs and driver. I highly recommend upgrading any legacy applications that use traditional DAQ.

If you are like me, you may have wondered why LabVIEW uses I32 for unsigned quantities such as loop iteration and count terminals, versus U32. The LabVIEW inventors intended I32 and DBL to be the prominent data types for integers and fractional numbers, respectively. Therefore, LabVIEW uses I32 and DBL to help promote their use.

Rule 6.14 Use automatic formatting unless a specific format is required

By default, LabVIEW formats floating-point numbers within numeric controls and indicators automatically. Specifically, LabVIEW displays the appropriate number of decimals, and switches between decimal and exponential notation, similar to a hand-held calculator. Sometimes it is desirable to display a specific number of decimal places all the time, to indicate the precision of a physical measurement, coefficient, or other parameter. Unless you have a good reason such as this, use the automatic formatting.

 Rule 6.15 Show radix for hex, octal, or binary data

Numeric controls with integer representation are formatted as decimal data, by default. **Hexadecimal**, **Octal**, and **Binary** formats can also be selected from the **Format and Precision** tab of the **Numeric Properties** Dialog. Select **Advanced Editing Mode** to make the **Numeric Format Codes** visible. When configuring numeric controls to represent data in these less common formats, always make the control's radix visible. This is done by selecting **Visible Items»Radix** from the shortcut menu. If the radix is not visible, the data is generally assumed to be decimal format.

Figure 6-7 shows four different controls that may be used to write a digital pattern to an 8-bit digital I/O port. They include an array of Boolean and three numeric controls configured for binary, decimal, and hexadecimal notation, respectively. Although each control represents the same value, the different formats for each numeric control would be interpreted much differently without a visible radix.

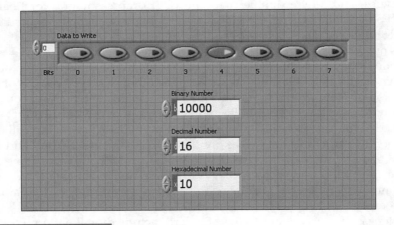

Figure 6-7
Four different controls contain a digital pattern for writing to an 8-bit digital I/O port. They include an array of Boolean, and three numeric controls formatted as binary, decimal, and hexadecimal. It is important to make the radix visible for numeric controls not configured for decimal notation.

6.2.3 Special Numeric

Four numeric data types that have special properties are the time stamp, reference number (refnum) and I/O name, text and menu rings (ring), and enumeration (enum). The **time stamp** data type stores absolute time with separate 8-byte fields for seconds and fractions of a second. Hence, the time stamp is extremely precise. **Refnums** and **I/O names** are references to a specific instance of an open resource, such as a file, instrument or device, network connection, image, LabVIEW application, VI, or control. They function similarly to a pointer to a data structure describing the resource. While the purpose of the two types of references is similar, the controls are substantially different. Specifically, I/O names display a value in an intuitive format, whereas refnums do not. VISA, IVI, and DAQmx Name Controls, for example, provide information about the specific hardware that they reference. ActiveX and .NET Refnums provide only a data type in a label. Because each type of resource has different requirements, the actual data type, memory use, and underlying data structure is different for each. The placement of refnums and I/O names in Tables 6-2 and 6-3, which list the data types ordered according to memory efficiency, only considers the reference portion of these data structures. It is important to note that this is not representative of the actual resource.

The ring and enum controls map text selections to numeric values. They are very useful for presenting a discrete number of selections that are intuitively described using text labels on the front panel, but the numeric is more functional on the diagram. **Enum** controls are always represented as unsigned integers, and the text selections are mapped to sequential numbers starting with 0. **Ring** controls may have any numeric representation, and the text selections are mapped to any value within the range allowed by the representation, including nonsequential, negative, and floating-point numeric values. The advantage of using enum and ring controls on the panel is that text selections can be more descriptive and meaningful than numbers, and selecting one item from a discrete list is very intuitive for the user. These controls play an important part in good programming style because they promote readability on both the front panel and the diagram.

On the diagram, enum and ring constants can display the text labels instead of or in addition to the numeric values they represent. Specifically, create a constant from the shortcut menu of a ring or enum terminal by selecting **Create»Constant** from the terminal's shortcut menu. The constant contains a pick list of the control's text selections, similar to the control on the panel. You can choose any of the text selections and show or hide the digital display. Additionally, when an enum is wired to the selector terminal of a Case structure, the text selections appear in the Case structure's selector area. This is why the enum is used for the State Machine design patterns discussed in Chapter 8, "Design Patterns."

Figure 6-8 shows an example of an instrument driver subVI that configures a digital multimeter. In Figure 6-8A, numeric controls are used to configure the **Function** and **Manual Resolution** parameters on the subVI's front panel. These parameters are programmed through terminals on the subVI's connector pane. The subVI call shown to the right uses numeric constants to configure these parameters. The value of **4** for **Function** is meaningless, and the value of **6.5** for **Manual Resolution** is risky. Specifically, the range of valid selections and the meaning of each selection is not very clear. In Figure 6-8B, text ring controls are used in place of the numeric controls. The subVI call contains text ring constants that were created from the terminals' context menus. These constants contain a discrete number of text labels that clearly describe each selection and limit the user to valid choices. Hence, the text rings are more functional, intuitive, and reliable than the numeric controls.

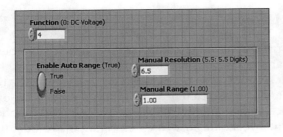 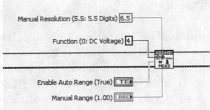

Figure 6-8A
The front panel of an instrument driver subVI contains numeric controls for programming the measurement function and resolution. The subVI call uses numeric constants for specifying these values, which are not meaningful and prone to error.

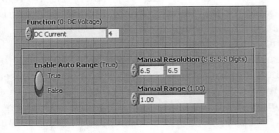 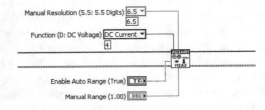

Figure 6-8B
The instrument driver subVI panel contains text ring controls instead of numeric controls. Constants created from the terminals' shortcut menus provide pull-down lists containing a discrete number of intuitively labeled choices.

 Rule 6.16   Use enums liberally throughout your applications

The text selections for enums and rings form a very important source of documentation on the panel and diagram that is built into these controls. I recommend enums over rings for maximum readability, including the case selector text. However, there are a few exceptions where a ring control must be used over an enum. Use a ring control when the text labels map to nonsequential, negative, or fractional values, or if localization is an issue. Also, use a ring control when creating an instrument driver or developer toolset intended for international distribution. The text embedded within an enum uniquely defines its data in the native language of the LabVIEW version in which it is created. The enum functions correctly when ported to localized versions, but the language of the enum text is not translated. For this reason, the NI Instrument Driver Development Guidelines recommends text ring controls and Booleans over enums. The text within the text ring gracefully ports to the localized language because the text is a symbolic mapping to the integer. Finally, the Strings[] property is read and write programmable for a ring but is readable and not writeable for an enum. A comparison between the enum and ring is summarized in Table 6-4.

Table 6-4 *A Functional Comparison Between Ring and Enum Data Types*

Data Type Behavior	Ring	Enum
Unsigned integer representations	✓	✓
Signed integer representations	✓	
Floating-point numeric representations	✓	
Nonsequential values	✓	
Text menu controls, indicators, and constants	✓	✓
Text menu case selector		✓
Strings[] property is editable	✓	
Strings are localizable	✓	

Rule 6.17   *Save enums as type definitions*

It is common to add and remove items from the enum controls many times during application development and maintenance. However, any clones you copy or constants you create do not maintain the same items as the original control after cloning, unless the original control is saved as a type definition. This is easy to overlook and a common source of misbehavior. Therefore, always save your enum controls and as type definitions. This is accomplished by choosing **Advanced»Customize** from the control's shortcut menu, select **Type Def.** or **Strict Type Def.** in the **Type Def. Status** ring of the Control Editor window, and save the control to a CTL file. Per Rule 6.6, choose the **Strict Type Def.** if the control has custom properties in addition to the text labels, and choose the **Type Def.** otherwise. Constants as well as controls that are formed from a type definition maintain an association so that when the type definition's items are edited, all instances are automatically updated as well. Note that a ring control must be saved as a strict type definition in order to provide similar behavior.

6.2.4   String, Path, and Picture

String, path, and picture are variable length data types stored in contiguous memory. String controls provide tremendous flexibility for entering or passing alphanumeric data. However, they lack filters or built-in methods to restrict the format, number of characters, case, or spelling of the data. Therefore, they provide the greatest opportunity for error.

Rule 6.18   *Avoid string controls on GUI VI panels unless required*

Avoid string controls on GUI VI panels except where free-form alphanumeric data is required. As developers, we should never trust the user to enter data in a specific manner, and the more invalid options we provide the user, the less robust our software is. As per Theorem 3.1, reliability is the developer's responsibility.

Two useful applications of string controls and indicators include entering descriptions and displaying instrument responses. If an operator needs the capability to enter free-form comments, such as a description of a one-of-a-kind test, or the cause of an alarm limit violation, a string control is appropriate. Instead of having a string control on the main GUI VI panel, create a dialog VI that prompts the user for the description and closes immediately afterward.

String indicators are invaluable for troubleshooting instrument communications. Instrument control is performed at a low level using calls to VISA Write and VISA Read. These functions write instrument commands and read responses as strings. Instrument drivers provide a layer of abstraction between the developer and the command and response strings, which are generally cryptic. However, when developing or troubleshooting an instrument driver, it is usually necessary to view these low-level messages in string indicators. Indeed, string indicators have some very useful display modes, including backslash (\) and hex. In **backslash** mode, we can view nondisplayable characters such as tabs, carriage returns, and line feeds via their backslash codes, such as \t, \r, and \n. In **hex** mode, we can view binary data in hexadecimal format.

Rule 6.19   Use enum, ring, and path controls in place of string controls where possible

Rule 6.20   Keep the Browse button visible for path controls on GUI VI panels

Use Table 6-1 to locate simpler alternatives to string controls. For example, enums and rings provide a discrete set of alphanumeric choices, as previously described. Always use path controls for entering or programming alphanumeric data representing a file path. LabVIEW formats the path using the standard syntax for the host's operating system. This ensures platform portability. The Browse button makes it simple for a user to interactively navigate and select a directory or file path, and should be kept visible on GUI panels. Configure the **Browse Options**, including **Selection Mode**, to restrict the user's choices and validate the data.

The picture data type stores a sequence of operation codes and data for drawing an image in a picture indicator. Operation codes are low-level instructions that tell the CPU what to draw. There are about 50 operation codes representing a wide variety of geometrical shapes and bitmap configurations. The data are the operands for each operation code. They include coordinates, colors, and other operation-specific data. The operation codes and data stored by the picture data type are transparent to the developer. Use the Picture VIs on the **Picture Functions** palette to create the pictures. Because the picture data type is stored in contiguous memory, it is a relatively simple data type, similar to string and numeric arrays.

6.3 Data Constructs

Data constructs are collections of one or more of the fundamental data types used to form a new data type. They include developer-defined constructs such as array, cluster, variant, variable, and queue, as well as built-in data types such as matrix, error, waveform, and dynamic. Arrays and clusters provide tremendous flexibility and utility, and are the primary means for organizing data in an application. This section presents style rules for data constructs, with emphasis on arrays and clusters.

6.3.1 Simple Arrays and Clusters

Array and cluster controls consist of shells or containers within which *any* data type or construct may be placed. The complexity of arrays and clusters depends entirely on their contents. This could be as simple as a single scalar Boolean or as complex as nested arrays and clusters. This section discusses simple arrays and clusters.

Arrays store data sets that have multiple values of the same data type. In memory, arrays consist of a heading containing the length of each dimension stored as 4-byte signed integers, followed by the data that comprises the array. Arrays containing one of the simple data types described in Section 6.2, "Simple Data Types," store the heading and data together in contiguous memory locations.

 Rule 6.21   Use arrays for multivalued data items; use clusters for grouping multiple distinct items

Most references consider arrays as collections of related data, very similar to clusters, except that the data elements have the same type. Indeed, this is correct. However, there are very important distinctions between arrays and clusters beyond the allowable data types. The elements of an array share the same properties, whereas each element of a cluster has its own unique properties. Specifically, an array control has only one owned label, control size, color, font, description, unit, caption, tip strip, and other properties that all elements must share. Consequently, the elements of an array all share the same *identity* and are distinguishable only by their index. As a result of this distinction, the elements of an array have a stronger association than the elements of a cluster. Arrays are used to store multivalued or vector quantities, whereas clusters are used for grouping multiple related but autonomous data items.

Array examples include the data samples of a waveform and the coefficients of a polynomial. These items may be thought of as single entities containing multiple elements. The index number satisfactorily identifies each element within the array as the sequential waveform sample number and polynomial coefficient number.

Cluster examples include the **error in** and **error out** clusters. These clusters contain a Boolean for the status, a numeric for the code, and a string for the source. Because the elements are related, they are combined into a cluster, which logically associates the elements. Because the element data types are dissimilar, an array is not an option. However, if we imagine for a moment that all elements are strings, then an array would be possible. However, the index would not be sufficient to uniquely identify the elements. Instead, separate controls with independent labels are needed. Hence, a cluster is required.

 Rule 6.22   Use arrays to store large or dynamic length data sets

Another distinction between arrays and clusters is that the number of elements in an array can be determined programmatically and resized as often as necessary, whereas the number of elements in a cluster is fixed. Also, it may be time and space prohibitive, in the literal sense, to create cluster controls containing a very large number of elements. Therefore, arrays are used instead of clusters if the quantity of data elements is large or may change during program execution and the elements share a common data type.

Let us review the Torque Hysteresis VI that was introduced in Section 6.1.3, "Create the Data Constructs." The clusters are implemented as shown in Figure 6-5. In the first step of the measurement sequence, UUT Information Dialog prompts the operator for some information regarding the unit under test, including the customer, model, production year, and more. This information is stored in a cluster named **UUT Information**. Here all the data is related and is of the same string data type. A cluster is chosen over an array because the data elements have unique identities requiring independent labels and descriptions.

Next, the application prompts the operator to enter some motion control parameters, including the velocity, angle, and torque limit. Because the data is related, it is stored in its own cluster, named **Motion Parameters**. Note that the **Motion Parameters** are weakly associated with the **UUT Information**. Specifically, the motion parameters are generally chosen to reflect the physical characteristics of the UUT. However, this is not a firm requirement. In Figure 6-5, the **UUT Information** and **Motion Parameters** are stored in separate clusters. If the application was larger and more complex, we could merge the elements from these two clusters into one cluster to minimize wire clutter and subVI terminals. The larger cluster could be renamed to indicate its more general scope, such as **Config Data**.

Next, the test runs, and torque and angle data is returned as a cluster containing two separate one-dimensional (1D) arrays. Arrays are used for torque and angle because they are each waveforms containing multiple samples acquired from two separate input channels. Note that a single two-dimensional (2D) array could have been chosen instead of the cluster of arrays. If the torque and angle data had been returned from the data acquisition VIs as a 2D array, I would continue the 2D array, for data type consistency as per Rule 6.3. However, in this case, the torque and angle are acquired from different subVIs within Run Test VI and are returned as separate 1D arrays. Bundling them into a cluster helps preserve their separate identities and is directly compatible with the XY graph indicator. Hence, the cluster of 1D arrays maintains consistent data types throughout the application, satisfying Rule 6.3.

Finally, the torque and angle data is processed and statistics are calculated and placed in a separate cluster, named **Statistics**. The elements in this cluster all have the same type, double-precision floating-point number, but a cluster is used to maintain their unique identities via independent labels.

Rule 6.23   Enter descriptions for array and cluster shells and control elements

Rule 6.24   Use alignment tools to keep clusters neat and compact

Arrays and clusters often form the data infrastructure for an application. When defined, they can proliferate throughout the application. As such, it is important that they are consistent and well documented. As discussed in Chapter 3, proper documentation includes succinct, intuitive labels that include the units in parentheses for any physical quantities. Additionally, include a description for each control element, as well as the array or cluster shell. Use the autosizing or distribute objects tools to keep the controls within a cluster neat and compact. I find that some clusters can really expand in number of elements as an application's requirements expand, so I prefer to make them compact, to minimize the front panel space. This includes use of the default properties for the controls that comprise the cluster, including size and font, and arrange or compress the contents. Specifically, arrange the elements in compact vertical form, in top-down order of the cluster order numbers, by selecting **Autosizing»Arrange Vertically** from the cluster's context menu. Alternatively, select and arrange any subset of elements in compressed vertical or horizontal order by selecting **Vertical Compress** or **Horizontal Compress** from the **Distribute Objects** toolbar menu. These techniques help reduce the front panel real estate occupied by large clusters.

Figure 6-9 shows the **Motion Parameters** cluster that is used throughout the Torque Hysteresis VI. The controls and labels within the cluster are independently aligned using the **Right Edges** and **Left Edges** alignment tools, respectively. The controls are evenly spaced and compressed using the **Vertical Compress** tool. Each control contains a description, including the **Velocity Amplitude**, as shown in the Context Help window.

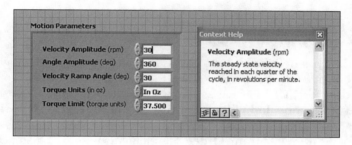

Figure 6-9
Motion Parameters cluster from Torque Hysteresis VI, with the Context Help window displaying the description for **Velocity Amplitude**. The cluster is neat, compact, and well documented.

 Rule 6.25 Save all clusters as type definitions

This rule cannot be overstated. Clusters form the data infrastructure of most applications and are highly subject to change. Save all clusters as type definitions or strict type definitions, and apply the type definition for each instance of the cluster. This ensures that any changes made to the type definition are automatically applied to every instance within every VI that uses it. This guarantees consistent cluster configurations, substantially streamlining application maintenance. Use the Control Editor window to create or edit the type definition, and use the **Type Def. Status** ring to specify the control type. Per Rule 6.6, choose the **Strict Type Def.** if the control has custom properties in addition to the element data types, and choose the **Type Def.** otherwise. Personally, I prefer strict type definitions over type definitions because I might want to edit the properties of the type definition after I have created multiple instances. Therefore, I can customize the appearance of the cluster or any element, and the properties are applied to all instances. Clone an instance of the type definition by choosing **Select a Control** from the Controls palette, or drag and drop a type definition from the Project Explorer window. Also, if you need to make a constant that is compatible with the cluster, select **Create»Constant** from the shortcut menu of one of the terminals of the type definition. Alternatively, select a type definition control file from the Project Explorer window or by using **Select a VI** from the functions palette, and drag and drop the type definition onto the diagram. Constants formed in this manner maintain compatibility with the type definition.

 Rule 6.26 Always use Bundle and Unbundle By Name

On the diagram, use the Bundle By Name and Unbundle By Name functions to access and replace the elements of a cluster. These functions provide more flexibility versus Bundle and Unbundle. Specifically, you can access or update any element of the cluster in any order using Bundle By Name and Unbundle By Name. Also, these operations do not break if your cluster changes, unless the specific elements that are being bundled or unbundled are affected. Hence, this rule works together with Rule 6.25 to streamline maintenance. Most important, the name labels uniquely identify the elements of the cluster and provide excellent documentation on the diagram. Because the Bundle and Unbundle By Name functions are sized according to the longest label, keep the control labels succinct and intuitive. Hence, Bundle and Unbundle By Name enhance the maintainability and readability of the diagram.

 Rule 6.27   Avoid clusters for interactive controls with Dialog VIs

Note that the strict type-defined clusters of the Torque Hysteresis VI shown in Figure 6-5A are not designed for interactive GUI behavior. Rather, the compression, default fonts, and properties make them less than ideal for GUI VIs such as dialogs. Additionally, there are some considerations with respect to tab key navigation. Individual controls that are not part of a cluster are configured for simple and immediate tab key navigation by default. The tab order of controls on the front panel can be edited by selecting **Edit»Set Tabbing Order**. The elements within a cluster can also be tabbed according to their cluster order, after the Key Focus property is applied to any element within the cluster. Subsequently, tabbing remains confined to the cluster until Key Focus is applied to a control outside the cluster. Therefore, tab key navigation requires additional programming for a cluster that is not necessary using the controls on the panel.

Figure 6-10 illustrates how a cluster can be used in conjunction with a dialog that prompts the user for the UUT information. The visible area of the panel simply contains string controls that have been customized for GUI interaction, as shown in Figure 6-10A. The **UUT Information** strict type-defined cluster is actually scrolled off to the side of the panel that is not visible, as shown in Figure 6-10B, along with the error clusters and **Cancelled?** indicator. Tab key navigation has been disabled for the noninteractive controls. This is accomplished by selecting **Advanced»Key Navigation»Tab Behavior»Skip this control when tabbing**. Notice that the type-defined cluster is much more compact than the corresponding controls on the panel. Hence, in this example, the controls on the panel are customized for best GUI behavior, and the type-defined cluster is optimized for subVI propagation.

The diagram of UUT Information Dialog bundles the data from the interactive GUI controls into the cluster when the **OK** button is pressed. In Figure 6-10C, we see that the terminals of the Bundle function are ambiguous. The developer has no way of knowing which terminal corresponds to which cluster control without examining the cluster order. Moreover, any changes to the cluster will create maintenance challenges for the developer using this approach. For example, the cluster order might change after adding and deleting or renaming string controls, without changing the number and data type of the elements. This would result in the same configuration as in Figure 6-10C, but an incorrect outcome. Figure 6-10D illustrates proper use of Bundle By Name for bundling the data into the type definition. As we can see, the Bundle By Name function's terminals are clearly labeled. Also, Bundle By Name is insensitive to the cluster order. If a cluster element is added, removed, or relabeled, the Bundle By Name and Unbundle By Name elements become invalid if they are affected. Therefore, Bundle By Name is much more readable and maintainable than Bundle.

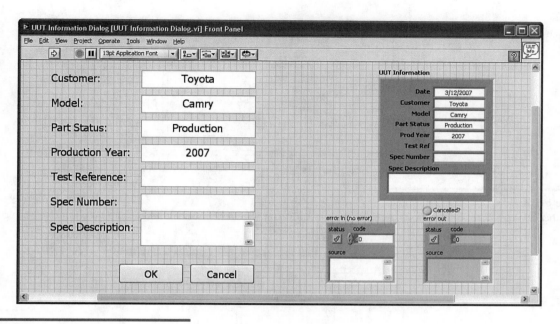

Figure 6-10A
The front panel of UUT Information Dialog contains string controls that are customized for GUI interaction.

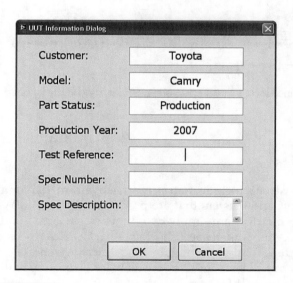

Figure 6-10B
The **UUT Information** type-defined cluster and other controls are scrolled off the visible area of the screen.

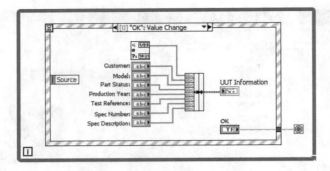

Figure 6-10C
The string controls are bundled into a cluster on the diagram. The terminals of the Bundle function are ambiguous.

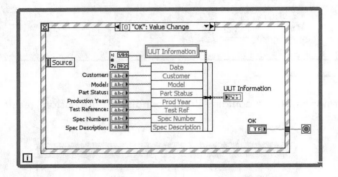

Figure 6-10D
Bundle By Name provides terminal labels that improve readability and maintenance.

6.3.2 Special Data Constructs

A few special data constructs include matrix, error, waveform, dynamic, and variant. The former four are variations of arrays and clusters that are predefined by LabVIEW and serve a specific purpose. The **real** and **complex matrices** are simply two-dimensional arrays of double-precision floating-point numbers and complex numbers, respectively. These arrays have been customized and saved as type definitions. They are used widely throughout the Linear Algebra VIs in the Math library. Likewise, the **error in** and **error out** clusters, used extensively for error propagation, are just clusters.

The **waveform data type** (WDT) can be considered a special type of cluster consisting of three elements plus attributes. The elements include a start timestamp, t_0; a time interval between data points, **dt**; and a one-dimensional array of numbers representing the samples of an analog or a digital waveform, **Y**. The attributes may contain any number of developer-specified name and value pairs. An example is a name or description for the waveform. In addition to the existence of attributes, WDT has conceptually different semantics with respect to math operations than standard clusters. Consequently, the WDT has a palette of dedicated functions for accessing and manipulating its data. Conversely, the standard cluster functions do not support the WDT.

Dynamic is a universal data type used with Express VIs. It was designed for novice developers as a very flexible data type that can store many types of data without learning about data types, data storage, and type conversion. You can wire the dynamic data type to any input terminal or indicator that accepts Boolean, numeric, or waveform data. LabVIEW automatically inserts the appropriate Convert to/from Dynamic Data function when required. In memory, the dynamic data type is represented as an array of analog double-precision floating-point waveforms.

A **variant** is a self-describing data type that encodes the data name, data type, data, and attributes or information about the data into a generic format. Conceptually, variant can be thought of as a wrapper that converts any data into a new format that is described in a universal manner. In LabVIEW, the variant data type is compatible with the standard used in Microsoft COM and .NET technologies. Specifically, variant is used with DataSocket and ActiveX communication protocol VIs. Additionally, variant is commonly used as a generic data type that passes data defined dynamically during runtime instead of edit time. It allows a data source or server to provide any type of data to a destination or client that can decode and use the data. The functional interface, such as a subVI connector terminal, maintains the variant data type, regardless of the variant's actual contents. This allows the client and server to be revised without affecting the interface.

The disadvantage of variant is that it is less memory and processor efficient than the conventional fixed data types, and may require additional programming to encode and decode the data within the client and server applications. In LabVIEW, variants are assembled using the To Variant and Set Variant Attribute functions. Alternatively, any LabVIEW data type is automatically converted to variant when wired to an input terminal of type variant. The conversion is denoted by a coercion dot.

6.3.3 Nested Data Structures

As noted previously, arrays and clusters are tremendously flexible data structures. For example, there is no practical limit to the number of dimensions arrays can have, although three or fewer is by far the most common. Also, you can create arrays containing clusters as elements, which can contain arrays and clusters, ad infinitum. Likewise, you can create clusters containing any combination of simple data types, arrays, and clusters, which may contain arrays and clusters. The only limitation is that arrays cannot directly contain other arrays, unless the contained arrays are bundled by a cluster. In any event, there is no practical limit to the number of layers of arrays and clusters within arrays and clusters that LabVIEW allows us to create.

Data constructs containing multiple layers of arrays and clusters are referred to as **nested data structures** and are considered complicated for two reasons. First, they are confusing for the developer to manipulate on the diagram. For example, changing a single element of data may require multiple calls to Index Array and Unbundle By Name to access the data element, followed by multiple calls to Bundle By Name and Replace Array Subset to rebuild the data structure. These cluster and array manipulation functions are frequently intermixed with nested looping structures used to index through the elements of each array layer. This becomes tedious and confusing after the first two layers.

The second reason nested data structures are complicated relates to how LabVIEW stores them in memory. Clusters are represented by a heading that describes the order and type of the data elements they contain, along with the values of any simple data types, in contiguous memory locations. The data of any arrays, strings, paths, and subclusters are not stored within the cluster. Rather, LabVIEW stores a handle containing the memory location where each multivalued data element is stored. Consequently, nested data structures may be stored as a network of handles and data that is distributed throughout the target computer's data memory space. The more layers, the more complicated this network can become.

Because LabVIEW manages memory automatically, the complication of a nested data structure's memory network is transparent to the developer. However, the developer should be concerned with memory efficiency. Specifically, each branch of the nested data structure requires additional memory operations to allocate and manipulate the data. First, most calls to Bundle By Name, Unbundle By Name, and Index Array cause a copy of the data provided at the output of each function to be created in memory. Additionally, as the size of the data contained in any array layer grows, the corresponding memory buffer may have to be reallocated, requiring more memory operations. The previous memory buffers fragment their memory blocks when new buffers are allocated in new locations. Hence, nested data structures use memory less efficiently than simple data structures. Per Theorem 6-1, memory access time is the principal latency in modern computing devices. Frequent operations involving nested data structures can reduce application performance.

Rule 6.28   Organize complex data using nested data structures

Rule 6.29   Avoid manipulating nested data during critical tasks

Despite these grim-sounding side effects, there are very practical uses for nested data structures that have limits to the number of layers and data size. Nested data structures help organize data in a hierarchical manner. If your target is a modern PC, you probably have plenty of memory resources at your disposal, and you can minimize the impact on performance if the nested data structures are applied properly. Specifically, try to manipulate the nested data at select locations in your application where performance is the least critical.

For example, a nested data structure can be designed to organize a mixture of configuration data, raw acquired data, and post-processed data. The configuration data may be queried from various locations and bundled or unbundled prior to initiating any resource-demanding operations. Accessing the nested data structure should be avoided during the execution of critical tasks, such as high speed data acquisition, deterministic control, and online data processing. Specifically, never bundle or unbundle continuously acquired measurement data into a nested data structure on-the-fly, as it is being acquired. As the size of the data increases, LabVIEW's memory manager may allocate new memory, causing a latency. Latencies affect the reliability of software-timed tasks, which may result in lost data or unacceptable jitter.

Rule 6.30   Limit the size of arrays by initializing to maximum length

Another method of preventing memory allocation latencies is to predetermine the maximum size of all arrays, including the array layers of a nested data structure, and initialize and maintain them at their maximum length. All array operations subsequent to the initialization can then be performed using the Index Array and Replace Array Subset functions. In this manner, the data structures maintain a constant size, and memory management is limited and predictable.

Use a specific uncommon initial value to help distinguish valid data from initialization data. The **NaN** constant, which stands for Not a Number, is a particularly useful constant for initializing floating-point numeric arrays. NaN can easily be identified and searched for using LabVIEW's array functions and cannot be confused with valid measurement data. Additionally, all of LabVIEW's graph indicators will skip drawing any data points containing NaN.

Figure 6-11 contains an alternative implementation of the Torque Hysteresis VI utilizing a nested data structure to provide enhanced functionality. The controls that comprise the data structure are shown in Figure 6-11A. The **UUT Information** and **Motion Parameters** clusters from Figure 6-5A have been combined to form a new cluster, along with an array of cluster that combines the **Torque vs Angle** and **Statistics** clusters. The latter array stores data from multiple test runs, reusing the same **UUT Information** and **Motion Parameters** for each run. Additionally, the larger cluster is contained within an array, storing data from multiple UUTs that comprise a lot. This data structure contains five layers of nesting, as follows:

1. **Torque vs Angle** and **Statistics** subclusters
2. A subcluster containing the latter clusters that stores the data acquired from a single run
3. The **Multiple Run Data** array, consisting of an array of single run subclusters
4. A cluster that combines the **Multiple Run Data** together with the **UUT Information** and **Motion Parameters**, containing all data for a single UUT
5. The **Multiple UUT Data** array, consisting of an array of the composite data for the multiple UUTs that comprise a lot

Figure 6-11B presents the enhanced top-level diagram that takes advantage of the nested data structure. There are now two additional looping structures. The inner For Loop allows multiple tests to run on a single UUT, and the corresponding statistics are computed after each run. The middle While Loop allows multiple UUTs to be tested consecutively. Shift registers and wires are used to propagate the nested data structure between each loop. Within the subVIs, the data from each UUT and each test run is appended to the data structure. The application writes the data to file after each UUT but maintains the data in the structure until the end of a lot, at which point a report is generated summarizing the data for the multiple UUTs that comprise the lot. In this example, the nested data structure reduces wire clutter on the top-level diagram by combining multiple structures into one and adds the flexibility to perform multiple test runs per UUT, by storing the data for all test runs performed on multiple UUTs.

Figure 6-11C illustrates one possible approach for implementing Run Test VI, in which the **Torque vs Angle** measurement data is acquired and combined with the nested data structure. However, this approach is not recommended because of the continuous resizing of the arrays using Build Array, and manipulation of the **Torque vs Angle** subcluster within a time-sensitive acquisition loop. Specifically, the output of each Build Array function has a different array size than the inputs, causing LabVIEW to periodically allocate new memory buffers. Latencies resulting from the reallocation of buffers may cause unacceptable jitter during acquisition. Unbundle and Bundle By Name are used to combine the data with the **Torque vs Angle** subcluster within each iteration of the While Loop, which is unnecessary additional overhead within the looping structure. Because the While Loop is where the torque and angle measurements are acquired, this is a critical portion of the application, and data manipulation should be minimized.

Figure 6-11D contains the diagram of Populate Data Structure VI, a routine that initializes the **Multiple Run Data** portion of the overall data structure. It generates a **Torque vs Angle** subcluster, prepopulated with arrays of the NaN constant. Specifically, the maximum number of torque and angle samples that may be acquired per test run is calculated as the product of the **Angle Amplitude (deg)** and **Max Rate (Sa/deg)** parameters. An array of NaN is initialized to this maximum length and assigned to both the **Torque (In Oz)** and **Angle(deg)** elements of the **Torque vs Angle** subcluster. Next, the **Torque vs Angle** subcluster is bundled into the corresponding element of the single run cluster. An array of single run clusters is initialized based on the specified **Number of Runs**, forming the **Multiple Run Data** array. The **Multiple Run Data** array is bundled into the corresponding element of the single UUT cluster. The single UUT cluster then replaces the element of the **Multiple UUT Data** array corresponding to the **UUT Number**. Initialization of the nested data structure for multiple test runs using one UUT is complete. Populate Data Structure VI is inserted into the top-level VI just prior to the For Loop, as shown in Figure 6-11E.

Figure 6-11F illustrates a reliable and efficient method of implementing Run Test VI utilizing the previously populated nested data structure. During the time-sensitive acquisition loop, the data manipulation is reduced to replacing the elements of the torque and angle arrays with real measurements, using the Replace Array Subset function. Because these arrays contain simple data types with constant size, no extra buffers are created and no unpredictable latencies will occur during acquisition. Likewise, Replace Array Subset is used to insert the single run data into the **Multiple Run Data** array, and the single UUT cluster into the **Multiple UUT Data** array. The Replace Array Subset function always performs its operation in place, without either moving or making copies of the input data array. This is possible because the data array size is not changed by this function. After the tests are completed, a post-processing routine can simply search for the NaN constant and resize the arrays to their actual length. The post-processing routine should be performed off-line, after any time-sensitive operations have completed, and we are less concerned with performance.

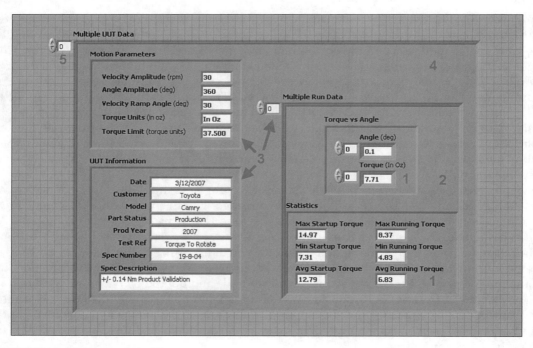

Figure 6-11A
An alternative data structure for the Torque Hysteresis VI has five layers of nesting.

Chapter 6 • Data Structures

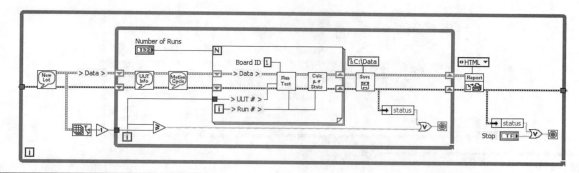

Figure 6-11B
The nested data structure propagates among all subVIs of the top-level diagram. The data structure and diagram enhancements facilitate multiple test runs per UUT, multiple UUTs, and organized data with fewer wires.

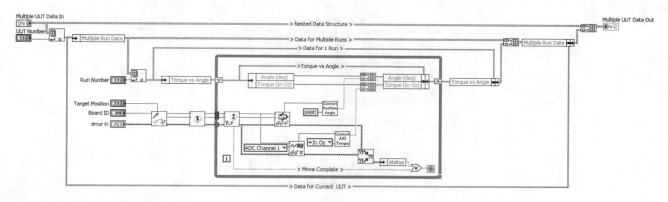

Figure 6-11C
One implementation of Run Test VI uses the Unbundle by Name, Bundle by Name, and Build Array functions to combine the newly acquired data with the **Torque vs Angle** subcluster. This implementation is inefficient because the Build Array function continuously resizes the arrays, allowing latencies due to memory allocations during the time-sensitive acquisition task.

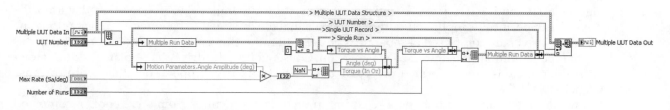

Figure 6-11D
Populate Data Structure VI is a subVI that initializes the nested data structure's **Multiple Run Data** array with **Torque vs Angle** data containing the **NAN** constant.

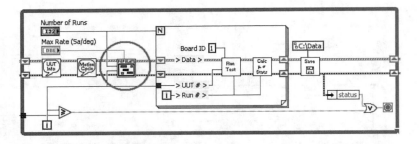

Figure 6-11E
Populate Data Structure VI is inserted into the top-level VI just before the For Loop that performs multiple test runs.

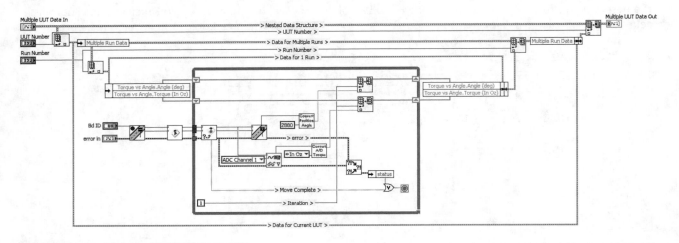

Figure 6-11F
Efficiency and reliability are improved within Run Test VI by replacing the data elements of each layer of the previously populated nested data structure using Replace Array Subset. The data structure remains a fixed size, and extra buffer allocations are avoided.

Astute readers may notice coercion dots on the input terminals of the motion control driver VIs in Figures 6-11C and 6-11F. These coercions are caused by dissimilar data types utilized within the driver VIs. Specifically, input and output terminals that are intended to be wired together among the VIs of the driver are comprised of separate type definitions for the controls and indicators. Wiring different type definitions together causes a coercion. Consequently, the driver VIs violate Rule 6.3. However, the coercions are performed on scalar data, which are trivial operations that will not cause latencies.

I would like to add one final note regarding nested data structures. Some references recommend avoiding nested data structures entirely. I disagree with this recommendation. As seen from the previous example, nested data structures can be effectively applied to organize data, increase functionality, reduce wire clutter, and improve maintenance. Performance is controlled by paying attention to where and how the data is manipulated. When applied properly, the organizational and functional benefits of a nested data structure offset the performance considerations. More examples are presented in Section 6.4, "Examples."

6.4 Examples

This section presents several more examples of proper and improper use of data structures.

6.4.1 Thermometer VI

Thermometer VI, shown in Figure 6-12A, is a subVI that acquires a voltage sample from a thermistor and converts it to engineering units. You may recognize it as a classic LabVIEW training exercise on how to create a subVI. It uses a vertical toggle switch labeled **Temp Scale** to select between units of degrees Fahrenheit and degrees Celsius. Although there are only two states, a Boolean control is not appropriate because the two states are not logical opposites, per Rule 6.8. Also, the diagram inefficiently computes both units all the time, regardless of the actual unit selected. Figure 6-12B replaces the Boolean with an enum. The enum improves maintenance over the Boolean because more units can be added in the future. Also, the diagram uses a Case structure so that only one unit is computed based on the **Temp Scale** selection. Additionally, the control and indicator are viewed as terminals, and the Express VIs are viewed as icons. SubVI calls for both versions are illustrated in Figure 6-12C. We see that the Boolean input of Thermometer VI is ambiguous from the caller VI's diagram. The enum is clearly superior to the Boolean in this example.

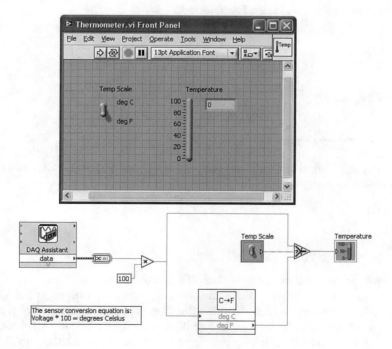

Figure 6-12A
Front panel and block diagram for Thermometer VI. The vertical toggle switch for selecting temperature units violates Rule 6.8 because the two states are not logical opposites.

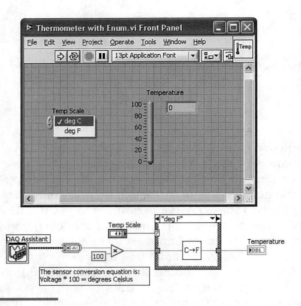

Figure 6-12B
Thermometer with Enum VI uses an enum to select the temperature units. The diagram is more efficient because the conversion to Fahrenheit is not made unless **deg F** is selected.

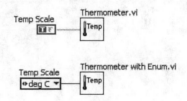

Figure 6-12C
The subVI calls for Thermometer VI and Thermometer with Enum VI illustrate that the Boolean is ambiguous from the caller VI's diagram, whereas the enum is not.

6.4.2 OpenG Variant

The OpenG[3] tools include an extensive set of VIs that use the variant data type. Many of these VIs are intended to operate on multiple LabVIEW data types, similar to polymorphic VIs. The LabVIEW data type is converted to a variant, which contains the data type, control name, and data. This information is used to increase the flexibility of the data that can be passed to a subVI through the connector terminals and the operations that can be performed on the data within the subVI. The OpenG tools use this concept to extend the capabilities of the LabVIEW libraries.

When a LabVIEW data type is wired directly to a variant input terminal, a coercion dot appears on the input terminal, and the conversion to variant is performed automatically. However, dissimilar data types are undesirable, per Rule 6.3. Ideally, coercion dots should be avoided for best style. However, the conversion from the LabVIEW data type to variant is unavoidable. As an alternative, the data can be converted to variant using the To Variant function, prior to wiring to the input terminal of the OpenG VI. This method replaces the coercion dot with an explicit function call. While the two methods are functionally equivalent, the explicit conversion conveys that the conversion is deliberate and not an oversight.

For example, in Figure 6-13, the OpenG Empty Array VI is used to determine whether the numeric array wired to its input terminal is empty. The Context Help window indicates a purple wire for the array input terminal, representing the variant data type. The illustration contains three functionally equivalent dataflow paths. The top path connects the array directly to the Empty Array VI, and a coercion dot appears on the VI's input terminal. The middle path inserts the To Variant function to remove the coercion dot. The bottom path uses the LabVIEW Empty Array? function for the same purpose. The LabVIEW function is polymorphic and does not require a conversion. In terms of performance, the top two paths are identical. The middle path is mildly preferred over the top path because it demonstrates that the type conversion is intended and eliminates the coercion dot. The bottom path is the most efficient overall because the polymorphic function maintains native LabVIEW data types and the type conversion is eliminated.

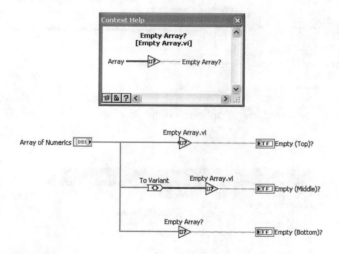

Figure 6-13
This diagram contains three methods of determining whether the input array is empty. The top dataflow path contains a coercion dot on the input terminal of Empty Array VI. The middle path removes the coercion by explicitly converting to the variant data type. The bottom path uses a polymorphic LabVIEW function to maintain consistent data types.

The OpenG Variant Configuration File VIs really showcase the power and flexibility of the variant data type. These VIs read and write data to text files in the standard configuration setting INI file format. They use clusters converted to variants to read or write multiple keys in one subVI call. Figure 6-14 illustrates the use of Write Section Cluster VI. The **Motion Parameters** cluster is converted to variant and wired to the **Section Formatted Cluster** input terminal. Write Section Cluster VI uses the data name, data type, and data information that the variant data type provides to create a section of an INI file. The elements of the cluster correspond to the section's keys, and the name of the cluster is the default section name when the section terminal is unwired. Figure 6-14A shows the source code in which five keys are written using one call to Write Section Cluster VI, and Figure 6-14B shows the resulting section of the INI file. Additionally, the OpenG Write INI Cluster VI populates an entire INI file containing multiple sections. As shown in Figure 6-14C, a nested cluster containing subclusters for each section is wired to the **INI Formatted Cluster** input terminal. The resulting INI file is shown in Figure 6-14D. In these examples, the variant data type is used to minimize development effort.

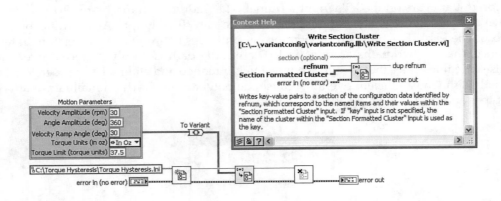

Figure 6-14A
The **Motion Parameters** cluster is converted to variant and written to a section of an INI file using Write Section Cluster VI.

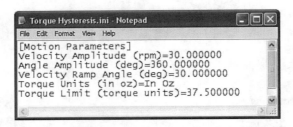

Figure 6-14B
The **Motion Parameters** section of the INI file resulting from the operation in Figure 6-14A

Chapter 6 • Data Structures

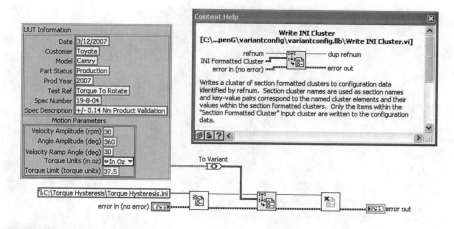

Figure 6-14C
A nested cluster populates multiple sections of an INI file using Write INI Cluster VI.

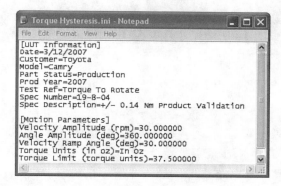

Figure 6-14D
The INI file formed from the operation in Figure 6-14C contains multiple sections.

6.4.3 Haphazard Data

Figure 6-15A contains a data structure that was thrown together rather haphazardly. It consists of an array of clusters containing a subcluster and three subarrays of clusters. It appears unkempt because it violates several of the front panel rules presented in Chapter 3. The alternative presented in Figure 6-15B contains several improvements, including transparent name labels, aligned and compressed controls, and boldface labels with plain-text units in parentheses. Additionally, these data structures contain more controls with fewer layers of nesting. Multiple two-dimensional arrays of clusters in Figure 6-15B are generally more memory efficient than one nested data structure of Figure 6-15A. However, the alternative increases the number of wires and subVI terminals required by a factor of four. The best alternative might be to maintain the single data structure of Figure 6-15A with the organization and neatness of Figure 6-15B.

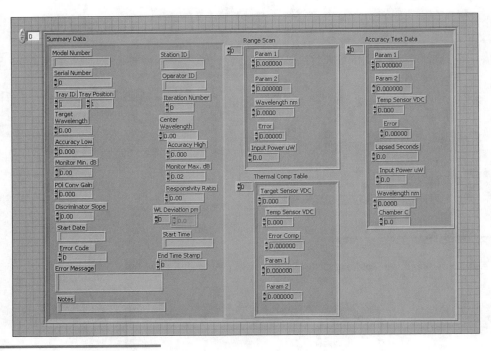

Figure 6-15A
This data structure is nested and the controls are haphazardly arranged.

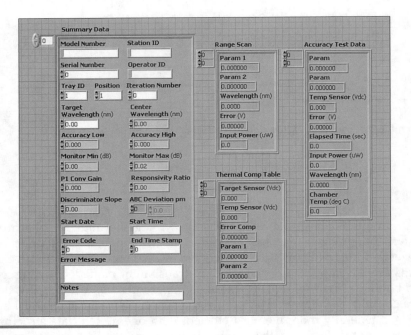

Figure 6-15B
The nested data structure has been divided into four data structures containing fewer layers of nesting, and front panel style rules have been applied to dramatically improve the appearance of each control.

6.4.4 Centrifuge DAQ VI

Centrifuge DAQ System VI, also discussed in Sections 3.5.4 and 4.4.6, is an application that performs dynamic measurements from soil samples in the payload of a centrifuge. The front panel shown in Figure 6-16A is highly configurable, allowing the user to configure a wide variety of signal conditioning modules. The hardware supports up to six SCXI chassis containing any combination of relay, accelerometer, LVDT, strain gage, voltage, and thermocouple conditioning modules. Any number of these modules can reside in any slots of any chassis. Each module type has a different set of programmable parameters, such as excitation voltage, signal amplification gain, filtering, bridge completion, scaling factor, and engineering units. These parameters are configured using a set of nested data structures. When the user selects a sensor type using the **Sensor Selection** listbox control on the left, one of seven data structures becomes visible. Each data structure is an array of cluster containing one element per SCXI module. This allows each module to be configured independently. Figure 6-16B shows the data structure used to configure strain gages. Its appearance is controlled programmatically using Property Nodes. Also, on the front panel, the array index control is normally not visible and the control's background color is transparent.

Figure 6-16C contains a code snippet of the application's configuration state. Multiple nested data structures containing the configuration data for each module type are bundled into another more complicated data structure. This larger structure's sole purpose is to reduce the number of terminals and associated wire clutter for the subVI that follows. The nesting of this data structure might be a concern if the data was accessed during a time-sensitive routine. However, this data structure contains only configuration data, which is accessed and programmed only before the critical data acquisition routines begin. Therefore, the complexity of this data structure does not affect the performance of the critical task.

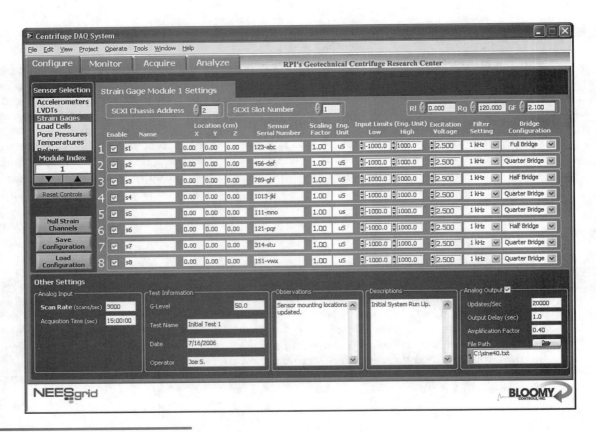

Figure 6-16A
The GUI VI panel for a highly configurable centrifuge data acquisition system

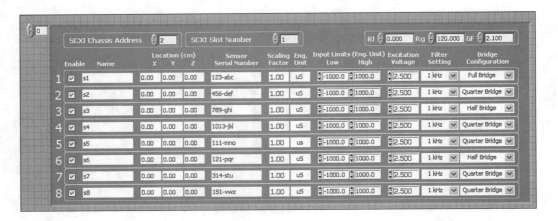

Figure 6-16B
This array of cluster containing strain gage configuration parameters is one of seven data structures used on the **Configure** tab to configure a corresponding SCXI module type. On the GUI VI panel, the array index control is not visible and the background is transparent.

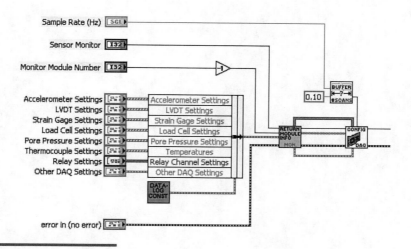

Figure 6-16C
The configuration routine bundles data structures to form a complex data structure that is passed into the subVI. Each data structure is an array of cluster containing configuration parameters for a module type, similar to Figure 6-16B.

Endnotes

1. The LabVIEW online help topic "VI Memory Usage" contains useful information concerning memory usage.
2. NI Instrument Driver Guidelines, Control/Indicator Naming and Data Representation.
3. OpenG is an organized community committed to the development and use of open source LabVIEW tools. The URL is www.openg.org.

Error Handling

7

A typical LabVIEW application controls instrumentation while acquiring, analyzing, displaying, and logging data—hence, the marketing slogan: *Acquire, Analyze, and Present*. Things can go wrong, however, regardless of the graphical programming methodology, application complexity level, and developer skill level. LabVIEW applications execute exactly as programmed and are not immune to bugs. The complexity of the applications LabVIEW is applied to is ever increasing. Also, in the dynamic world we live in, applications can begin misbehaving long after they are tested and deployed. For example, an instrument might hang up or become disconnected, or changes to the target computing device might cause a resource conflict, or a network directory path might change. Even if the application is completely bug-free at the time of deployment, it will not overcome the unexpected without an effective error handling scheme. Moreover, how quickly the developer and the application's users can identify and correct any problems that arise depends on how well the application incorporates error handling.

> ***Theorem 7.1:*** *Error handling is essential for troubleshooting applications, overcoming the unexpected, and implementing good style!*

Error handling is an *essential* part of *all* LabVIEW applications and serves multiple purposes, from development through deployment. First, error handling is imperative for debugging an application. When the application reports an error, we can quickly identify and correct many problems based on the error data. Error handling helps identify and describe errors long before we might otherwise observe the application misbehaving. Debugging an application without error handling is analogous to graphical programming wearing a blindfold. It is extraordinarily difficult to locate the source of most problems without good error handling.

After an application has been debugged, error handling helps test the limits of the application's capabilities. During testing, one may experiment with the maximum number of channels, fastest acquisition rate, most demanding analysis routine, fastest user interface update and disk streaming rates, or test the most unlikely combination of operator inputs. Error handling can tell us when something goes awry and pinpoint the operation that reports a problem. The application can be fine-tuned, such as setting the **Data Range** property of numeric controls, to help prevent invalid user inputs that may cause misbehavior.

When an application contains complete error handling, is thoroughly tested, and does not generate errors, we may rest assured that it is a reliable application. After deployment, the error handling must remain intact to report any new problems that may arise due to unforeseen circumstances. Many factors are beyond LabVIEW's control. Changes to the computing device's hardware, software, or network configuration are common sources of new misbehavior to an otherwise bug-free application. The presence of thorough error handling helps users and developers quickly identify and resolve new problems. Hence, thorough error handling serves the purpose of preventative maintenance in our applications.

Not only is error handling important for identifying problems, but error handling also is directly related to good LabVIEW programming style. As discussed in Chapter 4, "Block Diagram," data flow is the fundamental principle of LabVIEW, and propagating the error cluster is the primary means of establishing data flow. The error clusters are LabVIEW's most universally recognized data structure on a VI's front panel, connector pane, and diagram. Put them to good use!

7.1 Error Handling Basics

An **error** is a failure of a function or VI to complete its programmed task. Most nodes in LabVIEW propagate error data using the **error in** and **error out** clusters. The error clusters consist of a Boolean for the **status**, an integer for the **code**, and a string for the **source**, and are located in the **Array, Matrix & Cluster** controls palette. The **status** uniquely indicates whether an error has occurred. The **code** identifies the type of error and is used by the error reporting VIs to look up a corresponding description. The **source** identifies the function or VI that generates the error. The error cluster's default value consists of **status** = FALSE, **code** = 0, and **source** = <empty string>, indicating no error or warning. If the **status** is FALSE but the **code** is nonzero and the **source** is not empty, then a **warning** occurred. A warning is similar to but considered less severe than an error. For example, the node is successful at performing its programmed task, but there is something unusual about the input values or result.

Examples of the **error out** cluster containing no error, a warning, and an error are provided in Figure 7-1. The default no error value is shown on the left. The warning shown in the middle is commonly returned from a successful call to VISA Read. It indicates that the number of bytes transferred is equal to the requested input count, but more data might be available. The error cluster on the right contains error code 43, indicating that the user canceled the operation. This is a very common error that occurs when a file dialog is canceled. Most VIs have one **error in** control and one **error out** indicator, assigned to the lower left and right connector terminals, respectively.

Chapter 7 • Error Handling

Figure 7-1
The **error out** clusters indicate no error or warning on the left, a warning in the middle, and an error on the right.

 Rule 7.1   All VIs must <u>trap</u> and <u>report</u> the errors returned from error terminals

Error handling involves trapping and reporting the errors returned from the functions and VIs that have error terminals. Trapping is capturing the error returned from each node's **error out** terminal. Reporting involves displaying or logging the error information using a dialog or log file. Thorough trapping and reporting of errors is the best means by which errors are handled. This section presents error handling basics, including rules, techniques, and illustrations for trapping and reporting errors. Error codes and ranges are also discussed.

7.1.1   Trapping Errors

 Rule 7.2   Trap errors by propagating the error cluster among the error terminals

Errors are trapped by propagating the error cluster among the nodes that have error terminals, throughout every VI in an application. Begin by placing the **error in** control and **error out** indicator on the panel of each VI, or simply use a template that has been prepopulated with these controls. On the diagram, lay out the nodes into data or sequence dependent groups, positioned from left to right in the order they will execute. Use the vertical alignment tools in the **Align Objects** pull-down menu to align the icons and simplify wiring. Use the horizontal distribution tools from the **Distribute Objects** pull-down menu to space the nodes for adequate terminal access. Each group of sequential nodes comprises an **error chain**.

Initiate a wire from the **error in** control terminal to the **error in** connector terminal of the first node in each error chain. Wire the node's **error out** terminal to the next node's **error in** terminal, and repeat until all of the error terminals are wired, with the exception of the last node of each error chain's **error out** terminal. For a VI with a single error chain, wire the last node's **error out** terminal to the **error out** indicator terminal or to an appropriate error reporting subVI. If a VI has multiple error chains, merge the **error out** terminals from the last node in each chain using Merge Errors VI. Figure 7-2 illustrates a simple error chain made up of File I/O nodes. Figure 7-3 contains two parallel error chains, including DAQmx VIs and File I/O nodes.

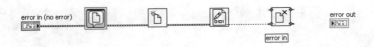

Figure 7-2
The error cluster propagates among several File I/O nodes to form an error chain.

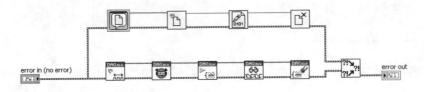

Figure 7-3
DAQmx VIs and File I/O nodes form two separate parallel error chains that are combined using Merge Errors VI.

Most functions and VIs that LabVIEW provides are designed to check the error status from the **error in** terminal and, if an error exists, skip their code and pass the same error information straight through their **error out** terminal. Hence, once an error occurs, the successive nodes of an error chain propagate the error information throughout the chain. This is desirable because the nodes of an error chain are generally related and interdependent. In the File I/O example of Figure 7-1, the file writing operation is dependent on a valid file reference number returned by Open/Create/Replace File, which itself is dependent on a valid file path returned by the File Dialog Express VI. An error generated by one prevents the subsequent nodes from succeeding. As long as all error terminals are properly wired, the first error that occurs is trapped within the chain of nodes and wires.

As discussed in Chapter 4, propagating the error cluster creates data dependency, which determines the order in which the nodes execute on the diagram. Data dependency is an underlying principle of data flow. Therefore, it is also important to wire the error cluster to nodes to specify the desired execution order. However, the most common mistake I observe in practice is incomplete error trapping. It seems that many developers propagate the error cluster for data dependency but fail to fully trap and report the errors. This is evident when the error cluster propagates between the functions requiring an order dependency but is not shared among the remaining functions or terminated appropriately.

 Rule 7.3 Trap errors from all iterations of loops

Special considerations are required with looping structures. It is important to examine the data flow and loop condition to see if the error is being trapped within all iterations. In Figure 7-4A, there is continuity between the **error in** control terminal, the error connector terminals on the File I/O nodes, and the **error out** indicator terminal. At first glance, the error appears to be trapped, but it is not. Instead, the loop input tunnel resets the error data within each new iteration of the loop. Also, the error data passed to the loop's output tunnel is overwritten in each successive iteration. Only the error data from the last iteration of the loop passes through the output tunnel. Hence, the error chain does not remember any errors that may have occurred in any of the iterations prior to the last. Two better alternatives are to either terminate the loop upon the first occurrence of an error or maintain the error data between successive iterations via a shift register. These alternatives are shown in Figures 7-4B and 7-4C. The former solution stops the loop immediately if an error occurs; the latter continues the loop but skips all successive File I/O operations. In both scenarios, the error is trapped and eventually passes through the While Loop to the **error out** indicator terminal upon loop completion.

Chapter 7 • Error Handling

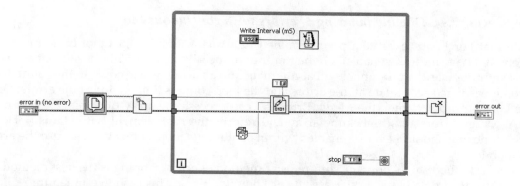

Figure 7-4A
Error data from the While Loop's input tunnel resets the error data, and error data from the file writing operation overwrites the output tunnel data within each successive iteration of the loop.

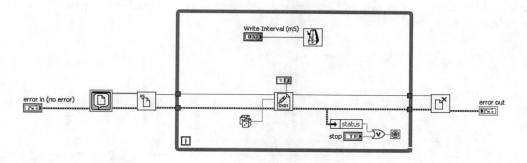

Figure 7-4B
The error **status** is checked within each iteration of the While Loop, and the loop is terminated if an error occurs. Any error is successfully trapped within the error chain and passes through the output tunnel to the **error out** indicator terminal.

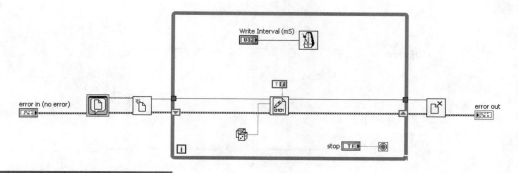

Figure 7-4C
A shift register is used to propagate the error data between successive iterations of the loop. If an error occurs, the loop continues but the file writing function skips its operation.

Rule 7.4 *Disable indexing of errors with continuous loops*

Note that For Loops execute a predetermined number of iterations, cannot be terminated prematurely, and have indexing enabled on their output tunnels by default. This means that an error cluster passed through an output tunnel will accumulate into an array of size equal to the count. If you have a good reason to accumulate all the errors and the loop count is limited, index the errors. Outside the For Loop's output tunnel, the Merge Errors VI can process the array of errors, as shown in Figure 7-5A. Merge Errors VI searches the array and returns the first element with **status** = TRUE, or no error. Indexing should never be enabled within continuous loops, however, because the array size is unbounded.

Similar to the While Loop, the shift register method of error trapping is the best method of propagating the error data between successive For Loop iterations. This is shown in Figure 7-5B. The shift register method maximizes efficiency in the event of an error because the functions and VIs that receive an error at their **error in** terminal skip their code. Also, the shift register method maximizes memory efficiency because it maintains the data from only one error cluster in memory. Moreover, the shift register extends a continuous error chain throughout the loop, whereas indexing resets the error chain in each new iteration. Consequently, the shift register method is generally preferred over the index array method.

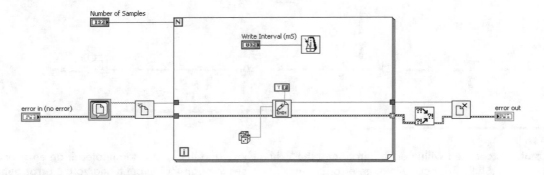

Figure 7-5A
A For Loop indexes an array of errors by default. Merge Errors VI returns the first element of the array with **status** = TRUE.

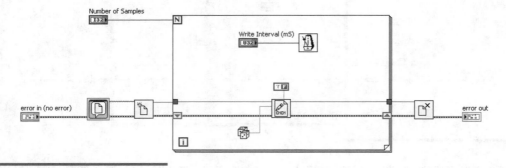

Figure 7-5B
A shift register extends the error chain between iterations of the loop. This approach is more efficient than array indexing.

Many VIs of medium or greater complexity benefit from parallel error chains, similar to what is illustrated in Figure 7-3. Parallel error chains are particularly useful when the parallel chains of nodes share common data, such as an I/O name or reference number. For example, in Figure 7-3 a task is propagated among the DAQmx VIs, and a file reference number is propagated among the File I/O functions. Positioning these groups of nodes in parallel helps facilitate the task and reference number propagation among the nodes, in addition to the error cluster. However, in some applications, the parallel nodes are functionally interdependent and should be combined into a common error chain. In this situation, the nodes can be arranged in parallel, to facilitate I/O name and reference number propagation, but a single error chain is shared by both groups of nodes.

For example, in Figure 7-6A, the DAQmx VIs acquire waveform data that is displayed to a waveform graph, while the File I/O functions log random data to file. These parallel nodes are functionally independent, and separate error chains are utilized. In Figure 7-6B, the DAQmx VIs acquire data that is logged to file using the File I/O functions. In this case, the parallel nodes are functionally *inter*dependent. Specifically, the Write to Binary File function depends on the DAQmx Read VI for its data. If DAQmx Read VI returns an error, the data it returns is not valid and should not be written to file. Likewise, the data acquisition task need not execute if an error occurs in selecting or creating the file. In this example, placing the nodes in parallel facilitates propagation of the task and reference number, and sharing a common error cluster creates error dependency while maximizing efficiency.

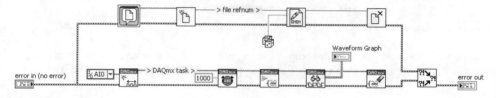

Figure 7-6A
DAQmx VIs and File I/O functions are independent, and parallel error chains are utilized.

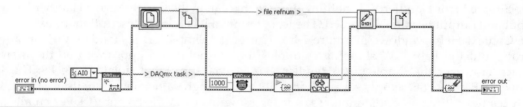

Figure 7-6B
DAQmx VIs and File I/O functions are interdependent. Sharing a common error chain facilitates error dependency and maximizes efficiency.

 Rule 7.5   Trap all errors from all nodes that have error terminals

For best results, trap *all* errors returned from *all* nodes that have error terminals. This maximizes the application's reliability. Some challenges and possible exceptions to this rule are explored in Section 7.3, "Prioritizing Errors." However, I generally recommend trapping errors from all error terminals to the maximum extent possible.

7.1.2 Reporting Errors

 Rule 7.6 Report errors using a dialog and/or log file

When an error is trapped, report the error using a dialog or log file. The dialog is the simplest and most popular method of error reporting. It consists of a call to the Simple or General Error Handler VI. These VIs, located on the **Dialog & User Interface** palette, examine the error passed to their **error in** terminal, look up an error description from LabVIEW's error code database, and generate a message describing the error. By default, they each open a dialog window that displays the error code, source, and description, which the user must acknowledge. An example is shown in Figure 7-7.

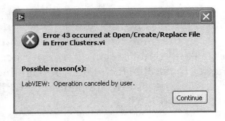

Figure 7-7
The Simple and General Error Handler VIs generate a dialog window that displays the error code, source, and description, which the user must acknowledge.

 Rule 7.7 Use General Error Handler VI over Simple Error Handler VI

The General Error Handler VI provides the flexibility to handle user-defined errors and exceptions. The Simple Error Handler VI is nothing more than a call to the General Error Handler VI, with fewer input and output terminals exposed. The icon, terminal labels, and description for each are shown in the Context Help windows in Figure 7-8A. Curiously, Simple Error Handler VI is more prevalent throughout the LabVIEW shipping examples than General and appears to be the more popular choice. However, the Simple Error Handler VI is a trivial subVI that violates Rule 4.9. The diagram is shown in Figure 7-8B. Simply stated, it does not add enough value to warrant its existence. Instead, it reduces flexibility and adds unnecessary processing overhead. Each subVI layer entails a nominal amount of processing time, including the call to Simple Error Handler VI that contains General Error Handler VI. Therefore, the General Error Handler VI is the more flexible and efficient choice for dialog error reporting.

Chapter 7 • Error Handling

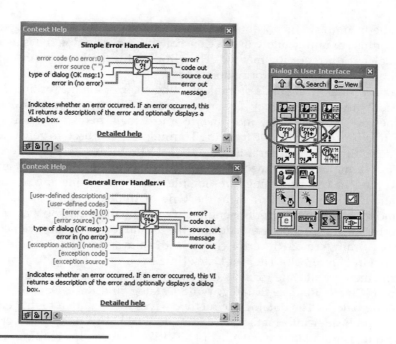

Figure 7-8A
The Simple and General Error Handler VIs are selected from the **Dialog & User Interface** palette. The primary difference is the number of terminals exposed.

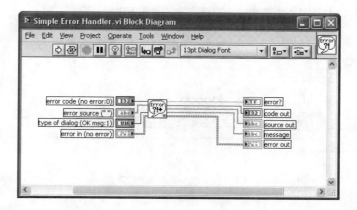

Figure 7-8B
The diagram of the Simple Error Handler VI contains a call to the General Error Handler VI. This is a trivial subVI that violates Rule 4.9.

 Rule 7.8 Implement an error log file for application deployment

After an application is deployed using the dialog error reporting method, it is often left up to the application's users to relay information regarding errors to the developer. Without explicit instructions, users may acknowledge an error dialog without recording the error, or the information they record might be incomplete or not thoroughly communicated to the developer. Additionally, error dialogs may confuse or annoy many users. Consequently, an error log file is used in addition to or in place of the dialog error reporting method. The error log file method simply consists of a routine for programmatically logging error information to file.

There are many possible methods of implementing an error log file routine. The primary considerations are readability and efficiency. For maximum readability, log a high-level error description, along with the code, source, date, and time, to a text file. This is shown in Figure 7-9A. The Get Date/Time String function, available from the **Timing** palette, is used to generate the date and time as ASCII character strings. The General Error Handler VI returns a message describing the error. The **message output** terminal is combined with the code, source, time, and date, and is written to a tab-delimited text file. The corresponding error log file can be read using any word processor or spreadsheet application. It is important to wire the **type of dialog** input terminal of the General Error Handler VI and to set the corresponding enumeration to **no dialog** if you want to suppress the dialog.

A second error log file method uses a datalog file format to optimize efficiency while sacrificing file readability. Specifically, bundle the error cluster with a time stamp and log the corresponding cluster to file, as shown in Figure 7-9B. The Get Date/Time in Seconds function, available from the **Timing** palette, returns the current time formatted using the time stamp data type. The datalog functions, including Open/Create/Replace Datalog, Write Datalog, and Read Datalog, are available from the **File I/O»Datalog** palette. The datalog file method saves the data in a highly efficient binary format, along with limited header information describing the data. This method is very simple to program. However, the data is not readable outside of LabVIEW. The datalog reading routine must consist of the Open/Create/Replace Datalog function with the cluster wired to the **record type** input terminal and the Read Datalog function. Use the datalog error log file if there is a possibility of many errors and efficiency is required, or if you want the file contents encrypted in binary format. However, text files are much more universal and are recommended otherwise.

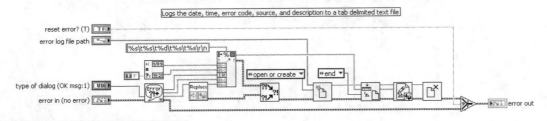

Figure 7-9A
The date and time are combined with the error message, source, and code, and are logged to a tab-delimited text file. Readability is maximized using this approach.

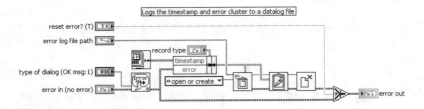

Figure 7-9B
Efficiency is maximized while readability is sacrificed by logging the error data and time stamp to a datalog file. The resulting file is not readable without a LabVIEW datalog file reading routine.

An important consideration with error log files is parsing the errors into files. In some circumstances, such as error reporting within continuous loops that do not terminate on error, appending every error to one file could result in a large or runaway file. Large error log files are difficult to load using a spreadsheet or word processor application and may contain many redundant errors. Similarly, creating a new file for each new error prevents large files but may generate large quantities of files. As an alternative, consider an intelligent error logging routine that limits the size and number of errors saved to the error log file. Specifically, the error logging routine can read the contents of the error log file and either append to file, replace the oldest entry, or skip its logging operation. The number of logged errors, file size, or existence of redundant error codes can be used as criteria for appending, replacing, or skipping.

It is important to notify the user or developer when an error occurs. For best results, combine an error dialog or front panel indicator with a log file. The dialog or indicator notifies the user that an error occurred, and the log file contains a complete record of the error details. An error indicator may be a Boolean that indicates a new error was logged, or a numeric that counts the number of errors that have occurred, or simply an error cluster displaying the most recent error data. Alternatively, LabVIEW can send an email that includes the error log file as an attachment, using the VIs on the **SMTP Email** palette.

Rule 7.9   Suppress dialog error reporting for unattended or remote operation

The dialog reporting method, either with or without an error log file, is not practical for applications designed to run unattended or remotely. When a dialog opens and nobody is present to acknowledge it, the application may suspend itself indefinitely. Additionally, beware that LabVIEW's web server cannot publish dialogs for remote clients accessing the application via web browser. In this case, the dialog will open on the server computer but is transparent to any remote clients. Avoid dialog error reporting in any of these situations.

It is important to carefully consider the placement of the error reporting routine. All error chains in an application must eventually terminate with a call to the error reporting VI. However, avoid calling the error reporting VI within continuous loops or within subVIs that might be called within continuous loops, unless corrective action is taken to gracefully recover from the error. Otherwise, the error reporting VI could cause the error dialog to repeatedly appear or the log file to grow very large. It is best to restrict the error reporting VIs to the top level, which helps consolidate error reporting and reduces the chances of reporting within a continuous loop. Additionally, it is helpful to merge all errors into one error reporting VI. The Merge Errors VI is a handy method of combining the errors from multiple error chains for reporting by a single VI.

The Torque Hysteresis VI's top-level diagram is shown in Figure 7-10. Observe that errors are trapped via propagation of the error cluster and are reported by the General Error Handler VI. If an error occurs the loop is terminated and the error is reported. The General Error Handler VI's placement outside of the loop ensures that the dialog will appear only once, even if the loop's termination logic changes. This example represents an extremely common error handling methodology applied to a simple application's architecture.

Rule 7.10   Avoid subVIs with built-in error reporting

LabVIEW's palettes contain several VIs that have error reporting built in. These VIs can be recognized on the palettes as high-level VIs, appearing in the top row of their subpalettes, that do not have error terminals. They include all VIs formerly known as the Easy I/O VIs, which are available from the traditional **Data Acquisition** subpalettes, and several of the high-level File I/O VIs. All such VIs are inflexible, providing only the dialog method of reporting errors. Additionally, the absence of the error terminals makes them more difficult to use in a dataflow sequence. Finally, these VIs are less efficient than the lower level VIs. Hence, I recommend avoiding them.

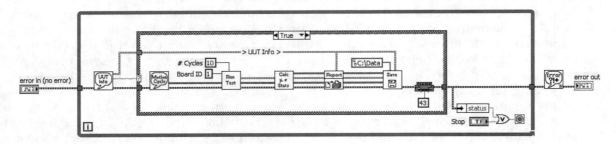

Figure 7-10
The Torque Hysteresis VI traps errors via propagation of the error cluster, terminates the loop on error, and reports the error using General Error Handler VI. Notice that the General Error Handler VI is called only once.

7.1.3 Error Codes

The error cluster contains a **code** control represented as a 32-bit signed integer. Codes are used to uniquely identify the errors that functions and VIs generate. LabVIEW maintains an internal database for which error codes are associated to descriptions. The categories and ranges of error codes, as of version 8.2 of LabVIEW, are shown in Table 7-1. This table relates the ranges of predefined error codes to the types of operations they are reserved for.

Table 7-1 *LabVIEW Error Codes*

Range	Error Code Table
−2147467263 through −1967390460	Networking
−1950679040 through −1950678977	Shared Variable
−1967362045, −1967362022 through −1967361997, and −1967345663 through −1967345609	Security
−1074003967 through −1074003950, −1074000001 and −1074000000	Instrument Driver
−1073807360 through −1073807192	VISA
−90165 through −90149 and −90111 through −90001	MathScript
−41005 through −41000 and 41000	Report Generation
−23096 through −23081	Formula Parsing

Range	Error Code Table
−23096 through −23000, −20141, −20140, and 20005	Mathematics
−20337 through −20301	Signal Processing
−20207 through −20201	Point By Point
−20119 through −20001	Signal Processing
−4644 through −4600	Regular Expression
−2983 through −2970, −2964 through −2960, −2955 through −2950, and −2929 through −2901	Source Control
−2586, −2585, −2583 through −2581, −2578, −2575, −2574, −2572 through −2550, −2526 through −2501, and 2552	Storage
−1821, −1820, −1817 through −1800	Waveform
−1719 through −1700	Apple Event
−1300 through −1210	Instrument Driver
−823 through −800 and 824	Timed Loop
−620 through −600	Windows Registry Access
0	Signal Processing, GPIB, Instrument Driver, Formula Parsing, VISA
1 through 20	GPIB
1 through 52	General
30 through 32, 40 through 41	GPIB
53 through 66	Networking
61 through 65	Serial
67 through 91	General
92 through 96	Windows Connectivity
97 through 100	General
102 and 103	Instrument Driver
108 through 121	Networking
116 through 118 and 122	General
1000 through 1045	General
1046 through 1050 and 1053	Script Nodes

Table 7-1 *Continued*

Range	Error Code Table
1051 through 1091	General
1101, 1114, 1115, 1132 through 1134, 1139 through 1143, and 1178 through 1185	Networking
1094 through 1157	General
1158 through 1169, 1318, 1404 and 1437	Run-Time Menu
1172, 1173, 1189, and 1199	Windows Connectivity
1174 through 1188, 1190 through 1194, and 1196 through 1198	General
1301 through 1321, 1357 through 1366, 1370, 1376 through 1378, 1380 through 1391, and 1395 through 1399	General
1325, 1375, 1386, and 1387	Windows Connectivity
1367, 1368, and 1379	Security
1337 through 1356, 1369, and 1383	Networking
1371, 1373, 1392 through 1394, 1400 through 1403, 1448, and 1486	LabVIEW Object-Oriented Programming
1430 through 1455, 1468 through 1470, 1483 through 1485, 1487 through 1493, and 1497 through 1500	General
1456 through 1467 and 1471 through 1482	3D Picture Control
1800 through 1809 and 1814	Waveform
4800 through 4806, 4810, 4811, and 4820 through 4823	General
16211 through 16554	SMTP
20001 through 20030, 20307, 20334, and 20351 through 20353	Signal Processing
56000 through 56005	General
180121602 through 180121604	Shared Variable
1073479937 through 1073479940	Instrument Driver
1073676290 through 1073676457	VISA

Additionally, LabVIEW reserves codes in the range of −8999 through −8000 and 5000 through 9999 for custom, user-defined errors. If your VIs need to flag a misbehavior that is not adequately described by one of the predefined error codes, define a new code in this range. The General Error Handler VI can be used to report user-defined errors. Simply specify the error codes and descriptions as arrays of integers and strings, and wire them to the associated terminals. This approach is very easy to implement if the application contains only one or a limited number of error reporting routines. However, it becomes more difficult if there are multiple instances of the General Error Handler VI, requiring multiple user-defined error codes. Specifically, the maintenance of the error codes and descriptions as arrays, along with propagating the data using wires, can be cumbersome.

Rule 7.11   Maintain user-defined error codes within an XML file

Instead, use an XML file to maintain multiple user-defined error codes and descriptions. Specifically, register custom error codes by creating an XML-based text file in the `LabVIEW\user.lib\errors` folder. This is accomplished using the Error Code File Editor that is launched by selecting **Tools»Advanced»Edit Error Codes**. The user-defined error codes are registered within LabVIEW, similar to a predefined error. This has several advantages over the diagram definition method. First, you need not wire additional inputs to the General Error Handler VI. Second, the user-defined error codes may be used within any VI that runs under the instance of LabVIEW that contains the XML file. Finally, the Explain Error feature recognizes and explains errors defined in this manner. **Explain Error** is a built-in utility that displays information about an error code in a dialog. It is invoked by selecting **Explain Error** from the shortcut menu from any error cluster, or by selecting **Help»Explain Error**. When deploying applications, be sure to create an installer or source distribution that includes the XML file.

Rule 7.12   Use negative codes for I/O device errors, and positive codes for warnings

With I/O devices such as data acquisition hardware, positive codes are used to represent warnings that may not directly affect the application's success, and negative codes represent significant errors that require corrective action or termination of the VI. As previously discussed, the error cluster's **status** Boolean uniquely distinguishes an error. All nonzero codes are warnings if **status** = FALSE. However, it is good practice to follow the well-established sign convention for I/O devices.

7.2 SubVI Error Handling

Most LabVIEW functions and VIs with **error in** terminals evaluate the **status** Boolean and skip their code if the value is TRUE. This aids the application's capability to recover from and report errors in a fast and efficient manner. For example, an I/O function such as VISA Read may wait 10 seconds for an instrument to respond to a query before returning a timeout error to the calling application. An error of this magnitude will likely degrade the application's performance. If similar instrument queries execute with similar results, the application appears to be hung up or unstable. Trapping the error and propagating it to all subsequent nodes in the error chain prevents this from happening. Instead, all subsequent instrument queries that receive the error are skipped, thereby accelerating the propagation of the error to the reporting routine and aiding the application in a graceful recovery.

Rule 7.13 Skip most subVI diagrams on error using an Error Case Structure

It is desirable for subVIs to mimic the error handling behavior of the LabVIEW functions and VIs. Specifically, most subVIs should skip their diagrams when an error is received at the **error in** terminal. This is implemented using an **Error Case Structure**, as follows: First, place the standard **error in** and **error out** clusters on the panel and assign them to the bottom left and right connector terminals, respectively. These standard connector assignments conform to Rule 5.26. Next, surround the completed subVI diagram with a Case structure and wire the **error in** cluster to the selector terminal. By default, the code will be contained by the **No Error** case. Use the Positioning tool to move the selector terminal near the bottom of the Case structure, and align the **error in**, case selector, and **error out** terminals vertically. Propagate the error cluster from the selector terminal to all nodes within the Case structure that have error terminals. Pass the error cluster through a tunnel on the right border to the **error out** indicator terminal. Inside the Case structure's **Error** case, pass the error cluster data from the selector terminal straight through the output tunnel on the Case structure's right border. The **Error** case is empty aside from the error cluster.

In the **No Error** case, you can also modify the error information, such as applying a user-defined error code and description if an error occurs. The error source can be provided by reading VI Server's **VI Name** property, or reading and indexing the call chain returned by the Call Chain function. The latter function returns the entire hierarchy of callers, which helps pinpoint the specific instance of a subVI call that generates an error. The Case structure ensures that all of the subVI's code will be skipped if an error is present. This maximizes the efficiency and responsiveness of your subVIs and the applications that call them.

Figure 7-11 illustrates subVI error handling incorporated by DAQmx Read VI. In Figure 7-11A, the panel contains the standard error clusters assigned to the lower left and right connector terminals. Figure 7-11B shows the diagram including **No Error** and **Error** cases of an Error Case Structure. As shown, the **No Error** case primarily consists of a call to a Call Library node that performs the read operation, as well as a call to a utility VI named DAQmx Fill In Error Info VI. The **Error** case is empty except for the error cluster passthrough and a constant. Figure 7-11C shows the contents of DAQmx Fill In Error Info VI. This utility VI checks if a new error has occurred and creates the error source information using a combination of data returned from Call Chain and any message passed to **extended error info**. Because the error occurred in the subVI's caller, the call chain is indexed by one element.

Chapter 7 • Error Handling

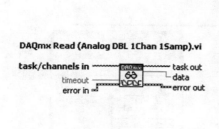
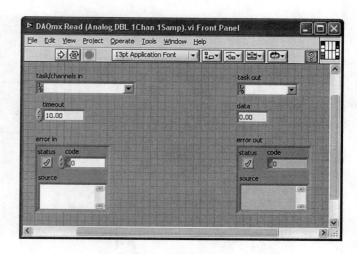

Figure 7-11A
The front panel of DAQmx Read VI contains the standard error clusters assigned to the lower left and right connector terminals.

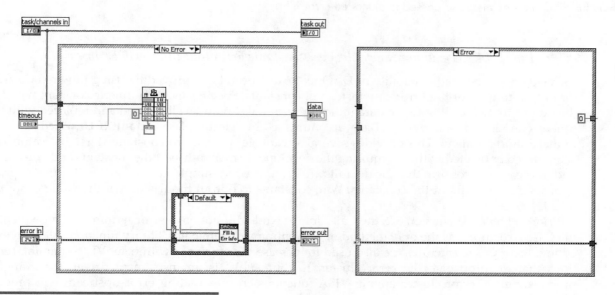

Figure 7-11B
The block diagram of DAQmx Read VI uses an Error Case Structure to skip its code if an error is present.

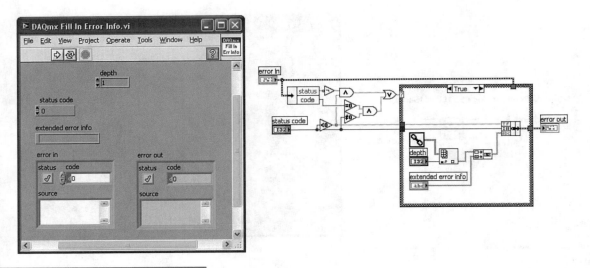

Figure 7-11C
DAQmx Fill In Error Info VI creates the error source information from a combination of data returned from the Call Chain function and any message passed to **extended error info**.

 Rule 7.14   Use unwired defaults over constants for output tunnels of Error case

As a side note, the maintainability of DAQmx Read VI can be improved by using unwired defaults for output tunnels of the **Error** case instead of constants. Specifically, the numeric constant wired to the output tunnel leading to the **data** indicator terminal should be eliminated and replaced with the tunnel's unwired default value. This is accomplished by selecting **Use Default If Unwired** from the tunnel's shortcut menu. This provides several advantages over wired constants. First, indicators can be modified or deleted without requiring the developer to manually edit the constants and wires. Second, it is easier to confirm that the defaults are being used by simple inspection. Finally, the diagram appears cleaner without the constants. Wire constants to output tunnels only if the value is not the data type's default.

Prior to LabVIEW version 8.2, the Call Library Node did not have error terminals. Consequently, the Case structure was the only method of preventing a dynamic link library function call from executing. Error Case Structures are also used to increase the efficiency within subVIs containing functions that do propagate the error cluster. If your subVI is an instrument driver, for example, propagating the error cluster among VISA functions ensures that an error upstream will prevent VISA functions downstream from executing. However, most instrument drivers perform substantial command string manipulation and response parsing, in addition to the instrument communications. Processing efficiency is optimized by enclosing these functions within the Case structure as well.

In Figure 7-12A, an instrument driver subVI for a digital multimeter assembles a command using several string manipulation functions and sends the command to the instrument using a call to VISA Write. This subVI contains incomplete error trapping. Specifically, the error cluster is passed to the VISA Write function only. The error terminals on three Format Into String functions are unwired. Consequently, these functions execute regardless of the error status of the **error in** cluster. In Figure 7-12B, an Error Case Structure encloses all the subVI's code, and the error cluster is shared among all nodes that have error terminals. Additionally, functions that do not have error terminals, such as In

Range and Coerce, Select, and Pick Line are now enclosed by the Error Case Structure and will not execute if an error is present. Because of the lack of error terminals, the Case structure is the only method that prevents these functions from executing. Hence, the Error Case Structure maximizes efficiency in the event of an error. Additionally, notice that the diagram nodes have been horizontally aligned about an axis defined by the error terminals.

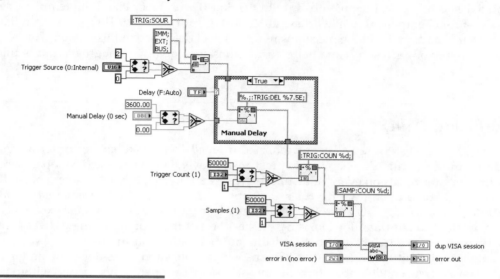

Figure 7-12A
An instrument driver subVI contains incomplete error trapping and inefficient recovery when an error is received.

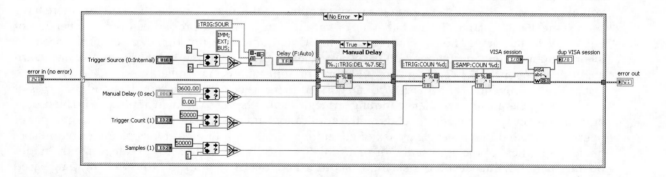

Figure 7-12B
The error cluster is shared among all functions that have error terminals, and an Error Case Structure maximizes efficiency in the event of an upstream error.

LabVIEW functions and subVIs that do not skip their code on error include most nodes that terminate a session to a resource. This includes Close Reference, VISA Close, Close File, and several nodes that have Destroy, Stop, Close, and Clear in their name. SubVIs that call these nodes should *not* skip them on error. Instead, confirm their behavior by carefully reading the **error in** terminal description in the node's detailed help or inspecting its diagram. If the description indicates the node runs normally if an error occurs before it, the node should be placed outside the subVI's Error Case Structure.

As an example, all LabVIEW Plug and Play instrument drivers include a Close VI that calls the VISA Close function. These subVIs do not contain an Error Case Structure.

 Rule 7.15   Use the SubVI with Error Handling template

If you like shortcuts, begin your subVIs with the SubVI with Error Handling template, selected from **New»VI»From Template»Frameworks»SubVI with Error Handling**. This template initiates a new VI that contains the error handling mechanisms for a subVI described in this section. This template saves a minute or two on each subVI and helps ensure consistent quality throughout your applications.

7.3 Prioritizing Errors

Per Rule 7.5, it is best to trap *all* errors returned from *all* nodes that have error terminals. This helps maximize the application's reliability. In practice, however, I know of few developers who rigorously wire the error terminals on every node of every diagram. In fact, there is a potential conflict between Rule 7.5 and some of the wiring rules in Section 4.2. In a dense diagram that contains parallel dataflow paths, wiring every error terminal may lead to overlapping wires, excessively large diagrams, or right-to-left data flow. Finding the best layout that optimizes these diverse requirements may become time consuming. What tradeoffs are permissible in the interest of saving time while maintaining reliability and good style? To answer this question, consider your understanding and comfort level of each node's operation, the importance of the operation's success within your application, and your tolerance for risk.

If you are like me and you learned to use the **Visible** property for a front panel control back in version 3.0 of LabVIEW, previously packaged as an Attribute Node that did not have error terminals, you might be able to justify skipping the error trapping on this one. However, beware that there have been some subtle changes in the behavior of various functions and properties from one version of LabVIEW to the next. The Synchronization and Communication functions, for example, seem to incorporate new enhancements with each new release. Error handling helps us identify and learn about the differences between versions and port our applications more reliably.

LabVIEW's functions and VIs can be categorized in terms of risk, to help understand the types of errors they generate and evaluate handling strategies with respect to a given application. The probability and severity of an error depends on the application and the system. However, different types of operations have different levels of risk of generating errors and different types of associated misbehavior, regardless of the application or platform. Specifically, I find it useful to categorize the nodes on the palettes in three levels of risk: high, medium, and low. Then prioritize the error trapping based on the perceived risk.

LabVIEW's **error out** terminal is the primary method by which LabVIEW errors are returned from nodes that generate errors. An exception includes nodes that return an error code as a scalar integer, such as the Mathematics VIs and many Call Library Function Nodes developed prior to LabVIEW version 8.2. The presence or absence of the **error out** or **error code** terminal in a node's connector pane is one indicator of the risk associated with that node. The basic mathematics functions available on the **Numeric** palette, for example, do not have error terminals and do not return errors. This is because LabVIEW gracefully handles all combinations of problems that may arise, and misbehavior is almost impossible. Division by zero, for example, returns the value **Inf**, which is a valid constant understood by all of LabVIEW's numeric functions, controls, and indicators. Therefore, simple mathematics functions do not require error handling. Likewise, all functions on the remaining palettes that do not contain error terminals do not generate errors. This includes many functions available from the **Structures, Array, Cluster &Variant, Boolean, String, Comparison**, and **Timing** palettes, as well as many other palettes.

At the opposite end of the risk spectrum are functions and VIs that perform input and output (I/O) operations. These nodes make calls to device drivers, DLLs or shared libraries, the operating system, or any application or resource that is external to the LabVIEW environment, including remote instances of LabVIEW. I/O operations include all of the nodes available on the following palettes: **File I/O, Measurements I/O, Instrument I/O, and Data Communication**. These nodes are risky because they rely on an external driver or a resource that may or may not be in a state that will respond appropriately to an application's requests. The external resources perform their I/O operations independent of LabVIEW, and there are many scenarios in which the operation could fail. Also, failed I/O operations are more likely to cause the calling application to misbehave, such as hang up waiting for a device to respond. Most I/O functions and VIs in LabVIEW return the error cluster. Trapping and handling the returned error information is an absolute necessity!

Medium-risk nodes have error terminals but are not calling resources external to the LabVIEW environment. These include Property Nodes associated with front panel controls, all functions in the **Synchronization** palette, VI Server when controlling the local instance of LabVIEW, most Express VIs (with the exception of the VIs on the **Express Input** and **Output** subpalettes), Scan from String and Format into String, and Mathematics VIs. Most of the errors generated by medium-risk nodes are well behaved and will not cause long latencies or crashes. Error handling is the best method of identifying problems with how these functions are being applied and is strongly recommended. For example, wiring the error returned from a Property Node helps determine that a programmed property value is invalid, which may otherwise go unnoticed.

Figure 7-13 provides source code samples for functions in the three categories of risk. Figure 7-13A contains a mathematical operation using numeric functions that do not have error terminals. None of the functions on the **Numeric** palette contain error terminals, and they have no risk of misbehaving. Figure 7-13B contains a Property Node and Scan from String function, with error terminals wired. These functions have medium level of risk, and error handling is recommended. Figure 7-13C contains VISA Write and Read functions, which are examples of high-risk I/O functions. Error handling is essential for I/O operations, including instrument communications with VISA.

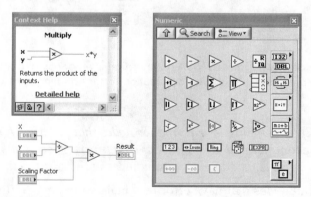

Figure 7-13A
The functions on the **Numeric** palette do not have error terminals, cannot generate errors, and do not misbehave. These functions are categorized as low risk.

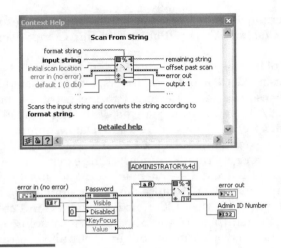

Figure 7-13B
Property Nodes associated with front panel controls and the Scan from String function are examples of medium risk nodes. Errors are usually not fatal, but error trapping is highly recommended.

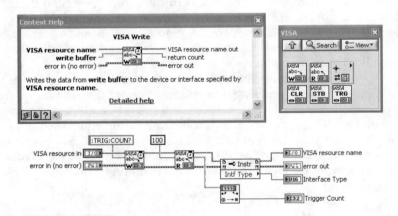

Figure 7-13C
Functions that call external device drivers such as VISA have a high level of risk of generating an error, and the corresponding errors are often detrimental to the calling application. Error handling is essential with I/O functions.

 *Rule 7.16 Error trapping is **required** for nodes that perform I/O operations, **recommended** for nodes that contain error terminals, and **optional** for diagrams that do not contain nodes with error terminals*

Trap all errors from all nodes that have error terminals, to the maximum extent possible. However, when style conflicts arise, prioritize error trapping according to the three classes of nodes just presented. A firm, minimum requirement is to positively handle any errors generated by nodes that perform I/O. As discussed, I/O functions and subVIs have a higher risk of generating errors and greater consequences when errors happen. Additionally, error trapping is highly recommended for the medium-risk nodes that have error terminals but do not perform I/O. Errors returned from these nodes help the developer understand how they work and help ensure reliable and robust applications. Finally, when creating routines that contain nodes without error terminals, error propagation via Error Case Structure or subVI with error terminals is optional. In the latter case, propagating the error cluster to the subVI is useful for data flow, maintains standard connector terminal patterns and assignments, and optimizes error reporting and recovery.

Figure 7-14A contains the front panel, block diagram, and Context Help window for Wait n mSec VI. This subVI simply inserts a delay within an error chain. The unique icon shape and connector pattern are discussed in Chapter 5, "Icon and Connector." The diagram contains the Wait (ms) function and proper subVI error handling. Note that the Wait (ms) function does not have error terminals and cannot generate an error. It is an example of a low-risk function. However, the subVI incorporates error propagation and an Error Case Structure for skipping the code on error, per the rules in Section 7.2. Moreover, the subVI's error terminals allow the calling VIs to use data dependency to specify when the delay occurs, and to skip the delay if an error is already present.

In Figure 7-14B, the Wait n mSec VI is applied to an instrument communications example. Specifically, the Wait n mSec VI is inserted between VISA Write and a Property Node that returns the Number of Bytes at Serial Port. This delay gives the instrument time to respond to the query before reading the response. Note that any errors returned prior to the Wait n mSec VI cause the delay to skip. Therefore, error handling with this subVI facilitates execution ordering via data dependency and optimizes the efficiency in the event of an error.

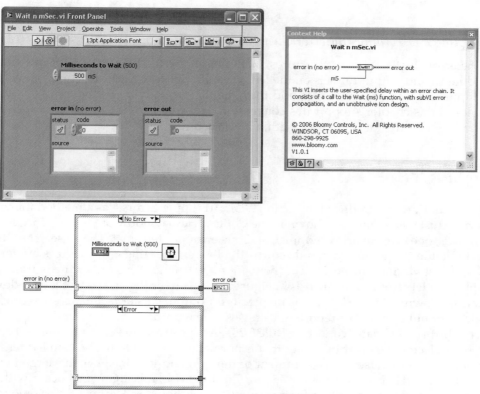

Figure 7-14A
Wait n mSec VI consists of the Wait (ms) function, a low-risk function without error terminals, and proper subVI error handling, including the Error Case Structure.

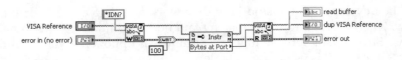

Figure 7-14B
Wait n mSec VI inserts a 100ms delay between the VISA Write function and a Property Node, allowing an instrument time to respond to a query.

7.4 Error Handling Tips

This section provides tips on error handling, including structure wiring, merging errors, clearing errors, and automatic error handling.

7.4.1 Structure Wiring

 Rule 7.17 Tunnel the error cluster near the bottom of structures

Wire the error cluster through structures, including loops, Case, Event, and Sequence, via vertically aligned input and output tunnels on the left and right borders. Each structure should have only one error input and output tunnel pair, and they should normally be the bottom-most tunnels. All the examples in this chapter support this convention. However, do not wire the error cluster through a multiframe structure if it is not used within the structure.

7.4.2 Merging Errors

Merge Errors VI was introduced in Section 7.1.1, "Trapping Errors," as a method for combining multiple error clusters into one. As shown in the Context Help window of Figure 7-15A, it has three **error in** terminals, one **error in array** terminal, and one **error out** terminal. The VI searches the error inputs in top to bottom terminal order and returns the first error it finds. If no errors are found, it returns either the first warning or no error if there are no warnings. This technique is used for combining multiple parallel error chains, such as in Figure 7-3, or converting an array of loop indexed errors into a scalar, as shown in Figure 7-5A. The merged error output cluster facilitates error propagation using a single **error out** terminal or reporting using a single reporting VI.

Merge Errors VI violates connector Rule 5.25, "Assign error clusters to bottom left and right terminals," by replacing the **error in** cluster that is normally assigned to the bottom left terminal, with the **error in array**. This causes a misalignment of the icon and terminals when Merge Errors VI is used with scalar error chains, as shown in Figure 7-3. Likewise, misalignments occur with loop indexed arrays of errors in parallel with reference numbers, as shown in Figure 7-5A. Moreover, the **error in**

array terminal has limited value. Arrays of error clusters are formed by indexing errors on loop tunnels or building an array that combines multiple error clusters. As discussed in Section 7.1.1, "Trapping Errors," indexing errors is not recommended. Instead, most loops check the error status and terminate on error, or propagate a scalar error cluster using a shift register. Also, the Merge Errors VI's merging operation defeats the purpose of forming the array, if one existed.

Figure 7-15B presents Merge Multiple Errors VI[1], an alternative to Merge Errors VI, containing five additional **error in** terminals and no **error in array** terminal. It uses the standard 4×2×2×4 connector pattern, with all left and middle terminals assigned to **error in** clusters. Hence, it merges up to eight scalar **error in** clusters into one scalar **error out** cluster. This increases the number of error chains that can be merged, while promoting good error handling, wiring, and connector style.

In Figure 7-15A, five parallel error chains are combined using the Merge Errors VI. The Build Array function is required to combine two of the error clusters into an array, and the subVI is positioned approximately near the middle of the pack of parallel error chains. In Figure 7-15B, the Merge Multiple Errors VI is used to combine all error chains. Because the lower left terminal conforms to Rule 5.26, this subVI is vertically aligned with the error terminals and subVI of the bottom error chain. Also, the Build Array function is eliminated, and the diagram contains fewer wire bends. Finally, a middle connector terminal is utilized to merge five error chains in this example.

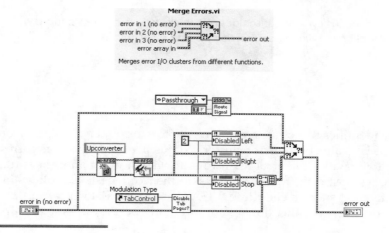

Figure 7-15A
Five parallel error chains are combined using Merge Errors VI. Because of the nonstandard **error in array** terminal at the bottom left, the subVI is not aligned with any of the other subVIs. Also, a Build Array function is required to merge five error chains.

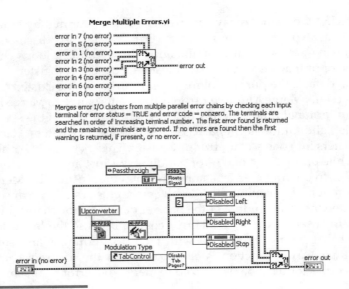

Figure 7-15B
Merge Multiple Errors VI combines up to eight scalar **error in** clusters using the standard 4×2×2×4 connector pattern. The subVI is vertically aligned with the error terminals and other subVIs.

7.4.3 Clearing Errors

Error handling identifies two classes of misbehavior: errors and warnings. Errors are uniquely distinguished by a TRUE value of the error cluster's status Boolean. Warnings correspond to any nonzero codes accompanying a FALSE error status. As discussed, when an error is trapped within an error chain, most nodes upstream from the error skip their code. The exceptions include most functions that close a reference, such as File Close, VISA Close, Close Reference, and Release Queue. In some situations, it is desirable to either downgrade or ignore specific errors so that upstream nodes execute normally.

As an example, consider a communications server application in which client connect and disconnect operations are routine tasks. The communications functions return an error when the client disconnects. The server program needs this information to clear its connection with the client and begin looking for new client connections. However, if the disconnect error is propagated throughout the server application, the communications functions will not execute, and any new client connection attempts will be thwarted. Instead, the server should process and clear the error associated with a client disconnect.

Figure 7-16 illustrates a utility subVI named Clear Error All or Specified VI[1]. Its purpose is to programmatically reset the error cluster with no error. Depending on the value of **Clear Mode**, it resets the error code always or just when the error code equals the programmed **Code to Clear**. Also, it can optionally display the error dialog before clearing the error. Notice the small icon size. This allows other wires on the calling VI's diagram to flow above the subVI without the icon obstructing their path. For example, any reference numbers running in parallel to the error cluster can pass without bending around the icon. The icon and connector for Clear Error All or Specified VI are discussed in Chapter 5.

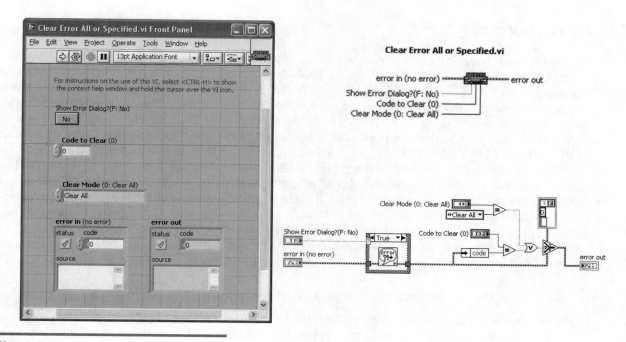

Figure 7-16
Clear Error All or Specified VI is a utility that selectively clears errors according to the values of **Clear Mode** and **Code to Clear**.

Figure 7-17 contains an illustration of Clear Error All or Specified VI applied to the Torque Hysteresis VI discussed in Chapter 6, "Data Structures." After each test, Save Data VI prompts the operator for a filename using a file dialog, and the test data is logged to file. However, if the user cancels out of the file dialog, the VI returns error code 43. This error causes the loop to terminate unless it is cleared from the error cluster before checking the error status. Consequently, Clear Error All or Specified VI is used to clear error code 43 and continue the application.

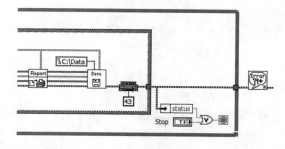

Figure 7-17
Clear Errors All or Specified VI clears error code 43 from Torque Hysteresis VI. This error naturally occurs when the user cancels a file dialog.

7.4.4 Automatic Error Handling

Automatic error handling is an alternate method for trapping errors that is transparent to the diagram. It consists of a VI property named **Enable automatic error handling**, selected from the **Execution** category of the **VI Properties** dialog. When an error occurs in any node with the **error out** terminal *not* wired, LabVIEW suspends execution, opens the diagram window, highlights the node, and displays an error dialog. If the node's **error out** terminal is wired, the error data propagates as programmed and no additional action is performed. Automatic error handling is disabled by default in all versions of LabVIEW except for 7.0, the version in which the feature was initially released.

It is important to note that automatic error handling does not ensure that errors are reported. If a node's **error out** terminal is wired but the error cluster terminates without a reporting routine, the error data is lost. Hence, automatic error handling helps only when one or more node's **error out** terminals are unwired, such as when nodes have been prioritized by risk, as discussed in the previous section. However, most error handling problems I have seen in practice are the result of incomplete error handling, in which many of the error terminals are wired but the error is not fully trapped or reported. Automatic error handling does not help in these situations.

 Rule 7.18   Leave automatic error handling disabled

If you are deploying an industrial application, the last thing you want is the application to suspend itself, open the diagram, and highlight a portion of the source code. This invites anyone who happens to be in front of the computer when the error occurs to tinker with the diagram. Hence, automatic error handling must be disabled before application deployment. Additionally, note that automatic error handling circumvents programmatic error handling. Per Theorem 7-1, programmatic error handling is essential for overcoming unforeseen problems before, during, and after deployment. Moreover, programmatic error handling is a principle method of establishing data flow and a key ingredient to good style. Therefore, use programmatic trapping and reporting of errors, and leave automatic error handling disabled.

7.5 Examples

This section contains more error handling examples, including loops, user-defined error, parallel error chains, and a state machine.

7.5.1 Continuous Acquire To File

Cont Acq&Graph Voltage-To File(Binary) VI is a shipping LabVIEW example that continuously acquires data and logs it to file using the DAQmx VIs and File I/O functions. The DAQmx VIs share a common task, and the File I/O functions share a common reference number and are placed in parallel to facilitate propagation of these data elements. Because the nodes are interdependent, a common error chain is shared between the VIs for error dependency and execution ordering. The loop terminates if an error occurs in any node, and an error message is reported with a dialog. The implementation shown in Figure 7-18A is located using NI Example Finder, as follows: **Hardware Input and Output»DAQmx»Analog Measurements»Voltage»Cont Acq&Graph Voltage-To File(Binary) VI**.

Chapter 7 • Error Handling

Figure 7-18B contains a few minor enhancements from the shipping example. First, the error cluster enters and exits the While Loop through vertically aligned tunnels located near the bottom of the loop. Additionally, the loop condition is placed in the bottom right corner of the loop, along with the code that unbundles the error status and evaluates the loop condition. Finally, the General Error Handler VI is used instead of the Simple Error Handler VI.

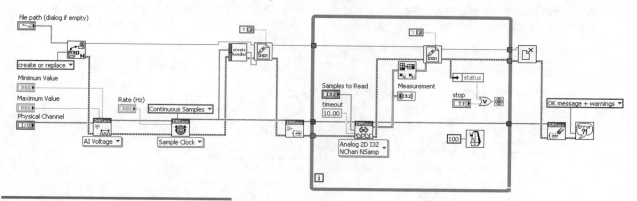

Figure 7-18A
The shipping example continuously acquires data and logs it to file. The error cluster is shared by the DAQmx VIs and File I/O functions because the nodes are interdependent. The loop terminates if an error occurs in any node, and it is reported in a dialog.

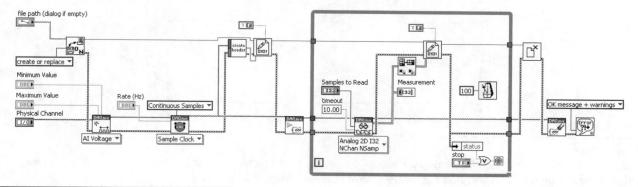

Figure 7-18B
Minor improvements to the shipping example include vertically aligned error tunnels at the bottom of the loop, loop condition at the bottom right, and General Error Handler VI.

7.5.2 Suss Interface Toolkit

Figure 7-19 contains the diagram of an instrument driver VI that controls a semiconductor wafer probe system. The diagram incorporates subVI error handling that includes error cluster terminals and an Error Case Structure. The **No Error** case of the Error Case Structure executes the Call Library Node and an error routine. The Call Library Function Node calls a function that resides within a dynamic link library (DLL). The function returns a parameter with integer data type named **Error Number**. This parameter returns a value of 0 if the function succeeds, or an integer in the range of 500 to 600 identifying any errors that occur within the DLL. The error routine evaluates the **Error Number** and generates an error cluster using Error Cluster from Error Code VI. The resulting LabVIEW error consists of a code that is derived from the original **Error Number** returned from the DLL but shifted into the range of LabVIEW user-defined errors (−8999 to −8000, 5000 to 9999). Specifically, an **Error Number** from 500 to 600 is shifted to an error code from −8500 to −8600. Per Rule 7.12, negative error codes are preferred for I/O device errors. Also, the error cluster's source is populated with an intuitive message consisting of the call chain and the response returned from the DLL.

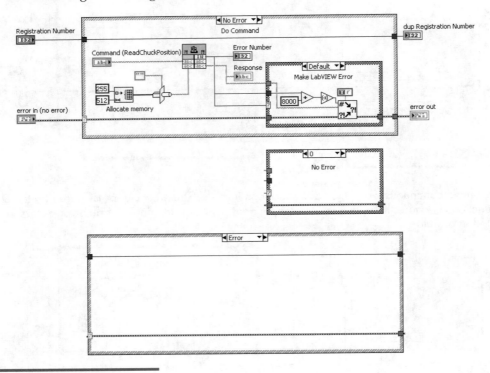

Figure 7-19
Instrument driver subVI uses a Call Library Function Node to call a function in a DLL. If an error is returned by the DLL, the **Error Number** is shifted into the range of user-defined errors.

7.5.3 Merge Parallel Errors

Figure 7-20 illustrates an initialization routine that reads configuration parameters from file using Config VIs, sets several control properties using Property Nodes, and initializes all control values to default using VI Server. These nodes form three parallel error chains that are merged prior to exiting the Case structure through an output tunnel. Merge Errors VI has a nonstandard **error in array** assigned to the lower left terminal. This causes a kink in the error cluster wire passed from the VI Server Close Reference function, as shown in Figure 7-20A. In Figure 7-20B, Merge Multiple Errors VI contains only scalar error terminals, including the bottom left terminal that is normally reserved for the **error in** terminal. This subVI eliminates the wire kink and promotes standard error terminal assignments.

Chapter 7 • Error Handling

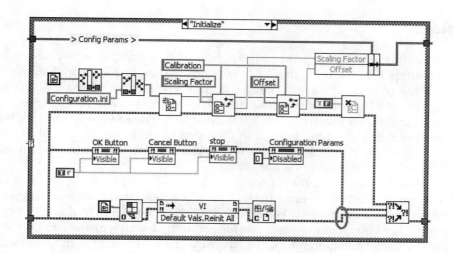

Figure 7-20A
Merge Errors VI has a nonstandard **error in array** terminal that causes a wire kink when used to merge parallel error clusters.

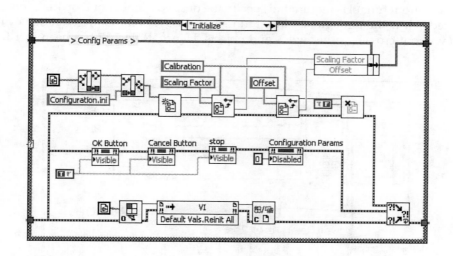

Figure 7-20B
Merge Multiple Errors VI has eight scalar **error in** terminals, which eliminates wire kinks and promotes good style.

7.5.4 Screw Inspection VI

Screw Inspection VI is an image processing application that was previously discussed in Chapter 4. The diagram snippet shown in Figure 7-21A contains two incomplete parallel error chains within a For Loop. The top error chain consists of three overlay VIs from the **Vision Utilities** palette. Mysteriously, the rightmost VI is omitted from the error chain. Also, the error is propagated among the Find Screw Ends VIs in the bottom error chain, but the first VI's **error in** terminal and the last VI's **error out** terminal are unwired. Hence, the error is not trapped in either error chain. Additionally, the Case structure containing the Wait (ms) function has no data dependency with the remaining nodes, and the order in which the delay occurs is not known.

Figure 7-21B illustrates proper error trapping for parallel error chains and For Loops. First, the error cluster is propagated to the **error in** terminal of all nodes that have error terminals. Second, the **error out** terminal of the last node in each chain is wired to the Merge Multiple Errors VI. Error propagation between For Loop iterations is accomplished using a shift register. Finally, the Wait n mSec VI replaces the Wait (ms) function and Case structure. This VI's error terminals facilitate execution ordering via natural data flow, and the icon's unique size and shape allow easy insertion within an error chain. The delay now reliably occurs at the beginning of the bottom error chain, before the four Find Screw Ends VIs. Figure 7-21C is similar to Figure 7-21B, except that a While Loop replaces the For Loop and a few nodes are rearranged. The While Loop allows the loop to terminate if an error occurs in any loop iteration. Nodes are rearranged so that the error cluster is the bottom-most output tunnel. A new input tunnel is required to specify the desired number of loop iterations. The input tunnel is located below the error cluster input tunnel, to pass it to the expression that evaluates the stop condition without crossing other wires and objects. Hence, Rule 7.17 has been sacrificed in favor of Rule 4.15.

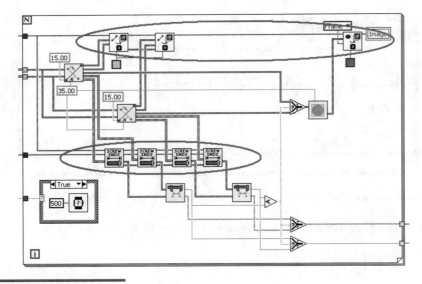

Figure 7-21A
Screw Inspection VI contains two incomplete parallel error chains within a For Loop, and a Wait (ms) function with unknown execution order.

Chapter 7 • Error Handling

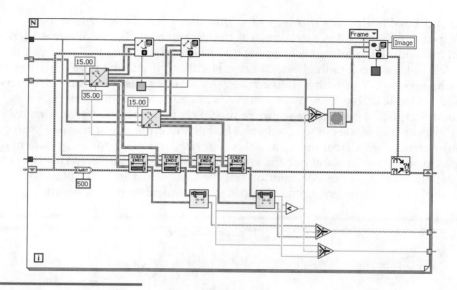

Figure 7-21B
Error trapping is completed by extending the error cluster to all error terminals, merging parallel error chains, and passing the error cluster between loop iterations using a shift register. The Wait n mSec VI uses data flow to order its delay before the Find Screw Ends VIs.

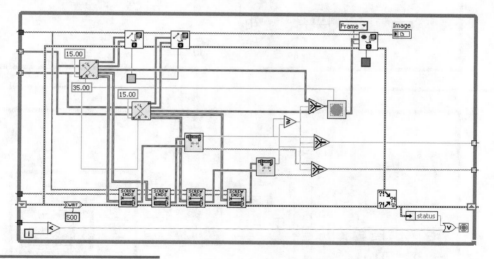

Figure 7-21C
A While Loop is used to terminate the loop on first error. Nodes are rearranged to eliminate output tunnels below the error cluster.

7.5.5 Test Executive VI

The application in Figure 7-22 is a test executive implemented using a variation of a state machine design pattern discussed in Chapter 8, "Design Patterns." Two cases of the primary Case structure,

Initialize and **Load Script**, are shown. In the **Initialize** case, the error terminals are unwired on all nodes, including two variables, nine Property Nodes, and a File Dialog Express VI. Also, a high-level VI is used for reading the test steps from file that has error handling built in. Specifically, Read Lines from File VI contains a call to the General Error Handler VI with the **type of dialog** input set to **continue or stop message**. This is shown in Figure 7-23. An error passed to this subVI generates a dialog, which may not be desirable if the test executive is intended to run unattended.

The **Load Script** case contains a sequence of subVIs that interact with a spreadsheet application. The error cluster propagates among all subVIs, but the error is not trapped or reported. The last subVI in the error chain's **error out** terminal is not wired, allowing the error data to disappear. Likewise, the error handling throughout the remaining cases is incomplete. This is evident at a glance because the error cluster does not enter or exit the Case structure or While Loop. Incomplete error trapping combined with no error reporting equates to nonexistent error handling.

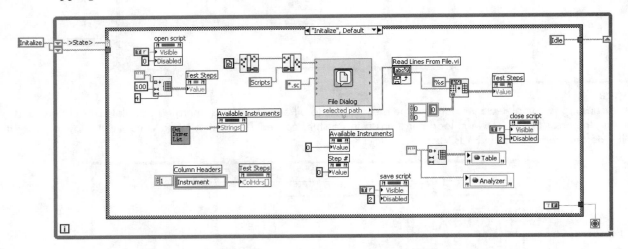

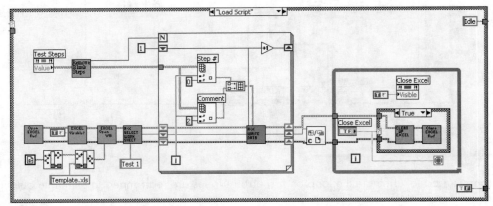

Figure 7-22
The **Initialize** and **Load Script** cases of a test executive are shown. The **Initialize** case has many nodes with unwired error terminals. The **Load Script** case propagates the error cluster between subVIs, but the **error out** terminal of the last subVI in the error chain is unwired, allowing the error to disappear.

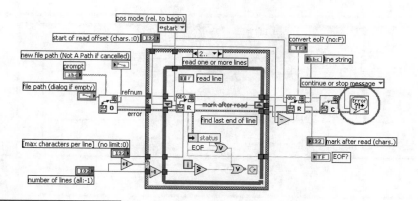

Figure 7-23
The diagram of Read Lines from File VI includes error handling via the General Error Handler VI. The error dialog is not desirable for unattended operation.

In Figure 7-24, Test Executive VI has been modified with proper error handling. First, the error cluster is read from the **error in** control terminal and initializes a shift register near the bottom of the While Loop. The error is read from the shift register and passes through a tunnel near the bottom of the Case structure. Each case of the Case structure, including **Initialize** and **Load Script**, propagates the error cluster among all nodes that have error terminals and returns the error cluster through the tunnel at the bottom right of the Case structure. Therefore, an error generated by any node in any case is trapped inside the error chain. Outside the Case structure, the error cluster's **status** is unbundled, and if an error exists, the **Error** is passed into the shift register on the right border and the **Error** case executes next. The **Error** case logs the error and time stamp to a datalog file, and clears the error from the error cluster.

Notice that the **Initialize** case now contains the intermediate level File I/O functions in place of Read Lines from File VI. This allows the error cluster to propagate in a manner that is consistent with the rest of the application and to avoid any dialogs from popping up during unattended operation. Finally, note that the **error in** control and **error out** indicator are not required in this example because it is a top-level VI. However, it is a good practice to write all VIs assuming that they might be used as a subVI and may require propagating the error cluster. For a GUI VI, the error clusters can be placed outside the visible area of the panel.

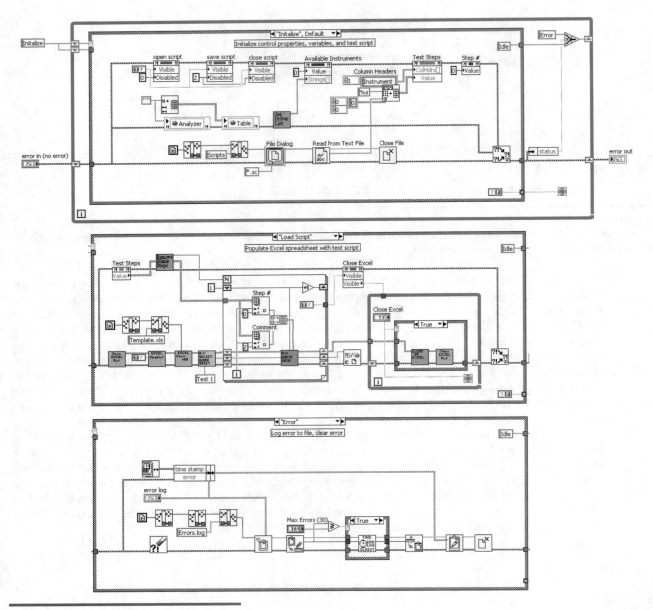

Figure 7-24
The test executive has been modified to implement proper error handling. Errors are trapped by propagating the error cluster to all nodes that have error terminals, and are reported by logging to a datalog file in the **Error** case.

Endnotes

1. Merge Multiple Errors VI and Clear Error All or Specified VI can be downloaded from www.bloomy.com/lvstyle.

Design Patterns

8

Design patterns are standard VI architectures that solve common software design problems. They consist of an arrangement of structures, functions, controls, and error handling that form a generic construct for common tasks such as looping, event handling, state transitioning, and data sharing and encapsulation. Design patterns are available as templates for promoting reuse, productivity, and quality. Also, one or more design patterns are combined with additional subVIs, controls, utilities, and libraries to form an application framework. An **application framework** is a more elaborate set of tools that advances the starting point for developing an application. Both the fundamental constructs and the higher-level frameworks are design patterns if they are common, recognizable, and reusable.

Design patterns promote good programming style. If all of the VIs in an application resemble well-known design patterns, it becomes easy for developers to recognize, understand, maintain, and reuse each other's code. Also, design patterns are ideal candidates for templates. Instead of creating every VI from scratch, start with a template and modify it to your needs. This saves time and helps ensure that your VIs employ common architectures consistently. Moreover, in a multideveloper organization, design patterns and templates promote standards that ensure consistent-quality software throughout the organization. As discussed in Chapter 1, "The Significance of Style," good style means good readability, maintainability, reliability, efficiency, robustness, simplicity, and performance.

In this chapter, multiple design patterns are discussed, with emphasis on the ones favored by the engineers at Bloomy Controls.[1] Because the topic is fairly lengthy, Table 8-1 provides a quick reference. For illustration purposes, the design patterns are applied to the Torque Hysteresis VI discussed in previous chapters. Before we begin, a few block diagram style rules presented in Chapter 4, "Block Diagram," are worth briefly reviewing here because they form the foundation upon which good design patterns are formed.

 Rule 4.6   Modularize top-level diagrams with subVIs

 Rule 4.27   Avoid Sequence structures unless required

 Rule 4.28   Avoid nesting beyond three layers

Table 8-1 *Summary of the Design Patterns and Application Frameworks Presented in This Chapter*

	SubVI	GUI VI	Top-Level VI	Simple Complexity	Medium Complexity	Advanced Complexity	Desktop OS	Real-Time OS	Event-Driven	Inefficient/Polling	Section	Page Number
Immediate SubVI	Y			Y			Y	Y			8.1.1	241
Functional Global	Y			Y			Y	Y			8.1.2	244
Continuous Loop	Y			Y			Y	Y		Y	8.1.3	246
Event-Handling Loop		Y			Y		Y		Y		8.1.4	250
Classic State Machine	Y			Y			Y	Y		Y	8.2.1	257
Queued State Machine	Y				Y		Y	Y	Y		8.2.2	260
Event-Driven State Machine		Y	Y		Y		Y		Y		8.2.3	262
Event Machine		Y	Y			Y	Y		Y		8.2.4	265
Parallel Loops			Y			Y	Y	Y	Y		8.3.1	269
Dynamic Framework		Y	Y			Y	Y		Y	Y	8.4.1	272
Multiple Loop Framework			Y			Y	Y		Y		8.4.2	278
Modular Multiple Loop Framework			Y			Y	Y	Y	Y		8.4.3	283

It is much easier to recognize a given design pattern when the diagram is modular. Always look for opportunities to combine groups of nodes that work together to perform a specific task into a subVI. There should be very few functions on the diagram of a top-level VI, outside of the functions that comprise the design pattern. Sequence structures are generally avoided because they circumvent the dataflow method of execution ordering and are inflexible. Ironically, one of the most accepted uses of Sequence structures is to help delineate and frame a design pattern. A Flat Sequence structure may be used when limited to the outermost structure of a top-level VI or initializing a variable or resource, but it should be avoided otherwise. The design patterns in this chapter use dataflow alternatives to Sequence structures. Avoiding Sequence structures helps prevent excessive nesting. The more layers of structures within structures there are on a diagram, the more difficult it is to comprehend the data flow. This equates to poor readability and maintainability. Most design patterns have three or fewer layers of nesting.

8.1 Simple Design Patterns

The simple design patterns are common and very easy to understand and implement. They include the Immediate SubVI, Functional Global, Continuous Loop, and Event-Handling Loop.

8.1.1 Immediate SubVI

The **Immediate SubVI** design pattern consists of the nodes that comprise the subVI and error trapping. There are no continuous loops, dialog windows, or GUI panels. Instead, an immediate subVI executes its code straight through to completion, in the order in which the error cluster propagates through the nodes on the diagram. Error trapping is incorporated per the subVI error handling rules presented in Chapter 7, "Error Handling." This includes an Error Case Structure for skipping the diagram if an error is detected, standard **error in** and **error out** clusters and connector assignments, and error propagation via error cluster wires. As the name implies, the Immediate SubVI design pattern is very common with subVIs and is the simplest design pattern.

You can use the SubVI with Error Handling template to develop immediate subVIs. This template is available from the **New** dialog by selecting **VI»From Template»Frameworks»SubVI with Error Handling**. It consists of error clusters, the standard 4×2×2×4 connector pattern, error terminal assignments, and a Case structure with **error in** terminal wired to the selector terminal. A variation is shown in Figure 8-1. It may save only 1 or 2 minutes of editing, but if you create an application containing 100 or more immediate subVIs, the minutes can add up to hours. Also, the template helps promote the subVI rules for front panels, connectors, and error handling discussed in Chapter 3, "Front Panel Style"; Chapter 5, "Icon and Connector"; and Chapter 7, "Error Handling." You can incorporate additional features to the template, such as standard documentation, icon, type definition controls, and more, to save more time.

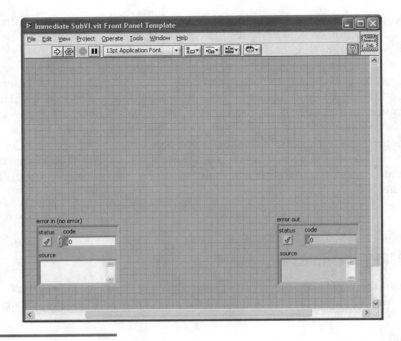

Figure 8-1A
The front panel of the Immediate SubVI template contains error clusters, the standard 4×2×2×4 connector pattern, and terminal assignments.

Figure 8-1B
The diagram contains a Case structure with the **error in** terminal wired to the case selector.

Figure 8-2 contains an example of the Immediate SubVI design pattern applied to a subVI called by the Torque Hysteresis VI. Specifically, Calculate Target Positions VI calculates several positions based on the user-specified motion parameters. The front panel layout, color, and text conform to the rules for a subVI discussed in Chapter 3. The block diagram contains the code within an Error Case Structure. Note that none of the functions that comprise the VI has error terminals. However, subVI error handling maximizes the efficiency in the presence of an error and helps facilitate data flow on the calling VI.

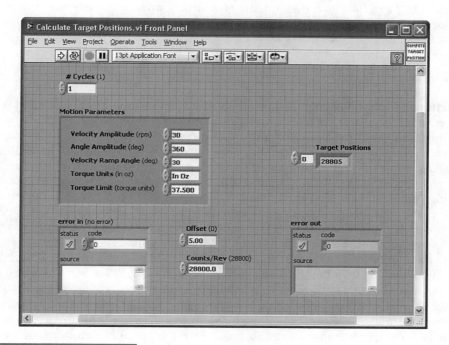

Figure 8-2A
The layout, color, and text of this Immediate SubVI panel conform to the front panel rules for a subVI discussed in Chapter 3.

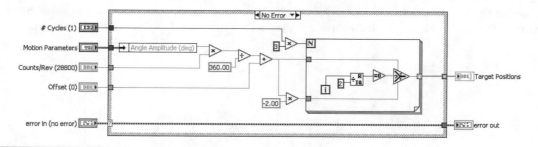

Figure 8-2B
SubVI error handling maximizes efficiency in the presence of an error and facilitates data flow on the calling VI.

If you are curious why the design pattern is called Immediate SubVI, although the LabVIEW template is called SubVI with Error Handling, allow me to explain. In the literal sense, both the terms *subVI* and *error handling* are misused. The complexity of today's LabVIEW applications is such that subVIs can have any level of complexity, from the principal components down to the low-level instrument drivers. The Immediate SubVI design pattern is only intended for very simple subVIs. Also, in Chapter 7, error handling is defined as the practice of trapping and reporting errors. The SubVI with Error Handling template performs error trapping but not error reporting. Therefore, an immediate subVI executes its code immediately, without any user interaction, continuous looping, or error reporting that might cause the calling application to wait.

8.1.2 Functional Global

The **Functional Global** design pattern consists of a subVI that contains a While Loop, a Case structure, an enumeration, and controls for reading and writing data. The While Loop contains one or more uninitialized shift registers, and a Boolean constant is wired to its conditional terminal, forcing it to stop after only one iteration. The loop's sole purpose is to store global data within the shift registers. The Case structure contains separate cases for writing and reading the shift register data. Additionally, you can add cases for initializing or resizing data structures such as arrays and strings, or performing calculations or data manipulation. The enumeration provides programmatic selection of the read, write, and other supported operations from the diagram of the calling VI. The subVI is applied to each location of an application that requires access to the data, similar to a global variable.

Functional globals are also known as **LabVIEW 2 style globals** because they were the only method of sharing data without data flow prior to the introduction of local and global variables in LabVIEW 3.0. However, functional globals have important uses today. They are more memory efficient and reliable than local and global variables. Each instance of a read local or global variable creates a copy of its data when read, allowing simultaneous write operations to occur without affecting the read data. This leads to extra memory buffers and race conditions. A **race condition** occurs when a write variable changes the data after a read variable has copied the previous value so that the read buffer data is different from the new variable data. A functional global has only one memory buffer per shift register and cannot be written to and read from simultaneously. Additionally, functional globals have diagrams that may incorporate expanded functionality beyond simple write and read operations.

Figure 8-3 shows the front panel and block diagram for a very simple functional global used for storing a numeric value. It supports only two operations, write and read, corresponding to the two cases of the Case structure shown. Notice that the functional global also meets the criteria for an immediate subVI. Hence, error trapping is applied as discussed in the previous section.

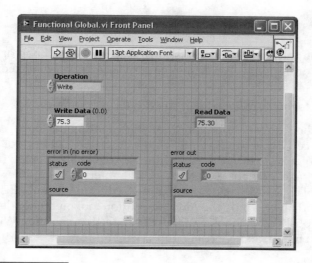

Figure 8-3A
The front panel of a functional global that stores a numeric value. It contains an enumeration for selecting the desired operation, a control for writing data, an indicator for reading data, and error clusters.

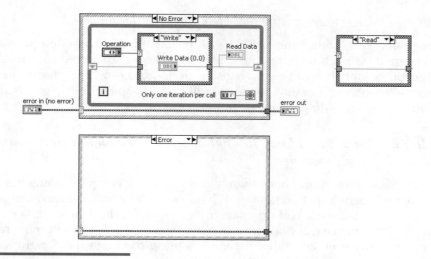

Figure 8-3B
The diagram of the functional global has a While Loop with a shift register for storing a numeric value, a Case structure with cases for reading and writing the data, and error propagation.

8.1.3 Continuous Loop

The **Continuous Loop** design pattern consists of a single While Loop or Timed Loop, shift registers, loop timing, and error trapping. The design pattern is equally applicable to top-level VIs and subVIs.

 Rule 8.1   Use multiple criteria for the loop condition

Carefully consider the stop/continue condition with continuous loops. Always make certain that the loop terminates gracefully, without requiring the Abort Execution tool. It is a good practice to use multiple criteria, which may include a user interface event such as the user clicking a **Stop** Boolean, an error occurrence, and perhaps a maximum time interval or number of iterations. Never use the Abort Execution tool as the means of terminating a continuous loop. Instead, use appropriate termination conditions and always hide the Abort Execution tool on the top-level VIs of deployed applications. This is accomplished by choosing **File»VI Properties**, selecting **Window Appearance** from the **Category** pull-down menu, and choosing **Top-level application**.

 Rule 4.35   Use shift registers over local and global variables

 Rule 4.36   Group most shift registers near the top of the loop

 Rule 4.37   Label wires exiting the left shift register terminal

Use shift registers with the looping structure of the Continuous Loop design pattern to store data locally and share it between loop iterations. Shift registers are preferred over local and global variables and functional globals because shift registers utilize data flow. Specifically, the data flow is readable, memory efficient, and immune to race conditions. Group the shift registers near the top of the loop, and label the wires exiting the left shift register terminal.

 Rule 8.2   Use a Timed Loop for highly precise or complex timing, and use a While Loop otherwise

A Timed Loop is used for applications with highly precise or complex timing requirements that are impractical to implement with the standard While Loop. These applications are typically deterministic applications intended for a real-time target. For example, the Timed Loop can use a hardware device or high-performance real-time processor to achieve 1 microsecond timing resolution. You can configure the period, offset, and other parameters using the input node, right data node, and VIs from the **Timed Structures** palette. Additionally, you can assign a priority to the execution of a Timed Loop that is independent of the calling VI's priority. However, the standard While Loop is simpler, requires less memory and processing overhead, and is appropriate for most desktop applications. In general, use a While Loop for most nondeterministic applications and a Timed Loop for more complicated and precise timing requirements.

Rule 8.3   Include a delay within continuous While Loops

Use the Wait (ms) function, Wait Until Next ms Multiple function, or Time Delay Express VI to incorporate a delay within continuous While Loops. Each of these nodes yields control of the execution thread back to the CPU and causes the loop to sleep for a number of milliseconds within each iteration. This allows parallel tasks to execute and improves the overall performance and responsiveness of the application. Note that LabVIEW is always multitasking, even if the diagram executes all nodes sequentially. LabVIEW contains multiple processes running in parallel, including the user interface and parallel diagram elements such as parallel loops and nodes. Additionally, the operating system may have other applications or background tasks that it must perform in parallel with LabVIEW. Including a delay in continuous While Loops allows these parallel tasks to run.

The Wait (ms) function and the Time Delay Express VI add a delay that is timed in parallel with the loop's code. The total loop execution time is equal to the larger of the time required to execute the code, or the specified wait or delay time. If the delay time is longer than the code execution time, the loop sleeps for the balance of the interval after the code completes. For example, if the loop code takes approximately 12ms to execute and you use a timing function with the wait or delay input set to 50ms, the total execution time is approximately 50ms per iteration, including 38ms of sleep. If the code takes more than 50ms to execute, the loop does not sleep. However, the wait or delay timing function forces the diagram's thread to yield control of the CPU, regardless of the delay time. Therefore, including the delay has benefits even if the delay time is very small.

The Wait Until Next ms Multiple function waits until the value of the system clock becomes a multiple of the specified **millisecond multiple**. You can use this function to synchronize the loop timing of multiple loops. Be sure that your code does not exceed the time interval that you are trying to maintain, or the loop timing may actually double. If a loop polls another resource, such as a communications port, instrument, or user interface object, and is paced by the resource instead of a time interval, use the Wait (ms) function. Set the **milliseconds to wait** input to the largest value that the loop can tolerate and still operate responsively. Note that delays are not necessary with event-driven constructs, including the Timed Loop and the Event structure. These structures sleep whenever no events are active.

The Continuous Loop design pattern template is shown in Figure 8-4. The front panel simply consists of the error clusters and a **Stop** Boolean control. The **Stop** Boolean contains a color and font scheme that makes it more legible for an industrial application than the default controls on the **Boolean** controls palette. Copy and paste the **Stop** Boolean to create a menu of Boolean controls. Edit the Boolean text to apply unique menu labels for each control. Rules for front panel text, including control labels and embedded text, are discussed in Chapter 3. The error clusters have been assigned to the lower left and right terminals of the 4×2×2×4 connector pattern, as recommended in Chapter 5. You can scroll these controls off the visible area of the panel, if desired. The diagram of the Continuous Loop design pattern template incorporates a While Loop with error handling, shift registers, delay, **Stop** Boolean, and termination conditions. Error handling consists of trapping the error by propagating the error cluster and reporting the error using the General Error Handler VI outside the loop. Three shift registers are positioned at the top of the loop. Replace the Boolean data with any data that needs to be stored between loop iterations. The Wait (ms) function is included for efficiency. The error status and the **Stop** Boolean control terminal are applied as the initial criteria for loop termination. The Compound Arithmetic function is used instead of the binary Or operation, for flexibility of adding terminals.

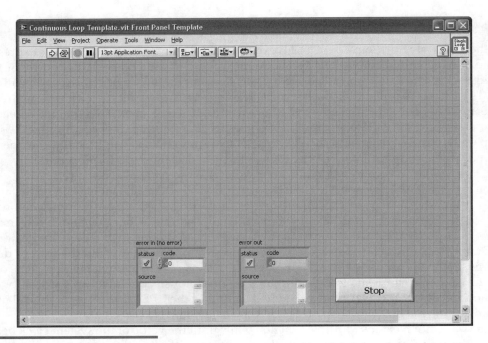

Figure 8-4A
The front panel of the Continuous Loop template contains error clusters and a customized **Stop** Boolean control.

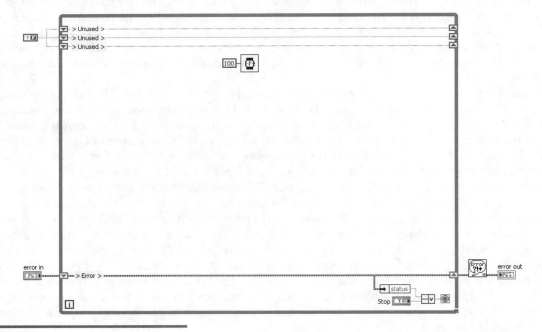

Figure 8-4B
The diagram of the Continuous Loop template consists of a While Loop with error handling, shift registers, timing, and several loop termination conditions.

Customize the Continuous Loop template for use as a subVI or top-level VI. For a subVI, delete the General Error Handler VI and Boolean controls, and surround the diagram with an Error Case Structure. For a top-level VI, hide the error clusters and customize the panel. Figure 8-5 shows a top-level implementation of the Torque Hysteresis VI using the Continuous Loop design pattern. The front panel contains a Boolean menu and graph indicator on a tab control. On the block diagram, the functionality has been divided into three Case structures. The first Case structure's **True** case contains dialog subVIs that prompt the operator for UUT information and motion parameters. This case is called when the user clicks a Boolean labeled **New UUT**. The UUT information and motion parameters are returned by the subVIs and maintained in shift registers. Also, a Boolean flag is set when the UUT information is validated, and the Boolean data propagates using a shift register. The middle Case structure runs the test and analyzes the data when the user clicks a Boolean labeled **Run Test** and the UUT information is valid. The acquired torque versus angle data and calculated statistics are stored in shift registers. The right Case structure prints a report and logs the test data to file when the user clicks a Boolean labeled **Log Data**. The loop terminates if the user clicks a Boolean labeled **Quit** or an error occurs. Note that the shift register data allows each task to execute on demand instead of sequentially by providing the most recently acquired data to the subVI terminals. In particular, it is common for an operator to configure the UUT and motion parameters once and to execute several test runs prior to logging data.

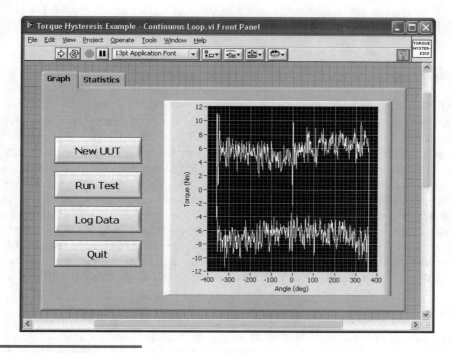

Figure 8-5A
The front panel of the Torque Hysteresis VI, based on the Continuous Loop template, contains a Boolean menu, XY graph, and tab control. The error clusters are scrolled off the visible area of the panel.

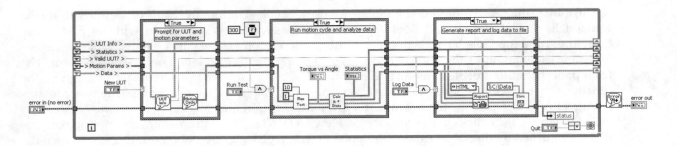

Figure 8-5B
The block diagram has been divided into three primary tasks that are executed on demand via the Boolean controls. Data is stored in shift registers.

 Rule 8.4   Avoid polling GUI objects

Polling a menu of Booleans within a continuous loop, as illustrated in the previous example, primarily applies to developers that do not have access to Event structures. This includes users of the LabVIEW Base package or versions of LabVIEW prior to 6.1, which introduced the Event structure. In general, GUI objects should not be polled. A more efficient alternative is presented next.

8.1.4   Event-Handling Loop

The **Event-Handling Loop** design pattern facilitates event-driven programming in LabVIEW. It consists of an Event structure within a While Loop. The Event structure contains an event case for each configured event that executes its subdiagram when the event fires. This is analogous to a Case structure with an event-driven selector. The Event structure maximizes the efficiency with which an application detects and responds to events on the graphical user interface, as well as any user-defined events. It goes to sleep, freeing the processor to work on other tasks, until a registered event fires.

The Event-Handling Loop design pattern is extremely useful for applications that involve user interface events of any variety. Almost any type of user interface activity can be captured and translated into an event. Additionally, user-defined events, such as events related to a test or measurement sequence, are handled by an Event structure using dynamically registered user events. Moreover, any application containing a graphical user interface benefits from the efficiency of an event-handling loop. The only caveat is that the Event structure is not available in the LabVIEW Base package.

 Rule 8.5   Use the Value Change event for most GUI controls

The **Value Change** event detects any control data changes by the user and is common to all types of controls. If a control's value changes, whether it is a Boolean command button, a numeric, a string, or another control type, the Value Change event fires and the corresponding subdiagram runs. For basic GUI applications, use the Value Change event to detect control value changes. This promotes consistency in the VI's performance, readability, and maintenance. Use other events, such as Key Down, Mouse Down, and events that are specific to a particular control, for special effects.

Rule 8.6   Place control terminals within their Value Change event case

Note that the Event structure allows you to read the new and old control values from the Event Data Node. Boolean controls configured with any type of latching action, however, will not reset their value to the default state until their data is read directly from the control terminal. Placing the control terminals within the Event structure's corresponding Value Change event case ensures that the terminals are read and latching Booleans are reset each time a value changes. It is a good practice to place all control terminals inside their corresponding Value Change event case unless they are needed elsewhere. This provides a handy reference point for the developers to locate the controls.

 Rule 8.7   Resize the Event Data Node to hide unused terminals

In most event cases, I find that I use only a couple, one, or none of the terminals from the Event Data Node. Conserve space and reduce clutter within the Event structure by resizing the Event Data Node to hide the unused terminals. In the examples in this chapter, the Event Data Node is reduced to the minimum single terminal and moved to the bottom left area of the Event structure.

 Rule 8.8   Avoid continuous timeout events

By default, the Event structure has a Timeout event case that becomes enabled whenever the Timeout terminal is wired. The Timeout event occurs if no other event fires within the specified timeout interval, and executes the corresponding Timeout event case. The Timeout event is frequently misused. It can be very tempting to place a continuous software-timed task inside the Timeout event because it saves the developer from creating an additional parallel loop and messaging scheme, reducing diagram complexity. However, note that the occurrence of any registered event will suspend Timeout events until the registered event is processed. The software has no control over how many statically registered events may fire or how often. Therefore, the Timeout event cannot be relied upon to perform a task at periodic intervals. Moreover, continuous Timeout events cause the Event structure to behave similarly to a traditional GUI-handling loop that polls the controls at timed intervals. The timeout code causes latencies in the response to other registered events and reduces the overall efficiency of the Event structure.

Figure 8-6 illustrates a template for the Event-Handling Loop design pattern. The front panel simply contains a tab control with a page containing several Boolean command buttons customized for an industrial GUI, a spare page, and a page containing error clusters. The diagram consists of the Event structure within a While Loop, shift registers, and a shutdown routine outside of the loop. The Event structure is preconfigured with cases for each Boolean command button's Value Change event, and cases for the **Panel Close?** and **Application Exit?** filter events. A two-button dialog confirms the user's decision to close or exit. Additionally, the Event structure is customized to promote good programming style for common applications. The control terminals are in their respective Value Change event cases, and the Timeout event case has been removed.

Note that three separate events trigger an application shutdown. In Figure 8-6C, the **Quit Value Change**, **Panel Close?**, and **Application Instance Close?** event cases are shown. All three events display a dialog prompting the user to confirm and then stop the loop, if confirmed. After the loop terminates, a shutdown routine outside of the While Loop runs. This prevents three separate instances of the same routine to appear in three different event cases. Configure the shutdown routine to close any open references, report errors, and exit LabVIEW, if desired. Because the **Panel Close?** and **Application Instance Close?** events normally cause the VI or LabVIEW to shut down immediately, these events must be discarded for the shutdown routine to run. This is achieved by wiring a TRUE Boolean constant to the **Discard?** filter event terminal.

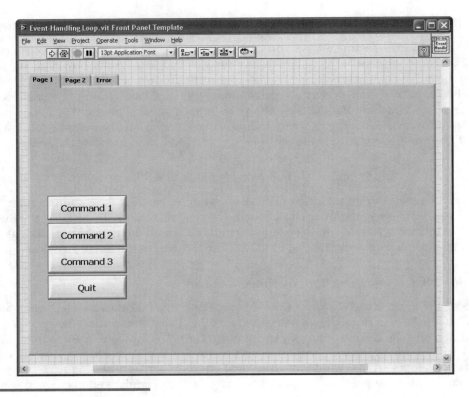

Figure 8-6A
The front panel of the Event-Handling Loop design pattern contains a tab control with a page containing several command buttons customized for use as an industrial GUI, a spare page, and a page containing error clusters.

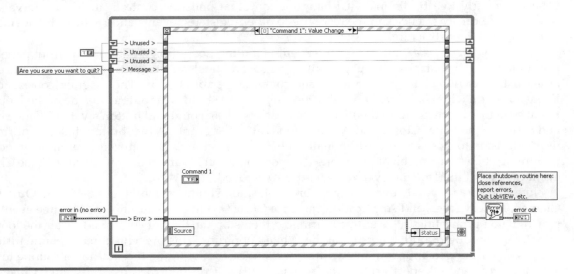

Figure 8-6B
The diagram consists of an Event structure within a While Loop containing shift registers, and a shutdown routine outside the loop. The control terminals are in their respective Value Change event cases.

Chapter 8 • Design Patterns

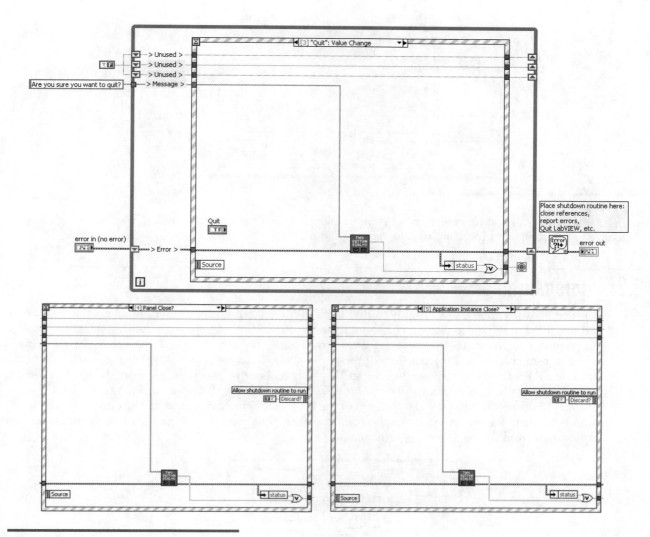

Figure 8-6C
Three separate events that trigger a shutdown include the **Application Instance Close?**, **Panel Close?**, and **Quit Value Change**. The former two events are discarded to allow the shutdown routine to run.

In Figure 8-7, the Torque Hysteresis VI has been implemented as an event-handling loop based on the template. The front panel is identical to Figure 8-5A and is not shown. Each of the controls from the Boolean menu have a corresponding Value Change event case that executes the same code contained in the Case structures in Figure 8-5B. The Event-Handling Loop is more compact, flexible, and efficient than the Continuous Loop implementation. A single Event structure with multiple event cases is more compact than the multiple Case structures. Also, it is easy to add more event cases or divide the existing event cases into multiple event cases without increasing the height and width of the diagram. Moreover, it is more efficient than the Continuous Loop because the diagram sleeps until a registered event fires, freeing the processor to work on other tasks.

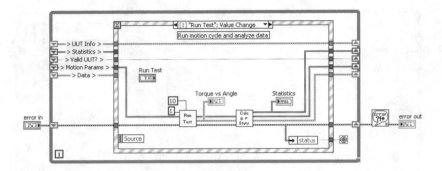

Figure 8-7
The Torque Hysteresis VI is implemented as an event-handling loop. This is a more compact, flexible, and efficient alternative to the Continuous Loop implementation in Figure 8-5.

8.2 State Machines

State machines are the most highly touted LabVIEW design patterns of all time. There are many variations, most of which consist of a Case structure within a While Loop, with a shift register or messaging construct wired to the case selector terminal. Each case of the Case structure contains a subdiagram corresponding to a state of the application. The case selector is an integer, string, or enumerated data type identifying the states. The shift register or messaging construct passes the next state selection from a previous case to the selector terminal in the next loop iteration. This is illustrated in Figure 8-8 using the classic form of the state machine. In a typical application, the state selection is determined by an event on the user interface, by a step in a sequential test or measurement routine, or from the result of a previous state.

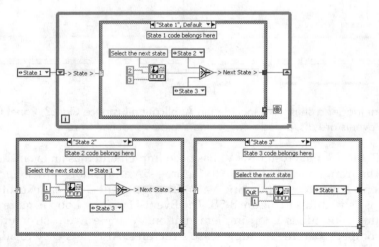

Figure 8-8
State machine design patterns consist of a Case structure within a While Loop and a shift register or messaging construct wired to the Case structure's selector terminal. Each case of the Case structure corresponds to a state of the application.

Rule 8.9   Use a state machine design pattern in most VIs of medium or greater complexity

State machines provide tremendous flexibility and expandability. The state machine's shift register or messaging construct allows the code to jump *from* any state *to* any state. Each state can determine the While Loop's run condition, allowing it to stop after any state. Also, you can easily add more code by simply adding cases to the Case structure, without increasing the overall size of the diagram. These powerful features make the state machine invaluable. For VIs of medium or greater complexity, you almost cannot go wrong using a state machine.

 Rule 8.10   Derive the application's primary states from the specification or design document

To use a state machine, the application must be divided into a series of states. This can be done during the planning phase using a properly delineated requirements specification or design document, or during the implementation phase using experience and intuition. As discussed in Chapter 2, "Prepare for Good Style," the more planning is performed before coding, the better the outcome of the application. The State Machine Diagram is a fundamental component of UML and an ideal method for defining the primary states of an application. Additionally, the State Diagram Toolkit is a utility available from NI that provides an editor for creating state diagrams, from which the corresponding LabVIEW code is generated automatically. However, several of the state machine implementations described in this chapter cannot be generated by the State Diagram Toolkit's code generator. Instead, you should use templates for initiating each design pattern.

 Rule 8.11   Divide the primary states into additional states

Sometimes an application's functional states do not translate perfectly into LabVIEW subdiagrams that fit appropriately within the state machine's case boundaries. As you develop the source code, implementation details are exposed that may lend themselves to a different delineation of states. For example, a given functional state may have a task in common with several other functional states, which warrants breaking out the task into its own separate case so that it can be called from multiple states. Also, when it becomes apparent that a state's code is too large for one case, create more cases as needed. Consequently, one functional state may ultimately be implemented using multiple subdiagrams in multiple cases. Remember to modularize your code into subVIs, and define the states for maximum reusability. A combination of planning, experience, and intuition is normally applied for best results.

 Rule 8.12   Use an enumerated type definition for the case selector

An enumeration saved as a type definition provides documentation and improves maintenance. When an enumeration is wired to a Case structure's selector terminal, the text items of the enumeration appear as the labels in the Case structure's selector area. Choose succinct and intuitive names for each state, and enter them as items of the enumeration. Any constants created from a type definition maintain synchronization of the item names with the type definition. Therefore, enumerated constants are used to specify the next state selection in each case. As items are added and removed from the enumerated type definition, all of the corresponding constants update themselves automatically. Enumerated type definitions are recommended for *all* variations of the state machine design pattern that use a Case structure.

Rule 8.13   Minimize code external to the Case structure

 Rule 8.14   Include states for Initialize, Idle, Shutdown, and Blank

Most applications require some general house-cleaning code, such as routines for initialization and shutdown, and perhaps a routine that runs when nothing else is running. For example, some control values and properties are initialized prior to the execution of any other states. Also, an event handler or resource-polling routine may need to run when no other states are being processed. A shutdown routine is required to gracefully shut down any open connections to hardware devices, data files, and remote applications. Create corresponding states for **Initialize**, **Idle**, and **Shutdown**. These routines may or may not have been considered during the design phase, but from my experience, they are always necessary. Maximize the utilization of the state machine by incorporating these routines as states within the state machine instead of external code. Finally, create a **Blank** state containing propagation of shift register wires for duplication purposes. Creating new states by duplicating the **Blank** state saves time.

For example, the diagram of Figure 8-9A contains initialization and shutdown code in sequence frames to the left and right of the state machine. Also, the state machine polls a **Stop** Boolean on the left of the Case structure and checks the error **status** on the right. This is an inefficient use of space. Figure 8-9B contains the equivalent code utilizing a state machine with states for **Initialize**, **Idle**, **Shutdown**, and **Blank**. The two-terminal shift register provides a buffer that stores the next two states. Each time the **Idle** state is called, it adds another **Idle** state to the end of the buffer, ensuring that **Idle** alternates with the other states. Additionally, the **Idle** state checks the error **status** and stops the While Loop if an error occurs. Consequently, the diagram of Figure 8-9B improves space efficiency by using states instead of external code.

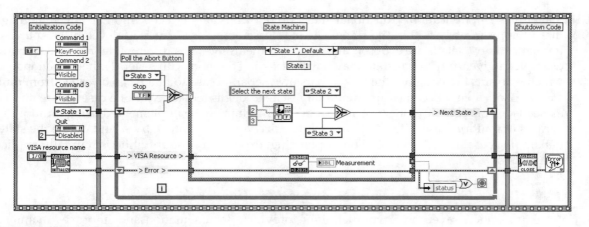

Figure 8-9A
A state machine design pattern is combined with a Flat Sequence structure to perform initialization, control polling, check the error status, and shutdown. The diagram uses excessive horizontal space.

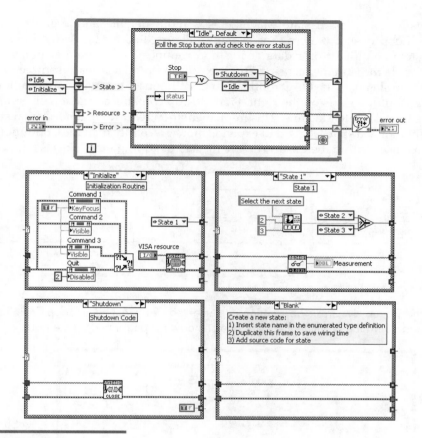

Figure 8-9B
Space efficiency is optimized by adding states implemented as cases in the Case structure for **Initialize**, **Idle**, **Shutdown**, and **Blank**. A two-element shift register ensures that the **Idle** state alternates with all other states.

Let us now discuss several popular state machine implementations.

8.2.1 Classic State Machine

The diagrams illustrated in Figures 8-8 and 8-9B are examples of the **Classic State Machine** design pattern, which uses a shift register for state transitioning. GUI activity, such as the state of one or more Boolean command buttons, is captured by polling the terminals within the **Idle** state. The Classic State Machine is appropriate for subVIs and routines of low to medium complexity. For example, it is a more functional and flexible alternative to the Sequence structure for executing a series of operations. The operations are implemented as states of the state machine instead of frames of the Sequence structure. Unlike a Sequence structure, the Classic State Machine can change its execution order on the fly or stop the sequence prior to completion. Also, the enumerated type definition helps ensure that each case is intuitively labeled. Finally, shift registers on the While Loop provide a graceful method of sharing data between states. An example of a classic state machine applied as a Flexible Sequencer was presented in Chapter 4.

Additionally, the Classic State Machine is a more flexible and organized alternative to the Continuous Loop design pattern. For example, Figure 8-10 contains two implementations of the Torque Hysteresis application's data acquisition routine. This routine acquires a finite number of torque and angle measurements using DAQmx VIs and publishes each sample to a shared variable. Figure 8-10A implements the DAQ routine as a Continuous Loop, with initialization VIs on the left of the loop and shutdown VIs on the right. Figure 8-10B contains a functionally similar routine implemented as a Classic State Machine. This implementation requires less horizontal space, contains better documentation via state labels, and expands more gracefully than the Continuous Loop by adding new states as cases within the Case structure instead of adding more objects horizontally.

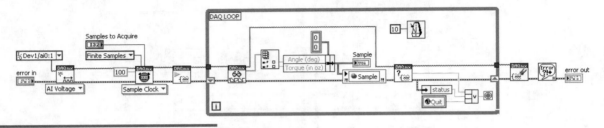

Figure 8-10A
The DAQ routine of the Torque Hysteresis application implemented as a Continuous Loop. Initialization code is on the left, continuous read operation is within the loop, and shutdown is performed on the right.

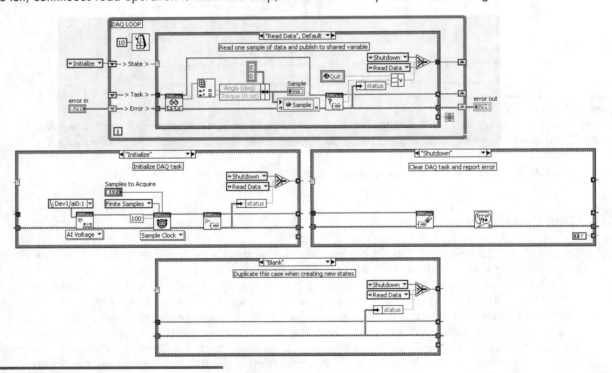

Figure 8-10B
The Classic State Machine implementation contains states for **Initialize**, **Read Data**, **Shutdown**, and **Blank**. This implementation requires less horizontal space, contains state labels, and expands more gracefully compared to the Continuous Loop.

Figure 8-11 provides a template for the Classic State Machine design pattern. In addition to initializing control values and properties, use the **Initialize** case to initialize instrumentation, read configuration parameters from file, and anything else required to set the application to its desired initial state. Use **Shutdown** for setting all hardware to a known state and disposing any instrument, data acquisition, or file references. The **Idle** case is normally the default case that is traditionally used to poll resources such as instruments or control values. The **Blank** case simply propagates all wires from the input tunnels straight through the output tunnels. Duplicate the **Blank** case to reduce edits when developing new states.

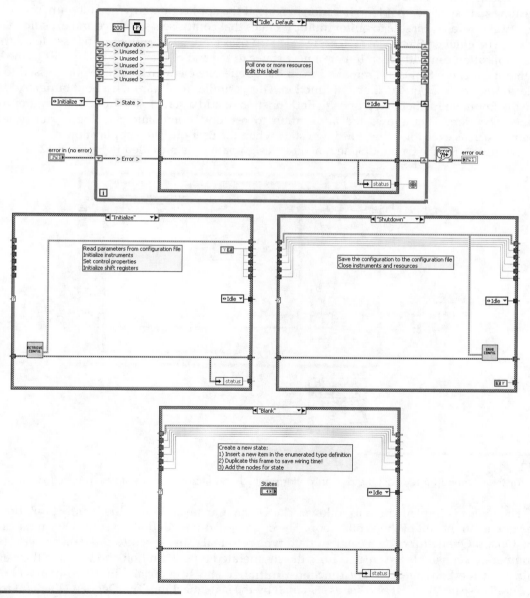

Figure 8-11
A template for the Classic State Machine design pattern contains states for **Initialize**, **Idle**, **Shutdown**, and **Blank**.

The Classic State Machine is *not* appropriate for complex VIs, top-level VIs, and GUI VIs. Alternative state machine implementations that utilize queues and Event structures are more functional and efficient for these applications.

8.2.2 Queued State Machine

The Classic State Machine design pattern's shift register method of state transitioning is limited to one new state specified per loop iteration or state of the application. In VIs of medium or greater complexity, it is often desirable to buffer multiple states that comprise a sequence and execute them one at a time. The Queued State Machine design pattern uses a queue to buffer multiple states. Any state of the application can add any number of new states to the end of the queue using calls to the Enqueue Element function. States are removed from the queue one element at a time and passed to the case selector using the Dequeue Element function. This is similar to a first-in, first-out buffer. Additionally, use the Enqueue Element At Opposite End function to add a state to the front of the buffer for immediate execution. This allows the application to respond immediately to high-priority actions or events, such as executing the Shutdown state when the user quits the application.

A template for the Queued State Machine design pattern is provided in Figure 8-12.

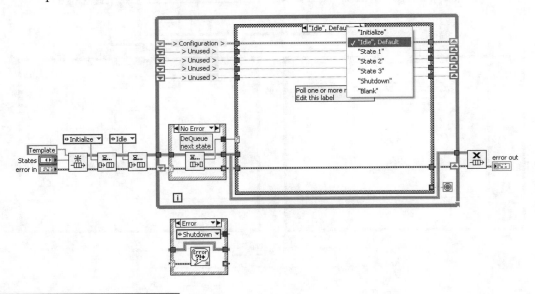

Figure 8-12
A template for the Queued State Machine design pattern, which maintains multiple states in a buffer.

The queue is implemented as follows: The Obtain Queue and Enqueue Element functions initiate the queue on the left of the While Loop. The enumerated type definition control terminal is wired to the Obtain Queue function's element data type terminal. This specifies the data type of the queue. Enumerated constants are created from the enumerated type definition and wired to the element terminals of the Enqueue Element functions. **Initialize** and **Idle** are the first two items added to the queue. These are the first two states executed by the state machine. The Dequeue Element function is placed within the **No Error** case of an Error Case Structure outside of the main Case structure. If no error is present in the error cluster, the next state is removed from the Queue and passed to the main Case structure's selector terminal. If an error is present, the error is reported using the General Error Handler VI and the **Shutdown** state executes. Within each case of the Case structure, additional states are added to the queue using the Enqueue Element function.

A variation of the Queued State Machine uses a queue element data type consisting of a cluster that contains the enumerated type definition bundled together with a variant. The enumeration contains the desired state for the case selector as normal. The variant is used to pass data from one state to another, using the queue functions instead of shift registers. For example, in Figure 8-13, the **Acquire Waveform** state acquires a waveform, converts it to variant, combines it with the state enumerated constant with **Update GUI** selected, and adds the cluster to the queue. In a subsequent loop iteration, the Dequeue Element function removes the cluster and unbundles the **State** and **Data** items. **Update GUI** state converts the variant back to a waveform and updates the waveform indicator. The variant data type maximizes the flexibility of the data that is passed between states. This approach involves fewer wires than the shift register method of data sharing. However, the data is stored only temporarily in the queue and is not accessible in subsequent states after dequeuing. By comparison, shift registers maintain and share the data among all states and use wires and dataflow. Therefore, shift registers are generally preferred over queues for data sharing between the states of a state machine.

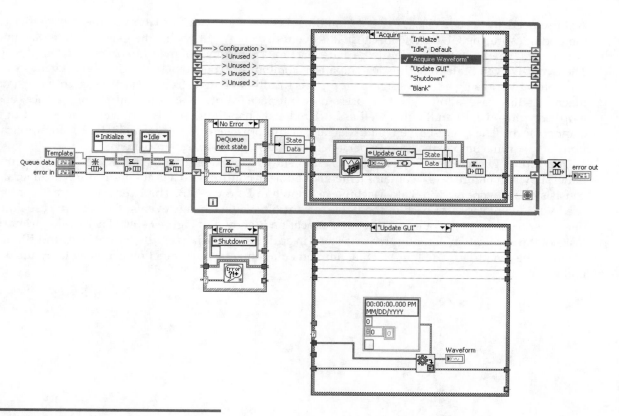

Figure 8-13
The queue is used to pass data between states via a cluster containing a variant and state enumeration. This reduces wires and shift registers, but the data is not accessible in subsequent states after dequeuing.

Rule 8.15 Avoid timeout with the Enqueue and Dequeue Element

Event-driven performance is achieved using the Queued State Machine design pattern by avoiding the **timeout** terminal of the Enqueue and Dequeue Element functions. Specifically, when the queue is empty and no states are active, the loop goes to sleep, allowing background tasks to run. As soon as an element is added to the queue, the Dequeue Element function immediately wakes up, removes the item, and runs the corresponding case. This is maximally efficient and represents a substantial advantage over the Classic State Machine, Continuous Loop, and other design patterns. However, whenever a timeout is specified, this advantage is eliminated. Therefore, avoid the **timeout** terminal, as shown throughout the examples in this chapter.

The Queued State Machine design pattern is appropriate for continuous routines and subVIs with medium or greater complexity. Similar to the Classic State Machine, the Queued State Machine is not appropriate for top-level and GUI VIs unless it is combined with an Event structure or a separate Event-Handling Loop. These design patterns are discussed in Section 8.2.3, "Event-Driven State Machine," and Section 8.3.1, "Parallel Loops," respectively.

8.2.3 Event-Driven State Machine

As discussed in Section 8.1.4, "Event-Handling Loop," the Event structure is the most efficient method of capturing GUI events because it is event-driven. Additionally, the Queued State Machine is a versatile method of implementing a state machine for routines of medium or greater complexity. The **Event-Driven State Machine** design pattern combines a Queued State Machine and Event structure into a hybrid single-loop design pattern. Specifically, an Event structure is inserted into the state machine's **Idle** case, allowing it to process user interface events in an efficient manner. Hence, one loop performs event handling as well as buffered state handling. This is a powerful construct in a compact form. The Event-Driven State Machine is appropriate for top-level and GUI VIs for applications of medium complexity. It is my personal favorite single-loop design pattern.

The Torque Hysteresis VI is rewritten as an Event-Driven State Machine in Figure 8-14. The front panel shown in Figure 8-14A contains a menu of Boolean controls and a graph indicator. The diagram consists of an Event Driven State Machine, as shown in Figure 8-14B. The Event structure is configured with Event cases for each of the Boolean controls' Value Change events. Each Event case adds one or more states to the queue. The state delineation is shown in Figure 8-14C. Each of the primary tasks of the application is configured as a state, in addition to **Initialize**, **Idle**, **Shutdown**, and **Blank**. Overall, the Event-Driven State Machine implementation is flexible, functional, efficient, organized, and compact.

Chapter 8 • Design Patterns

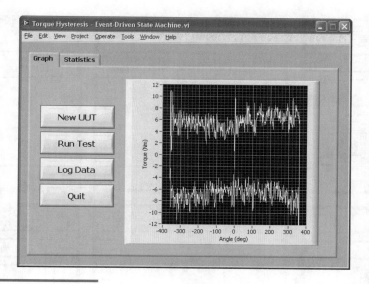

Figure 8-14A
The front panel of the Torque Hysteresis VI contains a menu of Boolean controls and a graph indicator.

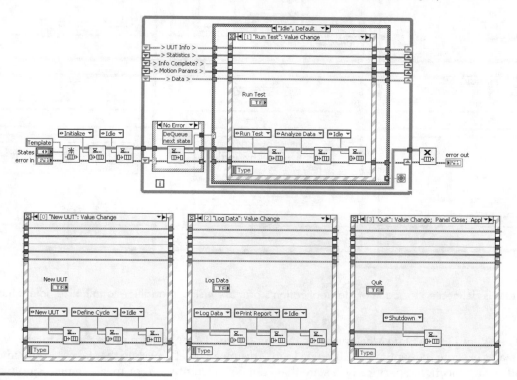

Figure 8-14B
The diagram of the Torque Hysteresis VI is implemented using the Event-Driven State Machine design pattern, comprised of a Queued State Machine with an Event structure in the **Idle** case. The Event structure is configured to buffer one or more states for each of the Boolean control's Value Change event.

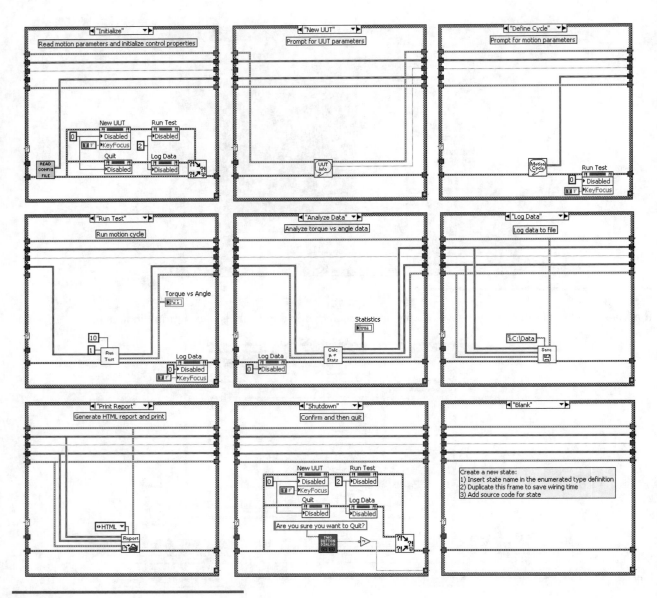

Figure 8-14C
A separate state is delineated for each of the application's primary tasks, in addition to **Initialize**, **Idle**, **Shutdown**, and **Blank**.

The Event-Driven State Machine is the most powerful design pattern presented so far. However, all of the states execute sequentially within one loop. This means that any state that takes a long time will suspend the application. Specifically, a state that contains a lengthy test or measurement step will prevent the application from running any other states, including **Idle**. This prevents the application from responding to GUI events during the lengthy state. If your application contains lengthy steps or parallel routines, proceed to Section 8.3, "Compound Design Patterns." Use the Event-Driven State Machine for the top-level VI of most applications that do not contain lengthy states or parallel processes.

8.2.4 Event Machine

The Event structure need not be limited to user interface event handling. Instead, you can configure custom user-defined events—known as **user events**—that are created, registered, and fired programmatically using the functions on the **Events** palette, located under **Programming»Dialog & User Interface»Events**. The **Event Machine** design pattern merges the functionality of the Case structure and Event structure of the Event-Driven State Machine into a new design pattern consisting of an Event structure within a While Loop. User events are defined for each state, and event cases are configured for any combination of user and GUI events. The Event structure handles user events in a similar manner to GUI events. Any registered user and GUI events that fire are buffered in first-in, first-out order until processed by the corresponding event case. The Torque Hysteresis VI is implemented as an Event Machine in Figure 8-15.

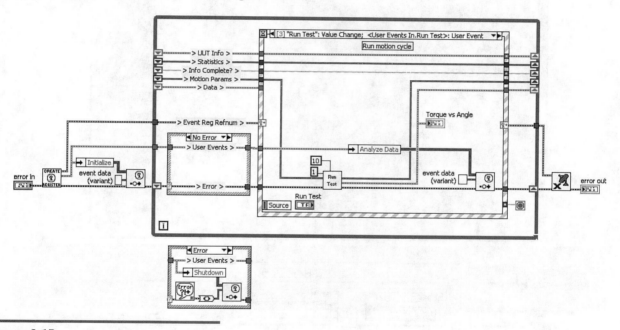

Figure 8-15
The Torque Hysteresis VI is implemented as an Event Machine design pattern. User events are used to merge the functionality of the Case and Event structures into an alternative design pattern with fewer constructs.

A user event is created and registered for each state of the Torque Hysteresis application using the Create & Register User Events VI, shown on the far left of the diagram in Figure 8-15. The subVI's front panel and diagram are shown in Figure 8-16. **User Event Data** is a cluster of variant controls containing one element per state, labeled using the state name. This cluster is saved as a type definition and is analogous to the enumeration of a conventional state machine, except that the data type of each of the cluster's elements defines the data type that is used to pass data into the event. Event data is useful if the routine that fires the event needs to pass data into the event. In particular, the event generation code can exist in locations where the data cannot be passed using wires. The implementation of Figure 8-15 uses shift registers to share data between event cases, similar to the previous state machine implementations. Shift registers provide the flexibility to share the data among all event cases, not just between the event generation routine and the corresponding event case. However, shift registers cannot be used if the event firing routine exists outside of the loop. Therefore, I recommend shift registers when the event data can be passed using wires, and event data otherwise.

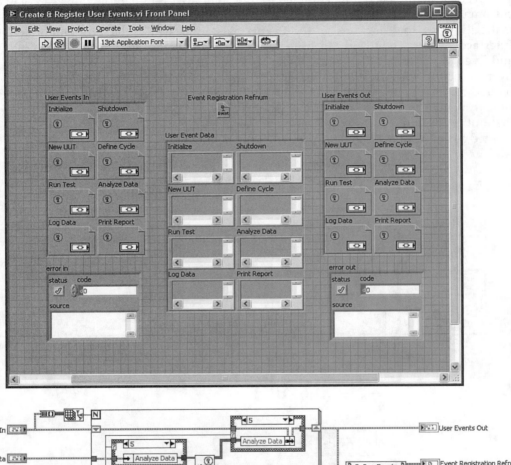

Figure 8-16
Create & Register User Events VI is a subVI that creates and registers multiple user events based on the labels and data types of the elements of the **User Event Data** cluster.

As shown in the diagram in Figure 8-16, the elements of the **User Event Data** cluster are unbundled and passed to the **user event data type** input terminal of the Create User Event function. This function defines the name and data type for each user event based on the label and data type of the data wired to the terminal. Create User Event returns a **user event** refnum that is bundled into the **User Events Out** cluster. These operations are repeated in the For Loop, generating a user event for each state of the application. The events are registered by passing the **User Events Out** cluster to the **event source** terminal of the Register for Events function. This function returns an **event registration refnum**, which is returned by the subVI and wired to the Event structure's **dynamic event** terminal, as shown in Figure 8-15. Finally, user events are fired anywhere in the application using the Generate User Event function. The user event refnum corresponding to the desired state is unbundled from the **User Events Out** cluster and passed to the corresponding input of the Generate User Event function, along with the event data. In the **Run Test** event case, the **Analyze Data** event is fired with an empty variant constant wired to the **event data** terminal.

The Event Machine design pattern is similar to the Event-Driven State Machine but maximizes the use of the Event structure. The primary advantage is that multiple Event structures can be configured to respond to the same user event, whereas each queue has a unique reference for enqueuing and dequeuing. This capability allows an application to have multiple Event Machines that all respond to a global command, similar to notifiers. However, there are several limitations of the Event Machine design pattern. First, it requires more maintenance than the Event-Driven State Machine. To add or remove an application state, both the user event data and user event refnum clusters must be modified to add or remove the corresponding elements. Also, corresponding calls to Create User Event must be added or removed, which also results in a new event registration refnum. This is in addition to the changes to the Event structure configuration, which is similar to maintaining an enumerated Case structure. Second, the Event Machine's event buffer is less functional than the Event-Driven State Machine's queue. Using a queue, you can preview, dequeue, or even flush queue elements without firing the corresponding state cases. Additionally, you can add new elements to the front as well as the back of the queue. I use this feature to enqueue the **Shutdown** state at the front of the queue whenever an immediate shutdown of the application is required. The Event Machine buffers events in a strict first-in, first-out manner, and all buffered events are executed until the While Loop stops. Therefore, I recommend the Event Machine only for complex GUI applications, where multiple Event structures are required.

Finally, a limitation common to all state machine design patterns discussed to this point is that each is capable of processing only one application state at a time. This causes any states containing time-consuming subdiagrams to prevent the application from responding quickly to other states or events. For example, the GUI will appear unresponsive to user interface activity during a time-consuming state or user event case. These limitations are addressed using the compound design patterns discussed next.

8.3 Compound Design Patterns

The design patterns discussed in the previous sections all have limitations when applied as the principle architecture for a large application. Immediate SubVI and Functional Global design patterns are strictly limited to subVIs. The Continuous Loop design pattern does not expand in a graceful, organized manner. The Event-Handling Loop and Classic State Machine design patterns are readily expandable through cases, but neither can perform the function of the other. The Event-Driven State Machine and Event Machine design patterns combine GUI event handling and state transitioning, but their performance is limited by the serial processing of a single loop. Instead, consider each single-loop design pattern as a building block upon which a large application is constructed. Most applications of advanced complexity are best implemented using a combination of multiple single-loop design patterns, to take advantage of the benefits of each while optimizing the overall performance.

Compound design patterns are created using a combination of two or more of the single-loop design patterns and adding a messaging scheme to pass data between loops. The messaging scheme adds complexity. Hence, good development tools, examples, and templates are beneficial. Develop applications employing compound design patterns, either using compound templates, or by merging the templates of the single-loop design patterns and adding the messaging. It is convenient and quite clever to have each of the single-loop design patterns available for selection from the **Functions** palette for fast and easy access, similar to the structures on the **Structures** subpalette. This is accomplished by creating a custom palette view and adding an icon for each single-loop design pattern template to the **Structures** subpalette. Configure the submenu icons to merge the contents of the corresponding design pattern template into the target VI instead of placing the icon as a subVI call. Developing compound design patterns becomes as easy as click, drag, and drop.

The custom palette view in Figure 8-17A contains icons representing four of the single-loop design patterns. Figure 8-17B shows the outline of a design pattern as it is being dragged and merged onto a diagram.

Figure 8-17A
Custom palette view contains several icons representing single-loop design patterns. The icons are configured to merge the corresponding design pattern onto the selected diagram.

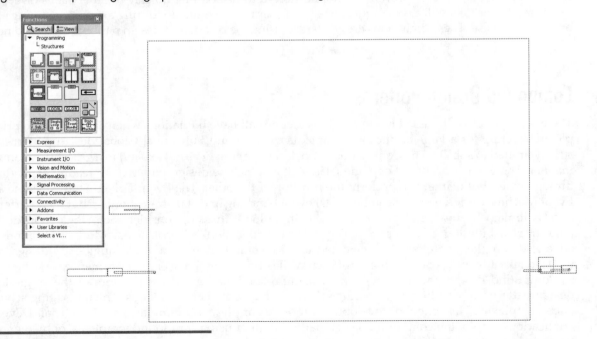

Figure 8-17B
Click, drag, and drop the design patterns from the **Functions** palette to merge them onto a diagram.

8.3.1 Parallel Loops

Parallel loops are required for applications that have multiple continuous tasks that must execute in parallel. For example, you might have a data acquisition application that acquires and logs data to file. Concurrently, the user needs to process and display the data on demand, in response to user interface events. However, the data processing and display tasks cannot interfere with the continuous acquisition and logging task. Consequently, you cannot put both tasks in separate cases of an Event-Handling Loop or state machine. Using any of the single-loop design patterns, any subVI, event, or state temporarily suspends the loop, preventing other subVIs, events, and states from executing until the code completes. Instead, the best way to architect this application is to divide the simultaneous tasks into separate parallel loops. Use a Continuous Loop exclusively for the continuous acquisition and logging task so that it operates without any interruptions from the user interface. Likewise, implement the user interface event handling and display in an Event-Handling Loop. Create a messaging scheme to share data between loops.

Rule 8.16   Use queues, shared variables, or RT FIFOs for parallel loop messaging

Messaging between loops can be accomplished using a variety of constructs, including local or global variables, functional globals, notifiers, user events queues, shared variables, and real-time first-in, first-out (RT FIFO) buffers. Local and global variables are the simplest but the least desirable in terms of performance and style. As discussed in Chapter 4, local and global variables undermine LabVIEW's dataflow principles. Functional globals offer performance and flexibility improvements over local and global variables, but they require some time and effort to develop and maintain. Notifiers are effective, but they are limited to storing and passing one message at a time. User events buffer the event data, but they are more difficult to configure and maintain, and the buffer is less functional than queues. I recommend using queues or shared variables for desktop applications and using RT FIFOs for real-time applications requiring programmatic configuration.

Queues and shared variables are built-in LabVIEW constructs that have evolved over time. Queues are available from LabVIEW's **Synchronization** palette, and shared variables are configured from the Project Explorer window. You can configure queues and shared variables to buffer more than one message or data element at a time. This is important if the loops run at different rates and you do not want to lose any data, or if the application transfers data in chunks instead of a single element at a time. Additionally, queues and shared variables are capable of synchronizing as well as sharing data between parallel loops. Queues and RT FIFOs are sized programmatically. Shared variables are sized during edit time but can transfer data over a network and incorporate a timestamp. Even if these features are not important for your application today, they provide more overall capability, in case it is required in the future.

Rule 8.17   Prioritize loops using delays or thread priorities

Parallel loops are very important for high-performance applications, including applications that are intended to run in real time. Each parallel loop is prioritized using a delay, Timed Loop Priority property, or subVI Execution properties. The simplest technique is to incorporate a delay within each While Loop using Wait (ms), Wait Until Next ms Multiple, or Time Delay Express VI. The loop with the smallest delay receives the highest priority. Alternatively, modularize each loop into a subVI, and configure the Priority and Preferred Execution System using the subVI's Execution properties, selected from **File»VI Properties»Execution**. This technique involves configuring threads, and a basic understanding of multithreading and hyperthreading is recommended[2]. The Priority property for a Timed Loop is configured using the input node's **priority** terminal. Many real-time applications have precise or complex timing requirements, and the advanced features of a Timed Loop are beneficial. For desktop applications, prioritize loops using delays for maximum simplicity. For real-time applications, use the Timed Loop's Priority property for precise performance.

 Rule 8.18   Size parallel loops to the same width and align vertically

 Rule 8.19   Minimize space between loops

 Rule 8.20   Label each loop in the top left corner

Here are a few rules that promote neat, organized diagrams containing parallel loops: Size each loop to the same width, align the loops vertically, and minimize the empty space between loops. In Chapter 4, we discussed limiting the size of our diagrams to one screen with 1280×1024 resolution. If a diagram contains multiple parallel loops, the one-screen rule is not feasible, unless the loops are modularized into subVIs. If you use parallel loops on the same diagram, be sure to minimize the space between loops, in an effort to maximize the diagram area that is visible on one screen. Finally, apply a free label to the top-left corner of each loop, identifying the primary task performed by the loop.

Consider the Torque Hysteresis application example. The primary objective is to acquire, analyze, and present torque versus angle data. In the single-loop implementations presented in the previous sections, the motion cycle is controlled while torque and angle data is acquired during the Run Test application state, using data acquisition routines within Run Test VI. When this subVI is called, it opens its panel and graphically displays live torque versus angle data throughout the motion cycle. When the test completes, the data is saved to file and a report is generated. However, it is desirable to acquire and display the torque and angle data continuously, throughout all states of the application. Specifically, live digital indicators for torque and angle on the front panel of the top-level VI, as shown in Figure 8-18A, allow the user to monitor these important parameters without initiating a test. This helps the operator set up the test fixture and unit under test, manually apply a torque and monitor the response, and confirm when the force is removed for safety purposes.

Continuous acquisition and display is accomplished by adding a second Continuous Loop on the top-level diagram, adjacent to the Event-Driven State Machine, as shown in Figure 8-18B. The data acquisition routine is extracted from the Run Test VI and runs in the parallel loop. This allows continuous operation while preventing the application states and user interface activity from interfering with the acquisition and display of the torque and angle data. Data is transferred between loops using a shared variable. The DAQ Loop incorporates a 10ms delay, and receives a higher priority than the Main Loop, which contains a 100ms delay.

Chapter 8 • Design Patterns

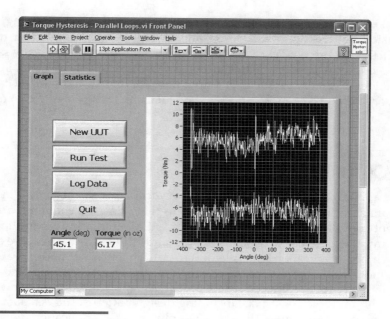

Figure 8-18A
The top-level front panel of Torque Hysteresis VI contains digital indicators with live display of the torque and angle data. This facilitates continuous monitoring of these important parameters.

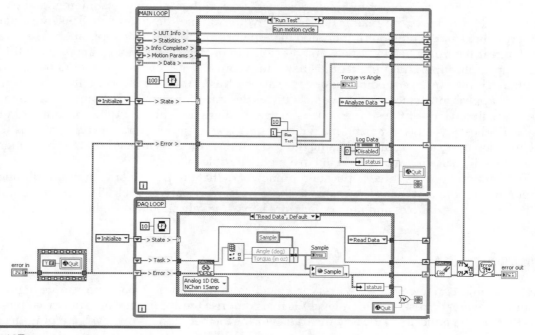

Figure 8-18B
Data acquisition and state transitioning are performed in separate parallel asynchronous loops. The DAQ Loop performs continuous acquisition of torque and angle data, independent of the Main Loop.

An application that uses parallel loops is quickly constructed by merging two or more of the single-loop design patterns onto a diagram and incorporating data sharing using queues, shared variables, or RT FIFOs. Alternatively, you can construct a multiloop template by saving a VI containing multiple single-loop design patterns and messaging as type VIT, and maintain it in the `LabVIEW\Templates\` folder. VI templates are loaded from the **New** dialog, and selected by choosing **New** from the **File** menu, or selecting **New** from a node's shortcut menu in the ProjectExplorer Window.

8.4 Complex Application Frameworks

Many complex applications share several common functional elements or routines, or are constructed in a similar manner. For example, most applications that I develop contain routines for data acquisition, instrument control, analysis, user interface event handling, state transitioning, and error handling. I have my own preferred design patterns and utilities that I reuse extensively. Instead of developing each new application from the design pattern templates and reusable utilities, I have an application framework that combines the common routines into a template and further advances the starting point for complex application development. An **application framework** is an elaborate template filled in with multiple constructs, including design patterns, subVIs, data structures, and messaging, as well as common routines. It serves as a blueprint for application development.

A crude example of an application framework is copying the entire application source code from a previous project and performing aggressive edits to satisfy the custom requirements of a new application. I refer to this approach as *hacking* and I strongly discourage it because the time saved by starting with a code base is offset by time consumed repeatedly searching, examining, and editing the source code. Moreover, no matter how closely you examine the code, it will likely contain some underpinnings of the previous application that do not apply to the new application. These remnants tend to rear their ugly heads at the worst possible opportunities. Imagine going through final checkout of an application, and a report or display screen contains the company name, address, and logo of a different company. Very embarrassing, to say the least! Additionally, code that has been hacked tends to appear much messier than code that was designed for its intended purpose from the outset. Hence, maintenance of the source code is generally more difficult and time consuming in this scenario. For best style, I recommend maintaining a *reusable* application framework that uses several of the functional elements common among the applications you encounter. Make the application framework generic, reusable, and architected in a highly flexible manner.

This section presents three application frameworks that are useful for complex applications. It is important to note that many variations to these frameworks are possible. The specific implementations, including examples and illustrations, are kept relatively simple to facilitate discussion.

8.4.1 Dynamic Framework

The **Dynamic Framework** uses VI server functions to dynamically load and run its high-level VIs from source files on disk. The high-level VIs that comprise the application are known as **components**. VIs that are dynamically loaded by the framework, whether high or low level, are **plug-ins**. Prior to the VI Call Configuration feature introduced in LabVIEW version 8.0, the Dynamic Framework was widely used to help improve the load time and memory efficiency of very large applications. This is also referred to as a **dynamic loader**. However, the VI Call Configuration dialog window facilitates dynamic loading, with the stipulation that the desired components are statically linked to the application as subVIs. The VI Call Configuration dialog window is opened by selecting **Call Setup** from the shortcut menu of a subVI call. Currently, the Dynamic Framework is required if the components are not linked to the application and are not identified until runtime. Specifically, the framework reads the available components from a dedicated directory on disk and loads them into memory on demand. This capability allows components to be developed and distributed after the application has been deployed, facilitating upgrades and maintenance.

A template for the Dynamic Framework is shown in Figure 8-19. The front panel simply consists of a tab control with pages for **Main** and **Error Info**. The **Main** page contains a listbox control, subpanel, and **Quit** Boolean. The listbox displays a list of the components read from disk, and the subpanel displays the front panel of the selected component. The component panel is considered a **child** panel when loaded into the subpanel. The child panel is fully interactive for as long as the component is running.

The diagram contains the Event-Driven State Machine design pattern without the queue, and three primary states shown: **Initialize**, **Idle**, and **Call Plugin**. The **Initialize** state reads the contents of a plug-in directory using the List Folder function and writes the VI filenames to the listbox control's Item Names property. The **Idle** state simply waits for an event on the graphical user interface. Specifically, when the user selects a component from the listbox, a Value Change event is fired, the path to the selected plug-in is created, and the **Call Plugin** state is selected as the next state to run. The **Call Plugin** state loads the component into memory using VI Server's Open VI Reference function and opens the front panel as a child of the subpanel using the Insert VI method. Data is passed to the component's connector terminals, and the component executes via a Call By Reference Node, similar to a subVI call. When the component stops running, it is unloaded from memory using Close Reference, and the child window closes from the subpanel using the Remove VI method.

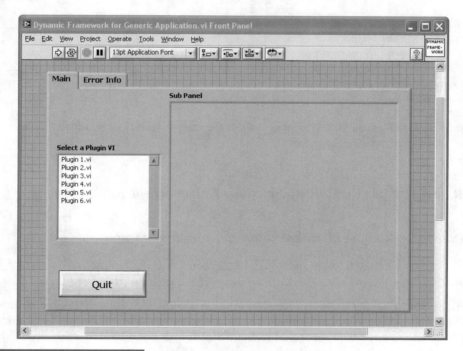

Figure 8-19A
The front panel of the Dynamic Framework VI Template contains a listbox control displaying a list of components, and a subpanel that displays the front panel of the selected component.

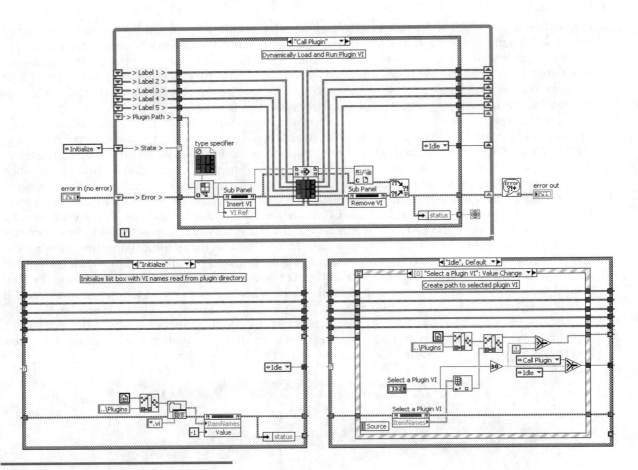

Figure 8-19B
The block diagram is an Event-Driven State Machine without the queue. Three of the primary states are shown. Data is passed between components via shift registers.

 Rule 8.21 Use the Call By Reference Node over the Run method

 Rule 8.22 Choose standard connector terminal assignments

 Rule 8.23 Maintain all components within a dedicated directory

An alternative to VI Server's Call By Reference Node is multiple calls to VI Server's Invoke Node with Set Control Value, Run, and Get Control Value methods. However, the Call By Reference Node executes faster and requires less source code than multiple Invoke Nodes. Also, the Call By Reference Node resembles a traditional subVI call, simplifying development and readability. Note that all components must share a common connector terminal pattern and assigned data types to use the Call By Reference Node. Also, maintaining the components in an appropriately named directory on disk helps simplify upgrades and maintenance. All of the examples in this section use a folder named \Plugins.

Rule 8.24 Pass data between components with shift registers

Design the plug-in components as independent stand-alone VIs that can execute in almost any order, and without much support from the framework. Limited data is passed between components using shift registers. The limitation is imposed by the developer's choice of connector pattern and assigned data structures. Careful consideration is required to specify data structures that are consistent among all components, including future add-ons. This is accomplished by examining the input and output requirements of each component, grouping related data into type-defined clusters, and specifying a maximum of four data structures that combine the *known* data requirements of *all* components. Duplicate these data structures as both controls and indicators on the front panel of each component. In this manner, each component will have read and write access to all data structures, whether they are used within the component or not. On the diagram, propagate the data structures from control terminals to any destination nodes, to the indicator terminals. Any unused data structures are wired directly from control to indicator terminals. Assign the controls and indicators to the 4×2×2×4 connector pattern, according to the connector rules presented in Chapter 5. This includes the **error in** and **error out** clusters assigned to the bottom left and right terminals.

 Rule 8.25 Assign one input and one output terminal as type variant

The controls and duplicate indicators of the prescribed four data structures, combined with the standard error clusters, occupy ten connector terminals. This leaves two spare terminals using the standard 4×2×2×4 connector pattern. However, spare terminals are useless with plug-in components. When the Open VI Reference type specifier is configured within the framework, based on the connector pattern of a representative plug-in, the connector assignments of all components must exactly match the type specifier. Hence, the terminal assignments cannot be added or removed after the framework is deployed. So why limit ourselves to a maximum of four data structures? To accommodate unknown data requirements in the future, assign one additional pair of input and duplicate output terminals as type variant, and propagate the variant between components using another shift register. As discussed in Chapter 6, "Data Structures," variant is a generic data type that is defined dynamically at runtime. The actual data contained by the variant is specified within the components. This approach maximizes flexibility for future expansion.

The diagram of the framework template shown in Figure 8-19B contains five shift registers for sharing data between components, and one shift register for passing the path to the desired component for dynamic loading. This allows all components to generate, modify, and propagate data. All data structures in the framework's template are type variant. Replace up to four of the variants with standard data structures required by the components for your specific application.

 Rule 8.26 Display plug-in panels within subpanels

The subpanel facilitates seamless integration of the component VI panels into the framework's graphical user interface. However, only one front panel is displayed in a subpanel at a time, whereas separate front panel windows can be opened and positioned according to the user's preferences. Because the Dynamic Framework loads, runs, and unloads one component at a time, the subpanel is mildly preferred over separate windows.

Consider how the Dynamic Framework is applied to the Torque Hysteresis application. The various design pattern implementations examined in the previous sections are modular, with each application state implemented as a subVI. Some of the subVIs are dialogs, such as UUT Information Prompt VI and Define Cycle VI. Dialogs are not complex and do not consume much memory. Other subVIs, such as Run Test VI, and Compute Statistics VI, are high-level VIs that contain a hierarchy of subVIs and large data structures. These are the high-level components of the application. The application's performance is improved by dynamic loading of the high-level components. This is easily accomplished using the VI Call Configuration dialog within any of the design patterns described in the earlier sections. Figure 8-20 shows a call configuration for Run Test VI. **Loading and retaining on first call** is the preferred selection because the Run Test VI normally runs many times.

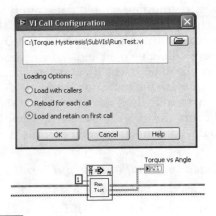

Figure 8-20
The performance of the Torque Hysteresis application is improved by dynamically loading the primary components, including Run Test VI. This is accomplished using the VI Call Configuration dialog window.

Improve the long-term maintenance and performance of the Torque Hysteresis application using the Dynamic Framework. In the implementation shown in Figure 8-21, all of the primary subVIs, including the dialogs as well as the high-level VIs, are implemented as components. This maximizes flexibility in terms of upgrades. The cluster of motion parameters is merged with miscellaneous parameters required by each component into a new cluster of configuration parameters. These parameters are initialized from a config file during the **Initialize** state. Each component has been modified to receive and return all data structures as inputs and outputs, whether they are required by the component or not. Finally, one pair of variant control and indicator terminals is included for future expansion. These enhancements maximize the flexibility of the functionality that exists within each component VI, while maintaining standard connector assignments for all components.

Chapter 8 • Design Patterns

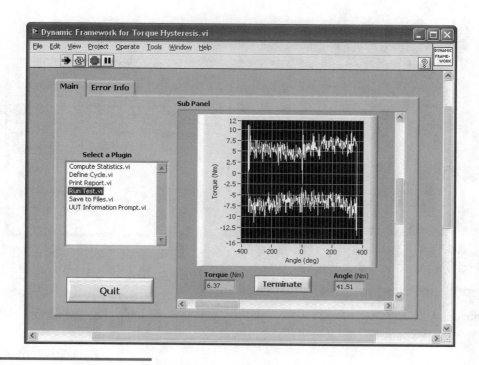

Figure 8-21A
The front panel of the Dynamic Framework applied to the Torque Hysteresis application. The component Run Test VI is loaded in the subpanel control.

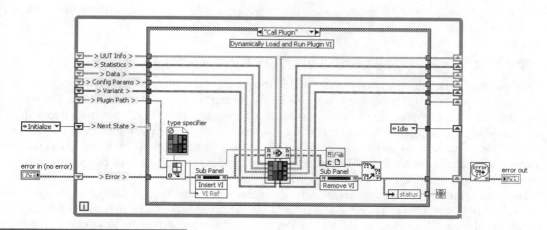

Figure 8-21B
The block diagram propagates the application's primary data structures using shift registers and passes them to the components in the **Call Plugin** state.

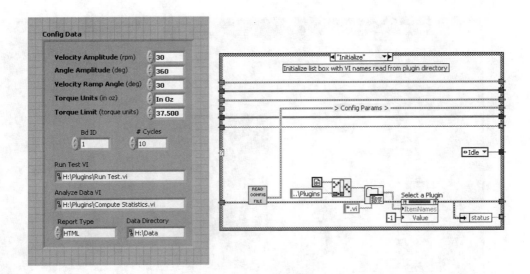

Figure 8-21C
The cluster of motion parameters is combined with other parameters into a single cluster of configuration parameters. The initial configuration settings are read from a config file within the **Initialize** state.

8.4.2 Multiple-Loop Application Framework

The Multiple-Loop Application Framework extends the compound design pattern concept to combine multiple single-loop design patterns, messaging, utilities, and controls into an application framework.

 Rule 8.27   Create a parallel loop for each cohesive parallel task

The flexibility and performance of an application framework is optimized by dividing the common functional elements into cohesive tasks and applying parallel loops for each task. Parallel loops provide the flexibility to execute multiple tasks simultaneously. LabVIEW spawns a separate thread for each parallel loop, and each thread can run on separate parallel processors, if available. Additionally, parallel loops allow you to specify and fine-tune the priority of each task using delays with While Loops and priorities with Timed Loops. Hence, an application framework based on multiple parallel loops helps optimize the application's performance.

The disadvantage of multiple parallel loops is the additional complexity of messaging required to share data between loops. Therefore, it is important to delineate tasks that are cohesive and not too interdependent with respect to data. The implementation presented in this section uses queues for loop messaging. The scope of this framework is a single PC, hence RT FIFOs and multiple process shared variables are not necessary. Queues are more functional and general purpose than local and global variables. Implementing each task as a parallel loop and using queues for data sharing and synchronization helps maximize the breadth of applications that the framework can be applied to.

Rule 8.28 Include loops for Event Handling, Main State Machine, Hardware I/O, and Error Handling

Most complex applications benefit from the following parallel loops: Event-Handling, Main State Machine, Hardware Input/Output (I/O), and Error Handling. The Event-Handling Loop is the most efficient method for capturing GUI events, maximizing the responsiveness of the GUI, and enhancing the user's experience. Always include an Event-Handling Loop with every complex application that contains a GUI. The Main State Machine facilitates state transitioning, organization, expandability, and maintainability. Implement any sequentially ordered tasks that are application-specific within the Main State Machine. Hardware I/O is often the most critical task for a LabVIEW application. Create a dedicated Hardware I/O Loop for each independent continuous I/O task so that they are appropriately prioritized with respect to the other tasks. Finally, an Error-Handling Loop provides a means of reporting all significant errors throughout the system using a single routine and recovering gracefully.

Event-driven performance is achieved by applying a queue or an Event structure for each continuous loop and avoiding timeouts. With Event structures, avoid continuous Timeout events or consider removing the Timeout event case, as per Rule 8.8. Additionally, avoid the **timeout** terminal of the Enqueue and Dequeue Element functions, per Rule 8.15. The unwired default prevents these functions from timing out. Apply this technique to each loop to ensure event-driven performance. On the surface, prioritizing loops using delays or Timed Loop priorities becomes less important when each loop is event driven. However, some loops might need to run at timed intervals and may still require loop timing. Also, delays prevent tight loops. A **tight loop** is a loop that monopolizes the processor due to its priority or the type of task it performs. Each call to a loop timing function such as Wait (ms), Wait Until Next ms Multiple, and the Time Delay Express VI forces the thread to yield control of the processor to allow other tasks to run. For best results, always use a delay within continuous While Loops.

Figure 8-22 contains a multiple-loop application framework developed by Greg Burroughs, Senior Project Engineer at Bloomy Controls. The main front panel contains a large invisible tab control and instructions for editing. The graphical user interface is comprised of multiple display screens implemented on tabs that are programmatically controlled from the diagram. The diagram consists of an Event-Handling Loop, a Main State Machine, an Error-Handling Loop, and a queue-based messaging scheme. Notice that each loop is controlled by a queue or an Event structure without timeouts, facilitating event-driven performance. Additionally, the framework contains multiple constructs for initialization, configuration, security, application control, error handling, documentation, and more. The built-in features are quite extensive and beyond the scope of this chapter. Multiple parallel loops are a key feature of this framework. The best single-loop design patterns are combined for flexibility, expandability, and performance.

This application framework and all subVIs have evolved over time and represent a reliable starting point for complex application development. Additionally, the framework and subVIs are all recognized standards within our organization that facilitate commonality, performance, maintainability, and good programming style. The framework is integrated into a LabVIEW project for file organization and management, similar to what is presented in Chapter 2.

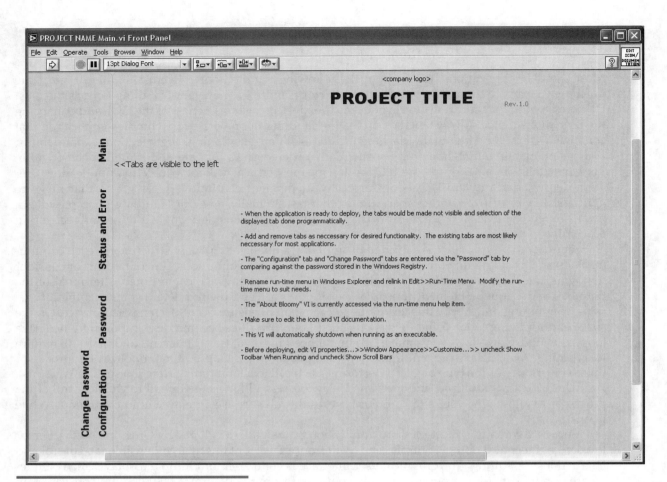

Figure 8-22A
The top-level VI front panel for a multiple-loop application framework consists of an invisible tab control and instructions for editing and use.

Chapter 8 • Design Patterns

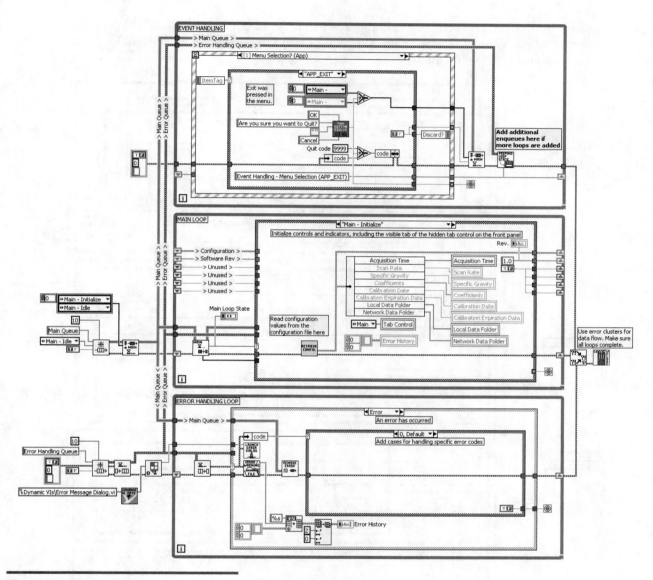

Figure 8-22B
The framework's top-level diagram is a compound design pattern that contains an Event-Handling Loop, a Main State Machine, and an Error Handling Loop. The framework is rich with features that have evolved over time.

In Section 8.3, "Compound Design Patterns," the performance of the Torque Hysteresis application was improved using parallel loops. In Figure 8-23, the application is further improved by applying the Multiple-Loop Application Framework. The separate Event-Handling Loop labeled **Event Handling** and the Main State Machine labeled **Main Loop** improve the response to GUI events and performance of state transitioning versus the Event-Driven State Machine design pattern. Data acquisition is performed in a dedicated loop labeled **DAQ Loop** for continuous monitoring of torque and angle data, similar to the parallel loops presented in Figure 8-18. Also, the loop labeled **Error Handling Loop** processes errors from any of the loops. This allows the application to report errors, gracefully recover, and continue running. Note that the Main State Machine, DAQ Loop, and Error-Handling Loops are all Queued State Machines. Delays are used to prevent tight loops, and messaging is accomplished using queues.

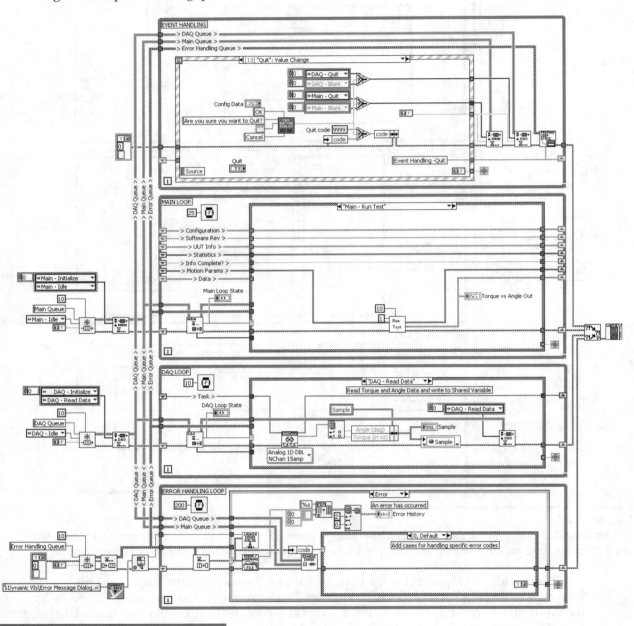

Figure 8-23
The top-level diagram of the Torque Hysteresis application contains separate loops for Event Handling, State Machine, Data Acquisition, and Error Handling.

The DAQ Loop is designed to add a new state to the queue after each read operation. As shown, the **DAQ—Read Data** case adds a new **DAQ—Read Data** element to the back of the queue each time this case runs. This causes this state to run continuously until the application ends or an error occurs. Consequently, a delay must be applied to this loop to prevent a tight loop. Likewise, delays are applied to the remaining While Loops as a precautionary measure.

8.4.3 Modular Multiple-Loop Application Framework

On the surface, the Multiple-Loop Application Framework seems scalable without limitations. Adding more functionality is as easy as adding cases to each single-loop construct, or adding more loops and queues for each new parallel task. A complex application appears quite impressive, as with Meticulous VI described in Chapter 1. However, the Multiple-Loop Application Framework violates several rules for good LabVIEW programming style. In Chapter 4, we presented rules for modularizing top-level diagrams with subVIs and limiting the diagram to one visible screen using 1280×1024 display resolution (see Rules 4.1, 4.4, 4.5, 4.6). Although the framework contains many subVIs, the top-level diagram vertically spans two or more screens, and the modularity index is often below 3.0. The Modular Multiple-Loop Application Framework is a more modular alternative to the Multiple-Loop Application Framework, in which the loops are modularized into subVIs.

 Rule 8.29   Pass control references into subVIs via a type-defined cluster

Because each of the Multiple Loop Application Framework's loops delineates a cohesive task, each loop is a candidate for subVI modularization. This would improve the modularity and reduce the size of the top-level diagram. However, within each loop are front panel terminals that need to be considered. Controls and indicators of the top-level VI are read from or written to within subVIs using control references and Property Nodes configured to read from or write to the control's Value property. However, because this is a template, many of the controls have not been created yet. So how can we modularize the loops of the application framework? Simply pass the top-level VI's control references using a cluster saved as a type definition, and add new control references to the type definition as new controls are added to the front panel. In this manner, you need to maintain only the type definition. The subVI connector assignments remain constant. Alternatively, consider modularizing the top-level VI's user interface using the subVI front panels and a subpanel control. This is similar to the Dynamic Framework discussed previously.

 Rule 8.30   Avoid subVI from selection with continuous loops

Let us refer to the construct that is comprised of a high-level loop modularized into a subVI as a **loop-subVI**. One might be inclined to expedite the modularization of loops into loop-subVIs using the SubVI from Selection utility. This is accomplished by selecting a loop and choosing **Edit»Create SubVI**. However, the SubVI from Selection utility does not handle front panel terminals gracefully. By default, it creates control reference constants for all control and indicator terminals within the selection and passes them into the subVI using individual connector terminals for each. Additionally, data written to indicators is returned through subVI connector terminals wired to the indicator terminals on the top-level VI. There are three issues with this approach:

1. If there are more than a couple front panel terminals within the loop, the subVI will have excessive terminals and wiring clutter inside and outside the subVI.

2. The indicator updates will not occur until the subVI completes, changing the behavior of continuous updates.
3. Maintenance is very tedious.

Specifically, adding and removing control references for each new control or indicator needed within the subVI entails terminal, connector, and wiring changes. Instead, always pass the control references using a cluster of control references saved as a type definition, and avoid the SubVI from Selection utility.

 Rule 8.31 Keep the Event-Handling Loop at the top level

 Rule 8.32 Keep high speed display updates at the top level

The Event-Handling Loop is naturally very tightly coupled to the graphical user interface. Indeed, many of the GUI VI's front panel terminals are maintained within this structure. Maintenance is reduced if this loop is maintained on the top-level diagram instead of modularizing into a loop-subVI. Additionally, Property Node read and write operations are several times slower than reading and writing directly to the control and indicator terminals. Therefore, performance is maximized and maintenance is minimized if the Event-Handling Loop is maintained on the diagram of the main GUI VI, normally the top-level VI. Likewise, maintain any loops that perform high-speed display updates at the top level, to optimize the update rates. Alternatively, consider distributing the top-level VI's user interface among the loop-subVI front panels, and load them into subpanel controls.

Figure 8-24A shows the top-level diagram of the Torque Hysteresis application implemented using the Modular Multiple-Loop Application Framework. Three queues and a type-defined cluster of control references are initialized on the left of the diagram. The queues are passed to the Event structure and loop-subVIs for messaging. The control references are passed to the loop-subVIs for reading from and writing to the controls of the top-level VI using Property Nodes. When the application ends, errors are merged and LabVIEW closes using the subVIs on the right.

The front panel and diagram of DAQ Loop VI are shown in Figure 8-24B. The front panel contains references for the **DAQ Queue**, **Error Queue**, and a type-defined cluster of control references. The **DAQ Queue** buffers the enumeration that drives the subVI's case selector. The **Error Queue** reports any errors to the Error Loop VI. The control references allow the diagram to update indicators on the panel of the GUI VI, including the **DAQ Loop State** enumeration and the **Sample** cluster. The diagram consists of a Queued State Machine. On the left of the Case structure, the diagram calls a subVI that removes the next state to run from the queue, and updates the **DAQ Loop State** indicator using a Property Node configured with the Value property just to the left of the Case structure. The corresponding indicator resides on a diagnostic tab of the GUI VI's tab control. It allows a user with appropriate privileges to monitor the active state of the DAQ LoopVI. In the **DAQ—Read Data** state, the subVI reads a sample of data, updates the **Sample** cluster indicator and shared variable, and adds another **DAQ—Read Data** state to the Queue. Loop timing is accomplished using the Wait Until Next mS Multiple function, based on the programmed loop rate.

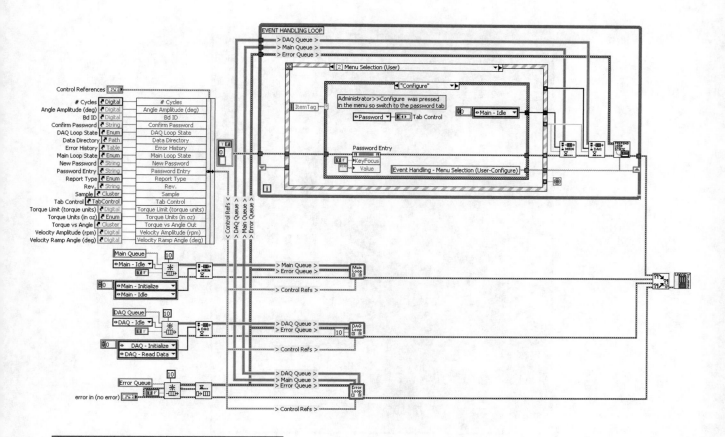

Figure 8-24A
The top-level diagram of the Torque Hysteresis application is modularized into three loop-subVIs plus an Event-Handling Loop.

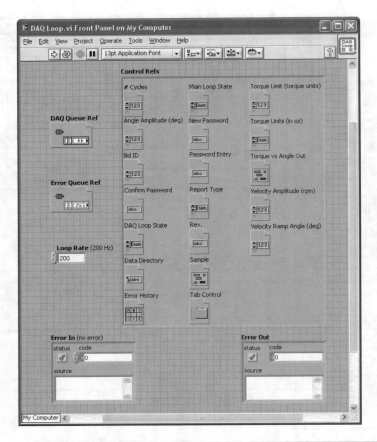

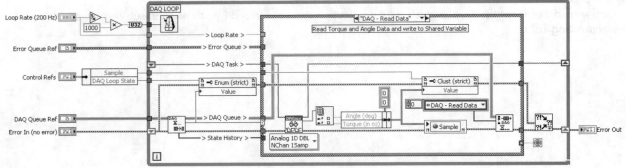

Figure 8-24B
DAQ Loop VI is a loop-subVI containing the data acquisition task. The front panel contains queue references and a type-defined cluster of control references. The diagram uses Property Nodes to continuously update the indicators on the top-level VI.

8.5 Examples

This section presents several more application examples of the design patterns and application frameworks.

8.5.1 Elapsed Time VI

Elapsed Time VI, shown in Figure 8-25, is a Functional Global that stores an initial time stamp in shift register memory and calculates the elapsed time with respect to the initial time stamp. When the **Mode (Update)** terminal is set to **Init**, the **Init** case of the primary Case structure calls the Tick Count (ms) function to acquire the initial time stamp and stores it in the shift register. Subsequent calls to the Elapsed Time VI with the **Mode** terminal set to **Update** select the **Update** case, which acquires new time stamps and calculates the elapsed time in the selected units. This VI is useful for monitoring or resetting the elapsed time from multiple locations in an application, or simply reducing clutter on the diagram of the calling VI.

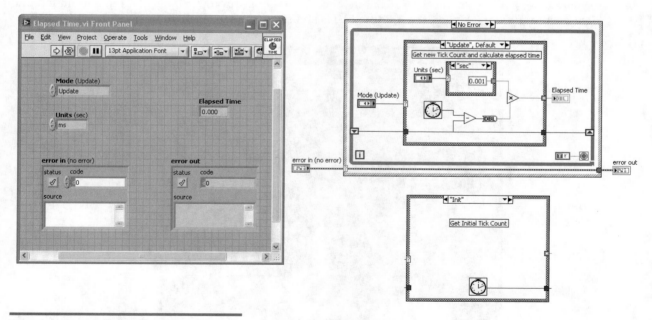

Figure 8-25
Elapsed Time VI is a Functional Global that maintains an initial time stamp in shift register memory and calculates the elapsed time for each subsequent call.

8.5.2 Poll Instrument Response VI

Poll Instrument Response VI is a subVI that reads a message from an instrument using a polling technique. This VI is as useful for serial instrument control applications as it is demonstrative as an example. The front panel in Figure 8-26A contains controls for the **resource name**, **bytes to read**, **timeout**, **response**, and **total bytes**, in addition to the standard error clusters. The diagram shown in Figure 8-26B monitors and reads the number of bytes available at serial port and builds a response string and byte count that are stored in shift registers. A 25ms delay is applied using the Wait (ms) function. The elapsed time is monitored and compared to the timeout condition using Tick Count (ms) and several arithmetic functions. The loop repeats until the byte count reaches the **bytes to read** or an error or timeout condition occurs.

Notice that this VI is a hybrid between the Continuous Loop and the Immediate SubVI design patterns. Specifically, it contains a While Loop with multiple termination conditions, including the timeout. Also, it has subVI error trapping, including error clusters and an Error Case Structure. Figure 8-26C presents an alternative diagram with some minor enhancements. Specifically, it uses Elapsed Time VI, the functional global introduced in the previous section, to replace the timing and arithmetic functions that were present at the bottom of the diagram in Figure 8-26B. Additionally, the error cluster wire enters and exits the Case structure and While Loop through tunnels on the bottom left and right vertical borders. Hence, the modified VI is stylistically superior than the original.

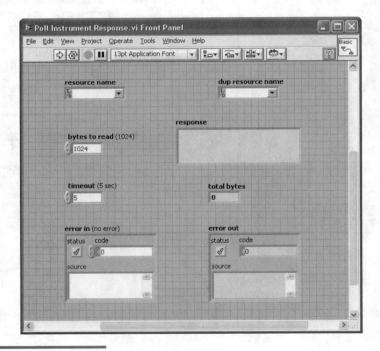

Figure 8-26A
The front panel of Poll Instrument Response VI, a subVI that polls a serial instrument for a response

Chapter 8 • Design Patterns

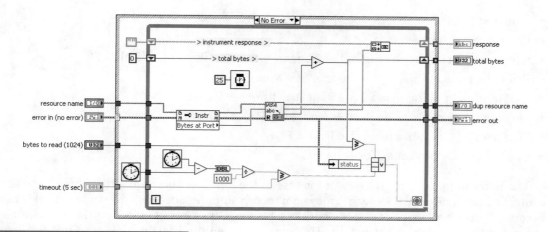

Figure 8-26B
The diagram is a hybrid between the Continuous Loop and Immediate SubVI design patterns.

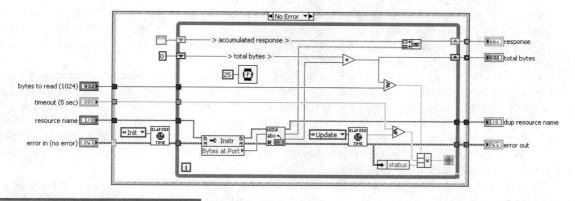

Figure 8-26C
The diagram of Poll Instrument Response VI is enhanced using the Elapsed Time VI functional global introduced in the previous section. Two calls to Elapsed Time VI replace several timing and arithmetic functions.

8.5.3 Unconventional State Machine

The VI illustrated in Figure 8-27 contains several features in common with a Classic State Machine design pattern, but the overall implementation is uncommon and includes multiple style rule violations. The common features include the While Loop containing a Case structure, a shift register, and a menu of Boolean controls. The diagram style violations are as follows:

- Wires enter structure through top horizontal border (see Rule 4.13)
- Right-to-left data flow (see Rule 4.22)
- No labels on wires exiting shift registers (see Rule 4.37)

- Function labels visible (see Rule 4.45)
- Errors are not trapped (see Rule 7.3)
- Simple Error Handler VI instead of General Error Handler VI (see Rule 7.7)
- Error cluster is not the bottom tunnel (see Rule 7.17)
- GUI terminal polling instead of Event structure (see Rule 8.4)
- No enumeration for case selectors (see Rule 8.12)
- Two parallel Case structures (see Rule 4.13)
- Control labels not visible (see Rule 9.1)

The Boolean menu is used to select the case of the top Case structure. The bottom Case structure is controlled by the **Run VI** Boolean. This is not readily apparent because the control label is not visible. Because both Case structures are controlled by Boolean controls, the two Case structures can be merged into a single Case structure. This entails appending the **Run VI** and **Quit** Booleans to the array that is searched for TRUE, and simply adding cases to the Case structure for the corresponding states.

Note that the shift register maintains the previous values of the Boolean menu and does not pass any state information. The Boolean menu alone controls the case selection of the top Case structure. Therefore, any code that executes within the Case structure cannot affect the selection of the next state. This is counter to the Classic State Machine design patterns, in which the desired next state is passed via shift register and wired to the case selector.

The first alternative presented in Figure 8-27C uses a Classic State Machine. Specifically, an enumerated type definition maintains the states and one Case structure contains a case for each state. The **Run VI** and **Quit** Boolean controls are appended to the Boolean menu and polled in the **Idle, Default** frame of the Case structure. The state selection is passed via shift register to the next loop iteration. Function labels are not visible, and free labels have been added to wires exiting shift registers. The state machine has states for **Initialize, Shutdown, Blank,** and **Run**, in addition to **Idle, Default**. Finally, all control labels are visible.

In this example, it is noteworthy that there is a one-to-one correspondence between Boolean controls and application states, with three exceptions: **Initialize, Blank,** and **Idle, Default**. Also, each state is called in response to a Boolean control's value change. Therefore, this VI can be implemented more efficiently using an Event-Handling Loop, as shown in Figure 8-27D. The Event structure is the most efficient method of processing user interface activity. Event cases configured to respond to each Boolean control's Value Change event contain the code from the corresponding cases of the Classic State Machine. The **Idle, Default** state that polls the Boolean controls within the Classic State Machine is eliminated by the Event-Handling Loop's Event structure. The **Initialize** state containing an initialization routine is replaced by the Event-Handling Loop's **Timeout** event case. Per Rule 8.8, continuous Timeout events are to be avoided. In Figure 8-27D, a shift register is wired to the Event structure's **Timeout** terminal. The shift register is initialized to 0, forcing the **Timeout** event to fire immediately. After the **Timeout** event case runs the initialization routine, the shift register is set to −1, disabling the **Timeout** event. Hence, the **Timeout** event case functions similar to the **Initialize** application state.

Chapter 8 • Design Patterns

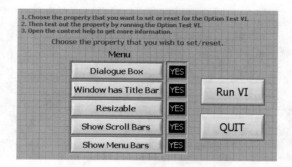

Figure 8-27A
The front panel window of Set Window Options VI consists of a Boolean menu and some LEDs.

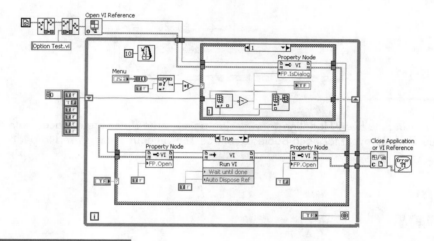

Figure 8-27B
The diagram contains several components of a State Machine design pattern but violates multiple style rules pertaining to State Machines and the block diagram in general.

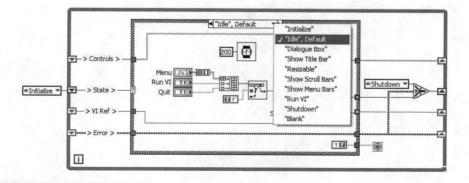

Figure 8-27C
This diagram contains the same functionality implemented with the Classic State Machine design pattern.

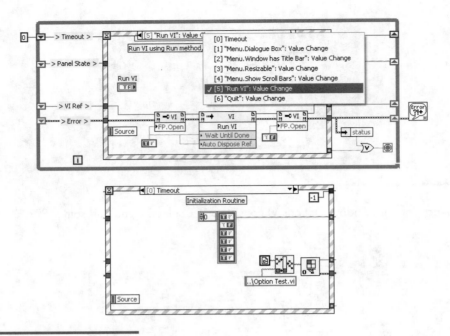

Figure 8-27D
The VI is implemented using the Event-Handling Loop design pattern. The states are replaced with event cases corresponding to the Value Change event for each Boolean control. The **Timeout** event is used to perform an initialization routine.

8.5.4 Centrifuge DAQ VI

The main loop of Centrifuge DAQ VI was presented in Chapter 4. The broader view shown in Figure 8-28 reveals a compound design pattern consisting of a Queued State Machine at the top, an Event-Handling Loop at the bottom, and a queue for messaging. The Event-Handling Loop violates several rules, including continuous Timeout events (Rule 8.8), dissimilar horizontal loop sizes (Rule 8.18), and incomplete error handling (see Chapter 7).

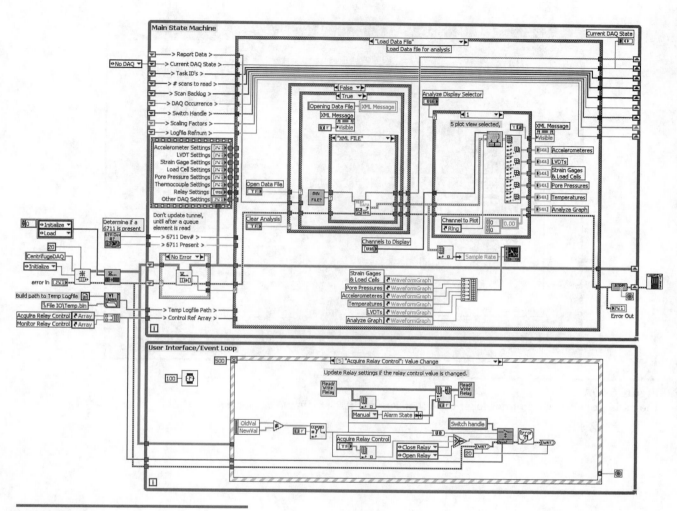

Figure 8-28
Centrifuge DAQ VI is a compound design pattern comprised of a Queued State Machine at the top, an Event-Handling Loop at the bottom, and a queue for messaging.

8.5.5 Transducer Control Utility

The application shown in Figure 8-29 is a full-featured instrument control utility that is bundled together with a driver for a digital pressure transducer. The control utility configures, acquires, and logs data, and provides full interactive control of any number of transducers. The diagram consists of a compound design pattern based on the Multiple-Loop Application Framework. It contains an Event-Handling Loop for processing GUI activity, a Queued State Machine for state transitioning, and a Continuous Loop for error handling, allowing multiple tasks to run in parallel. For example, several transducers can acquire and log data while another transducer is configured, without any interruption to the acquisition and logging.

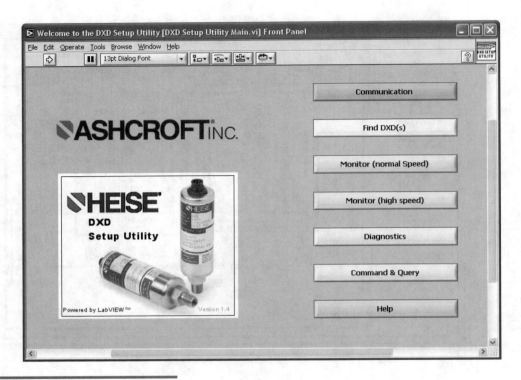

Figure 8-29A
The main front panel of a transducer control utility consists of a simple Boolean menu.

Chapter 8 • Design Patterns

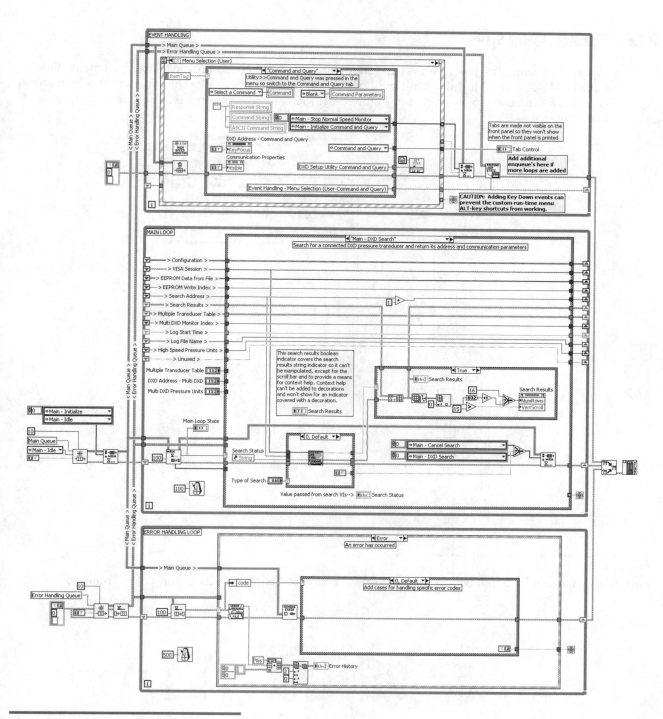

Figure 8-29B
The diagram uses the Multiple-Loop Application Framework.

8.5.6 Distributed Control System

Figure 8-30 contains the top-level diagram for a distributed control system implemented using LabVIEW RT. It is a Modular Multiple-Loop Application Framework. The diagram of a loop-subVI contains a Timed Loop, as shown in Figure 8-31. Messaging between VIs is accomplished using RT FIFO buffers.

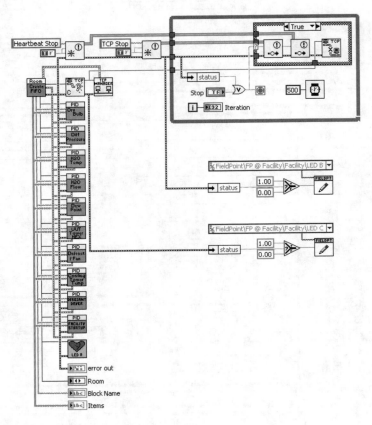

Figure 8-30
The top-level diagram for a distributed control system uses the Modular Multiple-Loop Application Framework.

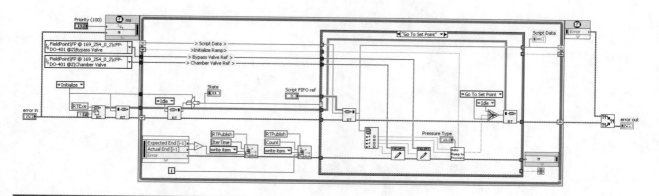

Figure 8-31
The diagram of one of the loop-subVIs consists of a Timed Loop. Messaging is accomplished using RT FIFO buffers.

Endnotes

1. Templates for several of the design patterns discussed in this chapter can be downloaded from www.bloomy.com/lvstyle.
2. "Using LabVIEW to Create Multithreaded Applications for Maximum Performance and Reliability" and "LabVIEW and Hyperthreading" are available for free download from NI Developer Zone (www.zone.ni.com).

Documentation

9

Documentation is any description of the structure, components, or operation of a system, application, or source code. Documentation includes specifications, design documents, source code documentation, user manuals, and online help. Specifications and design documents are discussed in Chapter 2, "Prepare for Good Style." Source code documentation consists of comments, descriptions, and text that are useful for the developers to understand the source code. A user manual is a document describing how to operate the software from the end user's perspective. Online help is documentation that is electronically integrated with the application.

All of these types of documentation apply to LabVIEW applications. Documentation requirements vary widely depending on the application type, developer, end user, organization, and industry. For example, a commercial application may require an extensive user manual and online help, whereas a one-of-a-kind lab experiment may require no documentation at all. Many organizations have a set of internal standards that all software and systems must adhere to. Government-regulated industries traditionally have stringent documentation requirements imposed by regulatory committees such as the Food and Drug Administration. This chapter focuses on LabVIEW source code documentation that comprises good LabVIEW development style, regardless of the developer, end user, organization, or industry.

> **Myth 9.1:** *LabVIEW diagrams are self-documenting.*

The graphical nature of LabVIEW presents a strong advantage over text-based programming languages in source code comprehension. A picture is worth a thousand words. However, it is a myth that LabVIEW source code is self-documenting. Because documentation is used to describe the source code, it cannot be exactly the same as the source code. Rather, LabVIEW diagrams are ultimately designed for graphical compilation, while documentation is explicitly designed for linguistic expression. Consequently, documentation is primarily implemented using text. Because LabVIEW source code is graphical, it is inherently not textual. Documentation serves to supplement and reinforce the readability of the graphical source code.

> ***Theorem 9.1:*** *Good development style reduces source code documentation effort.*

Readability is one of the key attributes of good style, as discussed throughout this book. The more human readable the source code implementation is, the less documentation effort is required. Indeed, some overlap exists between readable source code containing intuitive embedded text, and documentation. For example, in Chapter 3, "Front Panel Style," the importance of succinct and intuitive control labels is emphasized. In Section 9.2, the relationship between visible control labels and diagram readability is discussed. Multiple constructs in LabVIEW use text to facilitate readability, including the Bundle and Unbundle By Name functions, and the enumerated data type.

As another example, if your top-level diagram architecture is directly implemented using one of the standard design patterns discussed in Chapter 8, "Design Patterns," then you need not write a lengthy document describing how the design pattern works. Other references are available for this purpose. Standard design patterns are recommended for good readability, maintainability, and performance. However, if you decide to create your own custom VI architecture, then it will require documentation. The more unique or complex the implementation appears, the more documentation is required to explain it. The style rules presented throughout this book are designed to facilitate readability of the source code. Therefore, the better your development style, the less additional effort is required to document your source code. Moreover, when you use LabVIEW's Print VI Documentation features, your source code and documentation is automatically compiled into a document. Readable source code, good documentation practices, and built-in LabVIEW tools help to streamline the documentation process. Print VI Documentation is discussed in Section 9.4.

Documentation is directly related to readability, maintainability, and ease of use. Hence, documentation is integral to good style. Documentation is often neglected in the pursuit of fast development cycles. As discussed in Chapter 1, "The Significance of Style," shortcuts do not reduce development time when the entire life cycle of the application is considered. Rather, good development style reduces development time and effort.

This chapter presents style rules for source code documentation that is directly incorporated into the front panel, block diagram, icon, and VI and control properties. Additionally, we review style rules from previous chapters that directly relate to readability, such as appropriate use of text. If you master the rules from the previous chapters, your VIs are already readable and have an excellent foundation for streamlining the documentation process.

9.1 Front Panel Documentation

Front panel documentation consists of control labels, descriptions, tip strips, and free labels. These items are beneficial for users as well as developers. Style rules and examples pertaining to front panel text are presented in Chapters 3 and 6. The rules are repeated in this section for convenience, followed by an example.

 Rule 3.17   Minimize front panel text

 Rule 3.18   Delete template instructions immediately after edits are performed

Avoid entering lengthy comments as permanent labels on front panels. This is particularly important for GUI VIs. Intuitive GUIs are graphical, not textual. Long paragraphs of text should be limited to the user manuals and online help. SubVIs containing notes for the developer are an exception. Specifically, most template VIs contain instructions for editing and use within free labels. Always delete the instructions after each edit is performed. This helps signify that the edits have been completed.

 Rule 3.21   Use succinct, intuitive control labels and embedded text

 Rule 3.23   Provide default values and units in parentheses at the end of owned labels

Control labels are the most prominent type of text appearing throughout a LabVIEW application. It is extremely important to make them succinct and intuitive. Append the default value and units in parentheses for any controls representing physical parameters. Apply captions when long phrases are required or the desired GUI label differs from the control label. Captions are label alternatives that are visible on the front panel only when selected from the shortcut menu via **Visible Items»Caption**.

 Rule 6.5   Enter control descriptions

 Rule 6.23   Enter descriptions for array and cluster shells and control elements

Enter a one- or two-line description that further describes each control's purpose, data type, default value, and range, unless it is completely self explanatory based on the label. Arrays and clusters often form the data infrastructure for an application. It is useful to enter descriptions for the shells and elements before they proliferate throughout the application. Additionally, apply tip strips containing descriptive phrases to the controls of GUI VIs. Tip strips are abbreviated descriptions that appear when the mouse cursor is placed over the control. Tip strips, captions, and control descriptions help maintain succinct control labels and embedded text. Long phrases and explanations belong in the former locations instead of the labels.

 Rule 3.39   Always include a Help menu or button

Always include a **Help** menu or button for GUI VIs, with functional selections for providing the user help. This may include **Show Context Help**, as well as menu items for online documents or compiled help files. The **Show Context Help** menu item opens the Context Help window, which displays any descriptions available from the controls, indicators, and VI as the user scrolls the mouse over them. Additionally, LabVIEW can open online documents of a variety of formats using a browser, word processor, or reader application. Online documents are discussed in Section 9.4. Custom runtime menus are configured by selecting **Edit»Run-Time Menu**.

Menus are not appropriate with dialog VIs. Instead, provide Boolean command buttons for any required menu selections. This includes a Boolean for **Help** or **?**, which can launch the desired online documentation. Also create shortcut menus for any actions related to a specific control. Shortcut menus are created by selecting **Advanced»Run-Time Shortcut Menu»Edit** from a control's shortcut menu.

Figure 9-1A contains the front panel for niDMM Configure Measurement Digits VI, a subVI that configures a modular instrument. The Context Help window is open with the VI description, icon, and terminals visible. The control labels are succinct and intuitive. Only the **error in** cluster contains its default value in parentheses. The default value and range for **Resolution in Digits** and **Range** depend on the value of **Function** and, therefore, cannot be specified within the control label. However, the default value of **Function** is known and should be included in its control label. Default values and units in parentheses are particularly important with subVIs such as instrument drivers because they help the developer quickly decide whether to wire the corresponding terminals or leave as default.

Figure 9-1B contains the front panel for niDMM Configure Measurement Digits VI with the Context Help window displaying the control description for **Resolution in Digits**. The Context Help window updates with the description for the control that the cursor moves over. All of the controls of this subVI have descriptions. This helps developers understand the programmable functions of the instrument.

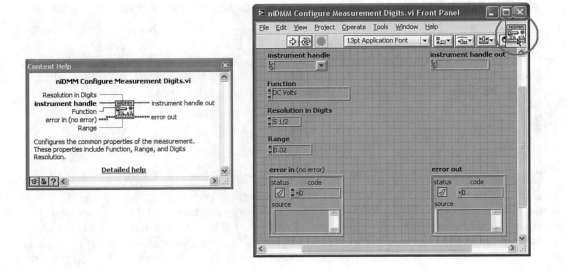

Figure 9-1A
niDMM Configure Measurement Digits VI is a subVI that configures a modular instrument. The controls have succinct, intuitive labels. The default values depend on the selected **Function** and are absent from the labels.

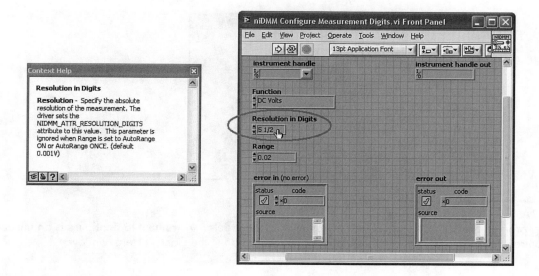

Figure 9-1B
The description of **Resolution in Digits** is shown in the Context Help window. All of the controls have descriptions.

Figure 9-2A contains the front panel of Configure Test Instrument VI, a dialog that prompts the user to configure a digital multimeter. The labels of the front panel controls are succinct and intuitive, with the exception of **Please Select Instrument to Configure**. This lengthy phrase should be a caption instead of a control label. Keep the labels succinct and intuitive, per Rule 3.21. Note that none of the control labels contain default values. Default values are not necessary for the labels of GUI VIs because the controls generally contain their default values when loaded, unless they are overwritten programmatically using local variables or Property Nodes. However, the default values are useful to have in the control descriptions. All of the controls on the panel of Configure Test Instrument VI contain descriptions and tip strips. However, the descriptions are not available to users, unless they know how to open the Context Help window via the **<Ctrl>+<H>** shortcut. Hence, this dialog requires familiarity with the LabVIEW environment to obtain help.

Figure 9-2B is the front panel of the same dialog as Figure 9-2A, with several improvements. The succinct control label **Instrument** replaces the lengthy phrase **Please Select Instrument to Configure**. Lengthy phrases, whether labels or captions, are usually not necessary with dialogs. The style of the panel and controls should be enough for the user to understand that the panel is interactive and action is required. Finally, a Boolean labeled **Help** has been added. This control is used to simply launch the Context Help window, allowing the user to view the VI and control descriptions. The diagram of Configure Test Instrument VI is presented in Section 9.2.

Figure 9-2A
This dialog prompts the user to configure a test instrument. **Please Select Instrument to Configure** is an unnecessarily lengthy control label. There is no obvious method for the user to open the Context Help window.

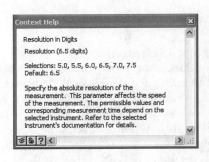 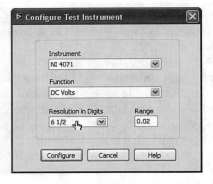

Figure 9-2B
Configure Test Instrument VI has been improved using succinct control labels and a **Help** button that launches the Context Help window for viewing the control descriptions.

9.2 Block Diagram

Block diagram documentation consists of text such as control labels, free label comments, and constructs containing text. This section covers block diagram documentation techniques.

 Rule 9.1 Keep front panel terminal labels visible on the diagram

The single most underrated source of documentation is the owned labels of the controls and indicators. In the previous section, the importance of creating succinct and intuitive labels for all controls and indicators is emphasized. These labels are a primary source of documentation on the diagram, but only if they are visible. A tremendous advantage of LabVIEW's dataflow paradigm is that you can trace the data path from source to destination. If you cannot identify the terminals, much of this advantage is neutralized. Observing the diagram entails toggling back and forth between the front panel and diagram windows, in an effort to identify each terminal and commit the labels to memory. On a diagram that conforms to good style, this exercise is completely unnecessary.

One might assume maintaining visible labels to be automatic. After all, owned labels are visible on the diagram by default. From what I have seen, however, many developers go out of their way to hide the labels. In some cases, the labels are even empty. This practice is particularly perplexing because LabVIEW assigns a default name when each control is created. I suspect that the developer does not want the labels taking up space on the diagram, and decides to hide or empty them. This is why it is important to make the labels succinct. Chapter 3 contains rules and examples for creating control labels that are succinct and intuitive, facilitating good documentation.

> *Rule 9.2   Apply free label comments in select locations*

> *Rule 4.18   Label long wires and wires from hidden source terminals*

> *Rule 4.37   Label wires exiting the left shift register terminal*

> **Rule 8.20   Label each loop in the top left corner**

> *Rule 9.3   Label every subdiagram of every multidiagram structure*

> *Rule 9.4   Label algorithms, constants, and Call Library Function Nodes*

> *Rule 9.5   Use default 13-point plain black application font for all diagram text*

Free labels are analogous to comments in a conventional text-based programming language. Apply them to select locations of your diagrams by single-clicking the labeling tool, or by double-clicking the auto tool in an empty area of the front panel or block diagram. One common rule is to apply a free label in every subdiagram of every multidiagram structure, with a few exceptions such as the enumerated Case structure that is described later. Additionally, it is useful to have a convention on the location of the label. At Bloomy Controls, our convention is to place the label in the top-left corner of most structures, as seen in many of the illustrations in this book. We use LabVIEW's default 13-point plain application font and a lightly shaded yellow background. Apply free labels on long wires. Use the greater-than sign to indicate the direction of data flow, and color the background white. Label wires exiting shift register terminals in a similar manner.

Label mathematical algorithms and constants. Note that algorithms are good candidates for subVIs. Constants may have owned labels, similar to controls and indicators. Simply right-click and choose **Visible Items»Label** from the shortcut menu, and enter the desired text. Likewise, the default icon and description of Call Library Function Nodes are not very informative. It is useful to label each Call Library Function Node with the name of the DLL and function that is being called. Alternatively, select **Names** format by choosing **Name Format»Names** from the shortcut menu. This resizes and populates the icon with the DLL's configured function name. Previously, developers were forced to navigate to the configuration dialog to identify the Call Library Node's function call.

 Rule 9.6   Leave notes for the development team

Another good use of free label comments is to leave temporary notes for the members of a multideveloper project team. Commercially available source code control tools are highly recommended for multideveloper projects. These tools normally prompt for comments when source files are checked in after editing. Additionally, LabVIEW's VI Differencing tool can identify source code changes between revisions. However, using free labels, you can enter comments directly on the diagrams, close to any source code that you want to draw attention to. This technique complements or extends the conventional source control and differencing techniques.

One scenario in which I use this technique is when the developers are sharing one computer, collaborating in series instead of parallel. This happens when the single computer has exclusive access to some critical instrumentation or equipment and is secured behind a firewall. Indeed, I have worked on projects that were supported by several developers across several shifts per day and weekends. Each developer writes notes describing any significant changes that were made in an area adjacent to the modifications. We might color-code the labels, varying the background shade to denote the shift. The author includes the date and appends her initials to each comment.

Some references recommend applying free labels liberally. I have deliberately used the term *selectively* instead of *liberally*. Free labels should never be so abundant that they dominate the diagram. In some situations, too much text can actually hinder readability. For example, if you implement a standard design pattern such as the Classic State Machine and choose to include paragraphs of text explaining how it works using multiple free labels, you may not recognize the design pattern among the comments. Instead, I recommend applying relatively succinct comments in select locations, as described. Longer explanations belong within the VI description, help files, or external documentation. Alternatively, if long comments on the diagram are necessary, either place the long comments in a location adjacent to the diagram or use string constants with scrollbars. String constants can be used in a similar manner as free labels, except that string constants can have vertical scrollbars that hide a portion of the text. From the shortcut menu, select **Visible Items»Vertical Scrollbar**.

 Rule 6.16   Use enums liberally throughout your applications

 Rule 9.7   Use enumerated data types with Case structures

Enumerated controls, or enums, map text selections to numeric values, similar to a text or menu ring control. However, unlike a ring control, when an enum is wired to a case selector, the Case structure replaces its selector labels with textual labels that match the items within the enumeration. Therefore, enums describe the function of each case diagram of the Case structure using intuitive text. Separate free labels for documenting every case of the Case structure, per Rule 9.3, are often not necessary when enums are utilized.

In the example illustrated in Figure 9-3, **Test to Run** is a numeric control that is used to select one of several tests to run. It is wired to the Case structure's selector terminal. Each case of the Case structure contains one or more subVIs that perform the selected test. In Figure 9-3A, **Test to Run** is a text ring control, and the corresponding selector labels are integers. Free labels are applied in each case to associate the case number with the actual test name. This requires effort to create and maintain the association of the free labels, test names, and case numbers. As the Case structure is modified, several items must be edited. In Figure 9-3B, an enum is used for the case selector. The selector labels contain the test names, and the free labels are not necessary. Additionally, the Case structure is maintained simply by editing the values in the enum control. Hence, the enum control is a central definition of the Case structure's cases and documentation.

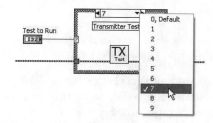

Figure 9-3A
This Case structure has a numeric selector, forcing the developer to translate between the numeric values and the test names.

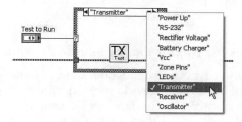

Figure 9-3B
The enumerated Case structure displays the test names in the selector area using intuitive text.

 Rule 6.17 Save enums as type definitions

It is common to use multiple instances of the enum on a diagram as constants as well as controls, and the text items are subject to change. Therefore, I recommend saving all enums as type definitions. This way, when edits are made to the items of the enum using the Control Editor window, all instances of the enum update automatically, maintaining identical items as the type definition. Ring controls must be saved as strict type definition to provide similar behavior. Enums, ring controls, and type defs are discussed in more detail in Chapter 6. Also, the enumerated Case structure is a fundamental element of the state machine design pattern that is discussed in Chapter 8.

LabVIEW provides a block diagram option called **Show subVI names when dropped**, available from **Tools»Options** and selecting **Block Diagram** from the **Category** list. I recommend disabling this selection, per the default value. All subVIs should be readily identified by a unique icon, and the name and path are available from the Context Help window. I recommend keeping the Context Help window open while editing because it contains valuable information. Specifically, the Context Help window contains the VI description, inputs, outputs, priorities, and data types, in addition to the VI's name. Hiding names of functions and subVIs reduces unnecessary text from the diagram, which saves space, reduces clutter, and improves readability of the graphics and data flow.

The name format of Property and Invoke Nodes affects the size of their icons on the diagram. Short Names are used by default and are specified from the Property or Invoke Node's shortcut menu by selecting **Name Format»Short Names**. Similar to a subVI, the long name of a property or method can be determined from the Context Help window. Therefore, use the short name format for Property and Invoke Nodes to reduce the space that they consume on the diagram, without eliminating the name entirely.

Rule 6.21   Use arrays for multivalued data items; use clusters for grouping multiple distinct items

When a multivalued data type is required, a cluster is often more appropriate than an array. Arrays are normally used when all of the elements of a data structure have the same data type, whereas clusters are used for defining data structures composed of mixed data types. However, you cannot document the individual elements of an array. On the diagram, all of the array manipulation functions access the array elements by index number only. Consider that each element of a cluster has an independent owned label and description. Also, the cluster wire is bundled or unbundled by name, using the corresponding diagram functions. If an index number sufficiently identifies the data you are accessing, then an array is appropriate. However, if unique identifiers and descriptions for each item are desired, then use a cluster. Array and cluster considerations are discussed in greater detail in Chapter 6.

Rule 6.26   Always use Bundle and Unbundle By Name

On the diagram, use the Bundle By Name and Unbundle By Name functions to access and replace the elements of a cluster. These functions provide more flexibility and readability versus Bundle and Unbundle. Specifically, you can access or update any element of the cluster in any order using Bundle By Name and Unbundle By Name. Also, these operations do not break the VI if your cluster changes, unless the specific elements that are being bundled or unbundled are affected. Most important, the name labels uniquely identify the elements of the cluster and provide useful documentation. This is further proof that control labels are an important source of documentation on the diagram. Because the Bundle and Unbundle By Name functions are sized according to the longest label, keep the control labels succinct and intuitive.

Configure Test Instrument dialog VI is continued in Figure 9-4. The block diagram in Figure 9-4A contains no documentation whatsoever. Specifically, all of the front panel terminal labels are not visible, and free label comments are missing from the Case structure's cases and wires exiting shift registers. Additionally, the Bundle function contains terminals that appear ambiguous. The diagram in Figure 9-4B contains multiple improvements versus Figure 9-4A. First, the front panel terminal labels are visible. Additionally, there are free labels on wires exiting shift registers, and free label comments in each case of the Case structure. Notice that the diagram in Figure 9-4B is significantly wider than the diagram in Figure 9-4A. This is due to the lengthy terminal labels. Perhaps this is the reason some developers hide their control labels on the diagram. Remember to keep the control labels succinct and intuitive. In particular, **Please Select Instrument to Configure** is an unnecessarily lengthy label. Additionally, notice that the terminal label style—raised versus transparent—is inconsistent.

Figure 9-4C contains the optimal documentation for this VI. The text ring control for **Please Select Instrument to Configure** has been replaced with an enumerated type definition labeled **Instrument**. The enum maintains the same instrument names for text items as the former text ring control. However, the instrument names now appear in the Case structure's selector area. This documents the Case structure's cases and eliminates the need for the free label comments in each case. All of the labels are succinct, which reduces the space they consume on the diagram. Bundle By Name replaces the Bundle function.

The terminal names simplify and reinforce the wiring assignments and reduce maintenance if the cluster order changes. Finally, all of the labels throughout the diagram have transparent backgrounds.

Figure 9-4D shows the revised panel and diagram, including the **Help** button, **Help Value Change** event case, and Context Help window. The user clicks the **Help** button to trigger the **Help Value Change** event. The Control Help Window function opens the Context Help window. The Context Help window displays the VI and control descriptions as the user moves the cursor over the objects on the panel.

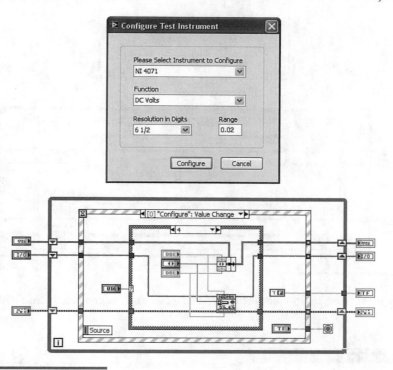

Figure 9-4A
Configure Test Instrument VI front panel and block diagram. Documentation is nonexistent on the diagram.

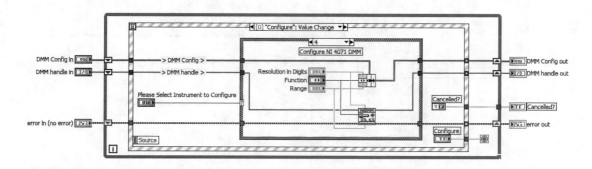

Figure 9-4B
This implementation contains documentation consisting of visible terminal labels, labels on wires exiting shift registers, and labels in each case. However, the diagram is wide due to the unnecessarily lengthy label.

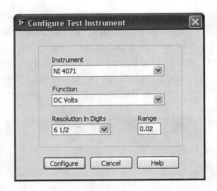

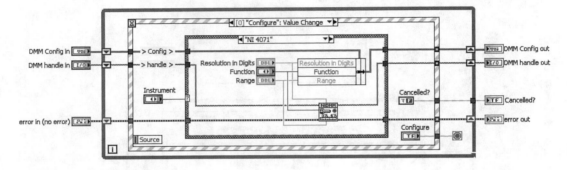

Figure 9-4C
Several improvements have been performed, including succinct labels, enumerated Case structure, and Bundle By Name function.

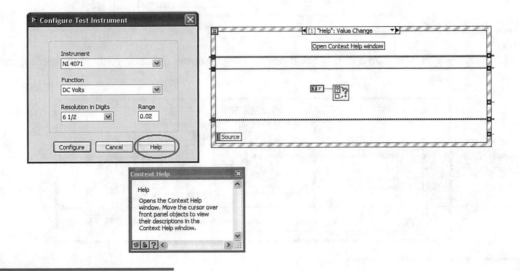

Figure 9-4D
The **Help** button fires the Event structure's **Help Value Change** event. The Context Help window opens, allowing the user to view the VI and control descriptions.

9.3 Icon and VI Description

Icons and VI descriptions comprise the most essential forms of documentation throughout LabVIEW. These critical elements reinforce the purpose of each VI using graphics and text. The icon graphically represents the VI as a subVI call on a diagram. For best results, create an icon that combines a universally recognizable glyph with color and text. The descriptions appear in the Context Help window when the user moves the cursor over the subVI's icon. Create a two- or three-sentence description, at a minimum, that summarizes the purpose of each subVI.

 Rule 4.10   Create a meaningful icon and cohesive description for every subVI

 Rule 9.8   Create the icons and descriptions as you develop your source code

Many developers consider developing VI icons and descriptions a lower priority than developing working source code. Most have good intentions, expecting to return to these elements at a later time, but procrastinate until another activity takes priority. As a result, LabVIEW's default icons are used instead of meaningful ones, and VI descriptions are skipped. Eventually, the software satisfies the operational requirements, and editing the icons and entering descriptions becomes an even lower priority compared to the next project that is in the pipeline. Some developers justify using LabVIEW's default icons and skipping the descriptions because they are uncertain as to whether the VIs will be included in the final application. Why waste time on icons and descriptions if they might end up in the recycle bin? In practice, the opposite is a much greater concern: beginning with default icons and empty descriptions but not having the time to go back and edit them. The readability and maintainability of such applications are sacrificed, as is the reusability of the subVIs. Moreover, it is much more time-consuming to develop meaningful icons and cohesive descriptions if you delay, versus immediately after you develop them. Therefore, create meaningful icons and cohesive descriptions *as you develop your source code*, while it is fresh in your mind. Furthermore, the icons and descriptions should be considered an integral part of your source code. Apply shortcuts to help expedite icon development, but never sacrifice unique, meaningful icons and cohesive VI descriptions.

The relationship between cohesive subVIs and VI descriptions is discussed in Chapter 4, "Block Diagram." An extensive discussion of icon style, including shortcuts and examples, is provided in Chapter 5, "Icon and Connector."

9.4 Online Documentation

Integrate stand-alone documents, such as help files and user manuals, into your LabVIEW applications as online documents. Online documents include HTML, portable document format (PDF), and compiled help (CHM) files. These documents can be programmatically launched from the **Help** menu of a top-level GUI VI, or from a Boolean control labeled **Help** or **?** from a dialog VI. Use an Event structure to programmatically launch the document when the menu is selected or when the **Help** button's value changes. This is very similar to launching the Context Help window, as discussed in the previous sections.

HyperText Markup Language (HTML) is the universal language for the creation of web pages. Today most word processors and many applications can create HTML documents. For example, LabVIEW's built-in printing features, as well as the VIs from the Report Generation toolkit, readily generate HTML. Create your online documents using the HTML format if you want to view the document in a web browser or if you want to provide network access to the documents via the Web. In particular, HTML is the easiest format to cross-reference multiple documents using hyperlinks. Also, if your documentation requires frequent updates, you can maintain the content as a website instead of shipping the documentation with the application. You can launch an HTML document in the host computer's default web browser using Open URL in Default Browser VI, located in the Help sub-palette under **Programming»Dialog & User Interface»Help**.

Portable Document Format (PDF) was created by Adobe Systems for representing two- and three-dimensional documents in a fixed-layout, cross-platform compatible file format. PDF files encode the exact look of a document in a device-independent manner. They are easily viewed or printed using any type of computer. Use PDF for documents with fixed content and format, such as a user manual. PDF documents can be viewed using Adobe Acrobat Reader, as well as most web browsers. Use the System Exec VI, located on the **Connectivity»Libraries & Executables** palette, to launch the document in the reader you specify. You must build the proper command line for the desired browser application and document path. Alternatively, use Open Acrobat Document VI, located in `<vi.lib>\Platform\browser.llb`. This VI automatically forms the command line for you using the computer's default reader. Unfortunately, this VI is not well documented or style conforming and is not made accessible from the palettes.

For commercial applications, consider a compiled help file. **Microsoft Compressed HTML Help (CHM)** is the standard help file format under Windows XP. CHM files are heavily indexed and include a hyperlinked table of contents that normally resides outside the main body text. First you create an HTML document using your favorite word processor or HTML editor, and then you compile it into a CHM file using a third-party help file compiler, such as Microsoft HTML Help Workshop or Adobe's Macromedia RoboHelp. From LabVIEW, you can programmatically launch the help file using the Control Online Help function. Wire the **String to search for** input to open the help file to a specific topic. The required content of a compiled help file varies widely based on application and is beyond the scope of this chapter.

Windows Vista utilizes Microsoft Assistance Markup Language (AML), a new generation of Help in which documents are defined by their context. As of this writing, the AML format is not available as an authoring platform. Therefore, CHM remains the preferred help file format for commercial applications.

Several varieties of online documentation are incorporated into the Torque Hysteresis VI shown in Figure 9-5. They include Context Help and a user manual in PDF, HTML, and CHM format. These resources are selected from the **Help** menu, as shown in Figure 9-5A. Figure 9-5B shows the diagram for handling menu selection events. Each menu selection corresponds to a case in the Case structure that launches the document using the appropriate browser. Specifically, **User Manual (HTML)** uses Open URL in Default Browser VI to launch the HTML document in the computer's default browser. **User Manual (PDF)** uses the System Exec VI to launch the PDF document using Adobe Acrobat Reader. **User Manual (CHM)** uses the Control Online Help function to launch the CHM file in the default browser. Finally, the **Context Help** menu selection opens the Context Help window for viewing the control and VI descriptions. This operation is built in, and no code is required on the diagram.

Chapter 9 • Documentation

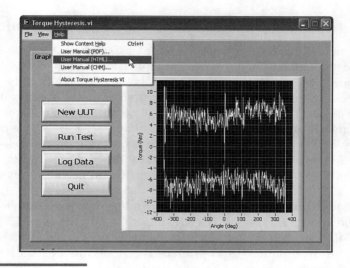

Figure 9-5A
The front panel of Torque Hysteresis VI contains **Help** menu selections for user manuals in PDF, HTML, and CHM format, in addition to Context Help.

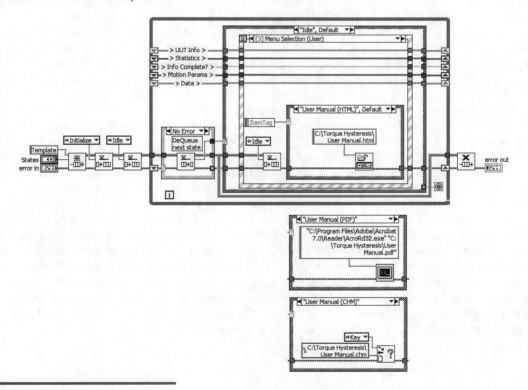

Figure 9-5B
The block diagram of Torque Hysteresis VI captures Menu Selection events using an Event structure and launches the corresponding user manual in the appropriate browser. Open URL in Default Browser VI is used to launch an HTML file, System Exec VI is used to launch PDF files, and the Control Online Help function is used to launch a CHM file.

Rule 9.9   Provide online documentation with deployed applications

Here is a hypothetical situation: You have just completed a project for a demanding customer. Multiple changes in the requirements have resulted in a much longer and more difficult development cycle than you envisioned. The project is well over budget and is significantly past due. You are going through final checkout and acceptance, and the customer asks, "Where is the documentation?" It is something that was never previously discussed as one of the requirements but is often expected as part of the software. Does this sound familiar? Do not panic! Your source code is well documented, per the style rules in this chapter. You started with a specification and design document, as per the rules in Chapter 2. You have detailed VI and control descriptions throughout your VIs. Your diagrams have appropriate labels, comments, and text. All you need to do is Print VI Documentation. Almost….

Print VI Documentation is a built-in tool for generating documentation for a VI. It extracts various pieces of documentation from the source code, depending on the user-selected fields, and creates a document in HTML, Rich Text Format (RTF), or Plain Text (TXT) file format. Alternatively, it prints a hardcopy of the documentation to a printer. This tool can be run interactively by selecting **File»Print»VI Documentation**, or programmatically using the VIs in the **VI Documentation** subpalette, located under **Programming»Report Generation**. Also, VI Server provides programmatic access to the properties of each VI, including elements of documentation.

There are a few considerations when using this tool. First, many of the selections are not relevant for a user's manual. If this is your objective, then you really need to select only documentation elements that apply to the graphical user interface (GUI), including the **VI Description**, **Front Panel**, **Controls**, **Control Descriptions**, and **Control Labels**. You can deselect everything else. Also, collect documentation only for the GUI VIs, including the top-level VI. Users may not need to see the inner workings of your subVIs, unless they will also maintain the source code or if they need to see the details of any mathematical algorithms. Also, there are some additional items that flow to the document that will not be useful for the end users. Specifically, Print VI Documentation prints the specified documentation for every control on the VI's front panel. These may include error clusters and any controls or indicators that you use during development for diagnostic purposes that are invisible or have been scrolled off the visible area of the screen. Finally, the control terminals are provided in the document, as they normally appear on the diagram. These items are not meaningful for typical end users. Consequently, you may need to perform significant manual edits to remove the documentation related to the error clusters, invisible controls, and terminals.

An alternative method of automating the documentation process involves utilizing Control References to obtain the images of the controls as they appear on the front panel, instead of the terminals, and developing a custom routine using VI Server to create the documentation. This approach is described in an article that appeared in LabVIEW Technical Resource Volume 10, Number 3.[1] The important point is to be sure to thoroughly document all of your VIs, and these automated techniques for generating documentation are available. I recommend always providing a user manual or other form of online documentation, accessible from the **Help** menu of the top-level VI, for all deployed applications. You will be one step ahead of your customers.

9.5 Examples

As discussed in the chapter's introduction, readability, maintainability, and ease of use are common goals of the style rules and examples provided throughout the book. Due to the close relationship of good style, readable source code, and documentation, there are related examples in every chapter. A few more examples provided here include SubVI from Selection VI, Filter Test VI, Meticulous Control Descriptions, and Temperature Profile Illustration.

9.5.1 SubVI from Selection VI

Figure 9-6 contains the Context Help window for SubVI from Selection VI, previously discussed in Chapters 3, 4, and 5. As discussed in the previous chapters, these VIs never conform to good style without significant manual cleanup. In addition to inappropriate terminal labels and icons, there is no VI description. These are violations of Rules 3.21, 4.10, 5.2, 5.3, and 9.7. SubVI from Selection w Cleanup VI, shown in Figure 9-6B, is an improved revision of the VI, containing appropriate terminal labels, and a meaningful icon and description. More importantly, there is no way of understanding the VI's purpose from the Context Help window or front panel. Always remember to finish any subVI from selections that you develop by applying the style rules.

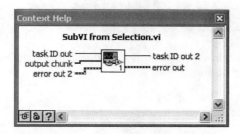

Figure 9-6A
The Context Help window for SubVI from Selection VI. This VI violates multiple style rules, including default icon, inappropriate terminal labels, and missing VI description.

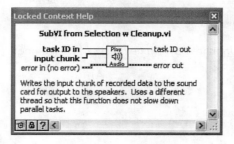

Figure 9-6B
SubVI from Selection w Cleanup VI contains a meaningful icon and VI description, and appropriate terminal labels.

9.5.2 Filter Test VI

The optical filter test application from Chapter 4 is continued in Figure 9-7. A routine that controls a laser utilizing clear wiring and good data flow but inadequate documentation is shown in Figure 9-7A. Specifically, a cluster is unbundled using the Unbundle function, an instrument driver that controls the laser contains generic icons, and there are no free labels. You cannot really understand by inspection what the VI is doing. In Figure 9-7B, the VI is improved using the Unbundle By Name function, meaningful icons, and several free labels. The laser driver's icon convention consists of a light beam glyph and text describing the function of each VI. Also, a free label is placed under each subVI describing its operation. Unbundle By Name contains terminals revealing the intuitive labels of the controls that comprise the cluster. The number of terminals is reduced to only those that are wired. Additionally, there are free labels on the long cluster wire and above the arithmetic functions that calculate the number of wavelength steps. The VI's overall readability is substantially improved.

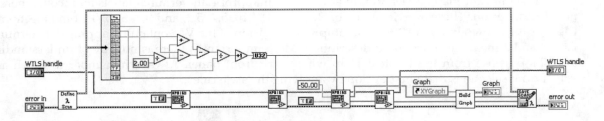

Figure 9-7A
The optical filter measurement routine contains clear wiring and data flow, but inadequate documentation via the Unbundle function, generic icons, and no free labels.

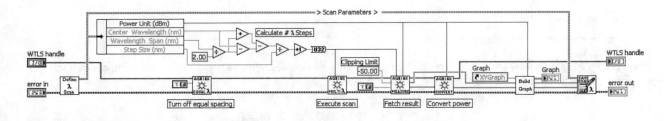

Figure 9-7B
Readability is improved utilizing the Unbundle by Name function, a meaningful icon convention, and free labels.

9.5.3 Meticulous Control Descriptions

Figure 9-8 contains a cluster of controls that are used to configure serial communications with one or more pressure transducers. Each control is very well documented with labels and descriptions conforming to all of the style rules. First, each control has a succinct and intuitive label with the default value in parentheses. Additionally, there are descriptions for the cluster shell as well as each control. The control descriptions all contain the control type, valid range, default value, and purpose.

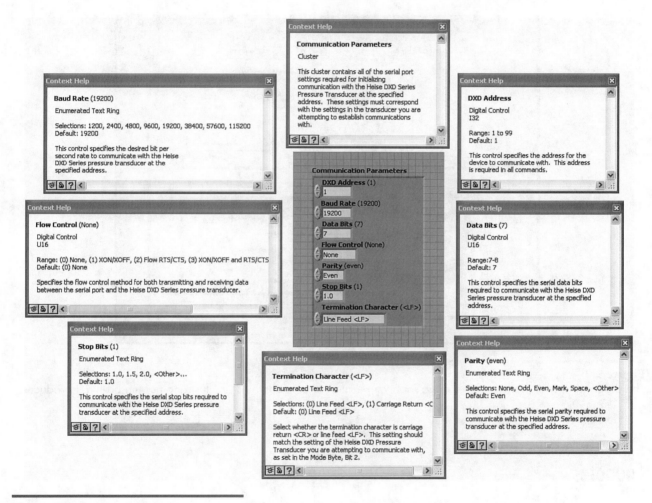

Figure 9-8
This cluster contains controls that configure serial communications with a pressure transducer. Each of the controls contains meticulous descriptions, as seen in the Context Help windows.

9.5.4 Temperature Profile Illustration

Documentation need not be limited to text. Sometimes an illustration is more descriptive. Figure 9-9 contains the block diagram of a VI that controls a ramp-soak temperature profile. It contains an illustration of the temperature profile that it produces. The illustration was created using a drawing application and was saved to an image file. The image file is copied and pasted onto the diagram by choosing **Edit»Import Picture to Clipboard**, followed by **Edit»Paste**.

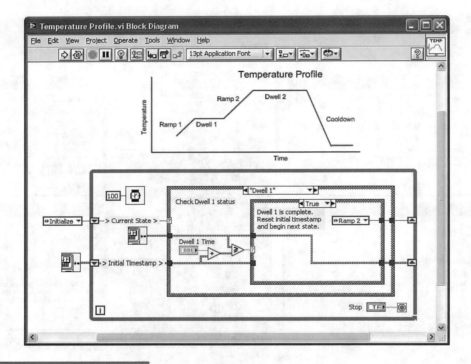

Figure 9-9
The diagram of a VI that controls a ramp-soak temperature profile contains an illustration of the profile it produces.

Endnotes

1. Chris Megandoff, "One-Click Documentation—Automating Your User Manual Development," LabVIEW Technical Resource 10 (3), 12-15.

Code Reviews

10

The LabVIEW Style Book provides many style rules that can help an organization establish a style convention. However, the rules, descriptions, and examples contained herein fall short of requiring their adoption and use. This chapter addresses how to enforce the standardization of a style convention within an organization. In practice, not everyone will agree with all the style rules, and even those who agree will not necessarily apply them consistently or effectively, without a system of checks and balances in place. **Code reviews** are systematic methods of reviewing source code quality, utilizing one or more independent perspectives or reference points. Code reviews have three primary benefits: identifying problems, exchanging knowledge and ideas that improve developer skill level, and promoting and evolving standards across an organization. Most importantly, code reviews help ensure high quality software that is readable, maintainable, and robust.

A secondary benefit of code reviews is that they introduce additional developers to your application. This provides some depth in the resources that are familiar with the application and its implementation. As a professional systems integrator, I find that resource options are often the difference between good customer service and excellent customer service. When changes are required in the future, it is very comforting to know that the code is high quality and that you have multiple resource options for further development. The chances of satisfying a new change request without interrupting other project schedules are dramatically increased.

Rule 10.1   *Enforce your organization's style convention using code reviews*

Rule 10.2   *Utilize a combination of self-reviews and peer reviews for best results*

Code reviews require one or more knowledgeable resources to provide objective feedback. Peer reviews involve coordinating a meeting with colleagues who are experienced with LabVIEW and the organization's style conventions. This is the traditional method of conducting code reviews. However, objective feedback can be obtained in more than one way. Self-reviews are performed utilizing alternate resources, such as the summary of style rules in Appendix B, "Style Rules Summary," and the LabVIEW VI Analyzer Toolkit. Self-reviews and peer reviews are complementary techniques that should both be used for best results.

10.1 Self-Reviews

As the name indicates, self-reviews are techniques for reviewing your source code by yourself. Automated and manual methods exist for conducting self-reviews. The automated method involves the LabVIEW VI Analyzer Toolkit, which is an add-on product from NI. Additionally, the style rules summarized in Appendix B can be utilized as a manual checklist. Simply go through the rules and check off whether your VIs conform to each rule.

Rule 10.3   *Perform a self-review prior to every peer review*

At a minimum, perform at least one self-review prior to each peer review. This helps you prepare for the peer review by considering your design decisions and identifying and correcting any style issues you discover before the peer review. This increases the productivity of the peer review.

One additional point regarding self-reviews: Many LabVIEW developers work by themselves. Perhaps you are the lone developer within your organization or are a self-employed consultant. In these circumstances, self-reviews are probably your only option. Self-reviews combined with some self-discipline should provide plenty of opportunity for maintaining good style.

10.1.1 VI Analyzer Toolkit

The LabVIEW **VI Analyzer Toolkit** is an add-on product from NI that automatically inspects VIs for style and performance. It contains both interactive and programmatic interfaces. The interactive interface enables you to configure the desired tests to run, save the configuration to file, run the tests, and view and save the results. Additionally, if you click on a failure in the Results window, the VI Analyzer opens the diagram and highlights the specific object or area containing the problem. This enables you to iteratively view and fix each problem. Additionally, the programmatic interface enables you to develop an application that performs the desired tests using the VIs on the **VI Analyzer** palette. For example, you can write an application that automatically inspects all the VIs in a source code repository and generates a report. You can configure the application to inspect the repository on a periodic basis and email or publish a report automatically.

The VI Analyzer contains more than 60 tests organized into the following categories: **Block Diagram**, **Documentation**, **Front Panel**, and **General**. The **Block Diagram** category tests several of the rules listed in Chapter 4, "Block Diagram," such as right-to-left data flow (backward wires), bends, and objects overlapping wires. Additionally, several tests identify more subtle issues that are not directly associated with style. These include unnecessary elements on the block diagram that cannot execute (referred to as "dead code") and hidden objects within structures. **Documentation** tests inspect several of the rules in Chapter 9, "Documentation," including VI and control descriptions, and comments in free labels. Additionally, the VI Analyzer can check the spelling of text on the front panel and block diagram. **Front Panel** tests inspect several of the rules in Chapter 3, "Front Panel Style," including control alignment and overlapping controls. Additionally, the VI Analyzer can check the default values of list items and arrays, which are common sources of inefficient memory usage. The **General** category contains tests for the icon and connector, file properties, and VI properties. The icon and connector tests correspond to several of the rules in Chapter 5, "Icon and Connector." File and VI properties include tests for the VI's saved version of LabVIEW and the compatibility of properties and methods with a built application.

The LabVIEW VI Analyzer Toolkit was introduced with LabVIEW version 7.0. By this time, Bloomy Controls had a well-established style convention and code review methodology, and we were pretty set in our ways. Not until I hired and trained a fresh young graduate did I become aware of the VI Analyzer's capabilities. Ordinarily, code reviews are very lengthy with new engineers because there are often many style issues to discuss. However, the quality of the code developed by one of my newest staff members significantly increased. Amazingly, all of her wiring was clear and nonoverlapping, with left-to-right data flow. Her code resembled that of much more experienced developers. The productivity of her code reviews was substantially increased, and the level of required training was reduced. Her secret was the VI Analyzer Toolkit. Several of my engineers now run the VI Analyzer on their source code prior to a code review.

Rule 10.4   Use the VI Analyzer to automate tedious inspections

The primary benefits of the VI Analyzer include the following:

1. It helps inspect and enforce good style.
2. It can be utilized by an individual.
3. It is automated and reduces tedium.

It is important to recognize that the VI Analyzer provides value that extends beyond testing a VI's style. Many of the tests inspect the likelihood of bugs and errors, similar to a secondary compilation process. Specifically, the VI Analyzer contains a **Warnings** subcategory of tests, including **Bundling Duplicate Names**, **Typedef Cluster Constants**, **Hidden Tunnels**, **Reentrant VI Issues**, and more. These are elusive sources of misbehavior that are difficult to identify through manual inspection.

Rule 10.5   Customize the VI Analyzer's test criteria

- Set the **Coercion Dots** maximum number on a single wire to **0**.
- Set the **Enabled Debugging** test to Fail Test if Debugging is **Disabled** during development.
- Set **Globals and Locals** maximum number to **0** each.
- Set **Comment Usage** to **Ensure subdiagrams contain at least one comment each**.

- Set **Control Alignment** for a **Pixel Tolerance** of **1**.
- Disable the **Dialog Controls** test for subVIs and industrial GUI VIs.
- Set **Connector Pane Pattern** to 32 and 0.

Utilize the VI Analyzer's **Select Tests** dialog to customize the test criteria according to your needs, and save the test configuration to file for repeated use. Specifically, set the maximum number of coercion dots on a wire to **0**, as shown in Figure 10-1. This test, configured under **Block Diagram»Performance»Coercion Dots**, enforces Rule 4.24, "Avoid array and cluster coercions," and Rule 6.3, "Choose controls and data types that facilitate consistent data structures throughout an application." The default value is 2, so it is necessary to customize it to identify every violation. Also, unless you are preparing to compile an executable, set the **Enabled Debugging** test to fail if debugging is **Disabled**. Most code reviews are performed during development, and debugging is desirable. During at least one pass through the code, set the **Block Diagram»Warnings»Globals and Locals»Maximum Number** for Write and Read Globals and Locals to **0** each. This helps you examine every instance of a variable to ensure that they are all necessary. If your VI is large and complex, and contains multiple variables and coercions, consider creating separate configuration files for examining only the variables and only the coercion dots, respectively. This enables you to focus on one issue at a time.

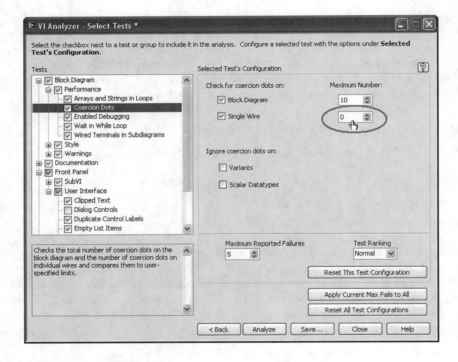

Figure 10-1
The VI Analyzer's interactive window is configured to run a test using customized settings. The **Coercion Dots** test is configured for a Maximum Number of **0** to identify every instance of this style rule violation.

Under **Front Panel»SubVI**, the **Control Alignment** test has a **Pixel Tolerance** parameter with a default value of **10**. If you use the alignment tools, as recommended in Chapter 3, you can reduce this to a value of **1**. Also, the VI Analyzer has a **Dialog Controls** test within the **User Interface** category that is selected by default. This test checks all the VI's controls to determine whether they are from the **System** palette. Dialog controls apply only to the GUI VIs of a desktop application that is designed to resemble the appearance of the operating system. Be sure to disable this test for all subVIs and industrial GUI VIs that utilize Modern or Classic style controls.

Additionally, the VI Analyzer has a **Comment Usage** test within the **Documentation»Developer** category. Select **Ensure subdiagrams contain at least one comment each** to enforce Rule 9.3, "Label every subdiagram of every multidiagram structure." Finally, under **General»Connector Pane Pattern**, choose the 4×2×2×4 connector pattern and the single terminal connector pattern, corresponding to pattern numbers **32** and **0**, respectively. Number 32 enforces Rule 5.21, "Use the 4×2×2×4 connector pattern for most VIs." Connector pattern number 0 prevents this test from flagging a failure when you test a top-level VI developed using a legacy version of LabVIEW. Prior to version 8.0, the single terminal was the default pattern, and most top-level VIs utilize the default.

Rule 10.6   Customize the rank for each test according to the priority of each rule

The test rank setting determines the order in which any violations appear in the Results window and the symbol that appears next to the test name. Set the tests corresponding to high-priority style rules to a rank of **High**, and set the normal-priority rules to **Medium**. Because the VI Analyzer is utilized to inspect performance issues as well as style, the default test rankings are significantly different than the priorities of the corresponding style rules provided in this book. For example, the **Error Cluster Wired** test has a default rank of **Low**. This test checks whether the **error out** terminals of nodes are wired, enforcing Rule 7.5, "Trap all errors from all nodes that have error terminals." As discussed in Chapter 7, "Error Handling," error trapping is an essential ingredient of error handling, and error handling facilitates robust applications. Customize this rule's test rank setting to **High** to reflect the priority of Rule 7.5. You can download a customized test configuration containing recommended analyzer test settings and rankings from www.bloomy.com/lvstyle.

Figure 10-2A provides a primitive version of the Torque Hysteresis VI discussed in previous chapters. This version was developed using a prior version of LabVIEW and a relaxed style convention. We refactor this VI in incremental steps throughout this chapter using code reviews. The VI Analyzer is configured to run a test utilizing custom settings recommended by Rules 10.5 and 10.6. In Figure 10-2B, the VI Analyzer's Results window displays the test results. The results are listed in top-down order of priority, with **!** denoting the high-ranking test failures, no symbol denoting the normal-ranking test failures, and **i** denoting the low-ranking test failures. Figures 10-2C through 10-2J contain the detailed results of specific tests, including all the failures identified within case 1 of the Case structure, along with modifications that eliminate the failures.

Figures 10-2C through 10-2G are the high-ranking test failures. In Figure 10-2C, the **Backwards Wires** test identifies a wire that flows data from right to left. In Figure 10-2D, the **Error Cluster Wired** test locates a function with unwired **error out** terminal. In Figure 10-2E, an unwired enumerated constant is identified by the **Unused Code** test. It is left over from the previous version of the File Dialog function that had an additional input terminal for selecting the operation type. In Figure 10-2F, the **Save** button does not contain a control description, flagging a VI Documentation failure. The cluster wire in Figure 10-2G originates within the Sequence structure on the left, passes through a tunnel, and briefly runs underneath the Sequence structure before connecting to the While Loop's shift register. This flags a **Wires Under Objects** failure.

Figures 10-2H through 10-2J are the medium-ranking test failures. In Figure 10-2H, the VI Analyzer identifies insufficient comments. A brief comment is added within each case of the Case structure using free labels. Additionally, two comments describing the function of the two primary structures are added, as shown in Figure 10-2K. In Figure 10-2I, the VI Analyzer's spell checker identifies the misspelling of velocity as **vel** in two control labels. Although this abbreviation was intended by the developer, abbreviating important terms within control labels on a GUI VI is not recommended. Hence, this problem is corrected by simply expanding the term to **velocity**. Similarly, each of the control labels within the **Statistics** cluster is expanded, as shown in Figure 10-2K. Figure 10-2J contains a violation of the **Wire Bends** test. A very short wire segment, known as a wire kink, is identified and corrected. Also, another wire contains four bends, exceeding the default maximum of three bends, which flags an additional occurrence of the **Wire Bends** test. These bends allow the wire to roughly follow the border of the Case structure, as the developer intended. Because wire bends are often deliberate, the **Wire Bends** test has a medium rank. However, in this particular example, the bends serve no useful purpose, and the wire is straightened in Figure 10-2K.

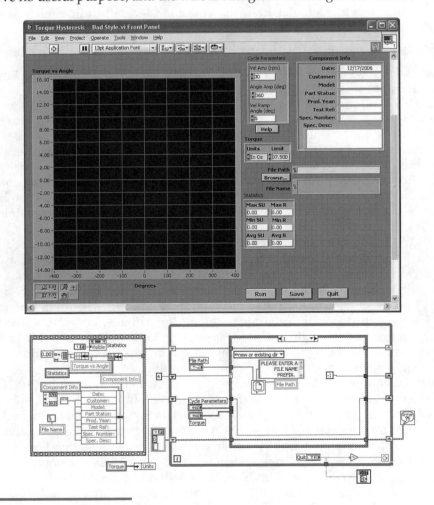

Figure 10-2A
Front panel and block diagram for a primitive version of Torque Hysteresis VI, prior to performing a code review.

Chapter 10 • Code Reviews

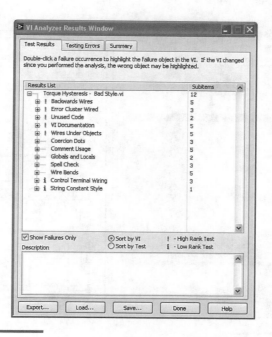

Figure 10-2B
The VI Analyzer Results Window displays the test results in top-down order of priority, with **!** denoting the high-rank test failures, no symbol for medium-rank test failures, and **i** denoting the low-rank test failures.

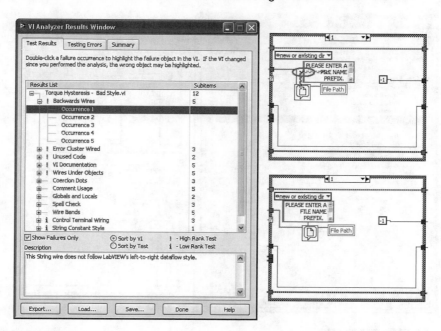

Figure 10-2C
The string constant is positioned above and slightly to the right of the File Dialog function, which causes right-to-left data flow. The VI Analyzer flags a **Backwards Wire** failure.

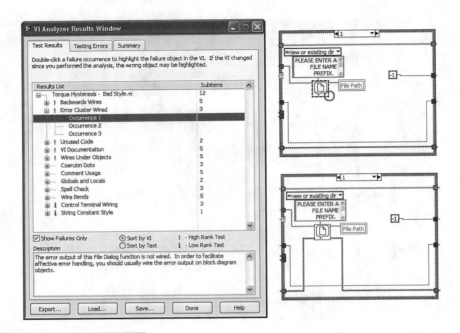

Figure 10-2D
The **error out** terminal of the File Dialog function is not wired, flagging the **Error Cluster Wired** failure.

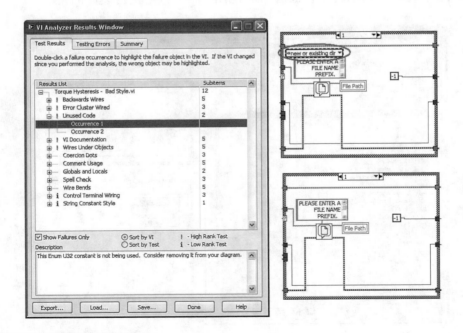

Figure 10-2E
Case 1 of the Case structure contains an enumerated constant that is not wired to anything. This flags an **Unused Code** failure.

Chapter 10 • Code Reviews

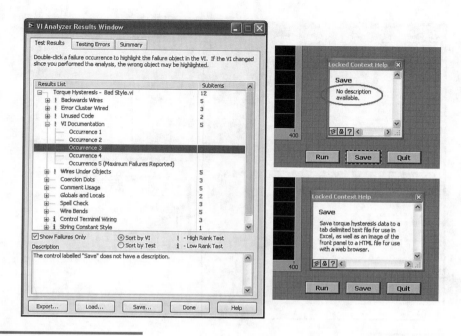

Figure 10-2F
Several controls do not have descriptions, including the **Save** button. This results in multiple **VI Documentation** failures.

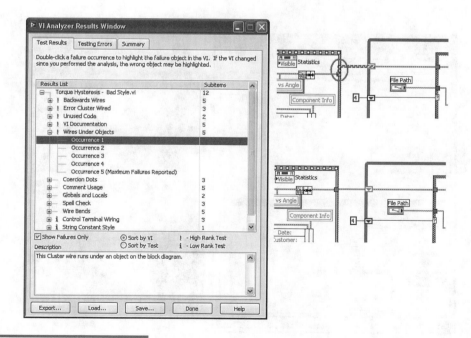

Figure 10-2G
A segment of the cluster wire runs underneath the Sequence structure, flagging a **Wires Under Objects** error.

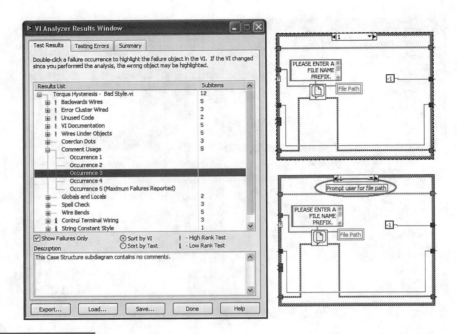

Figure 10-2H
The diagram contains no comments, including the subdiagrams of the Case structure, resulting in multiple **Comment Usage** errors. A brief comment is added to case 1.

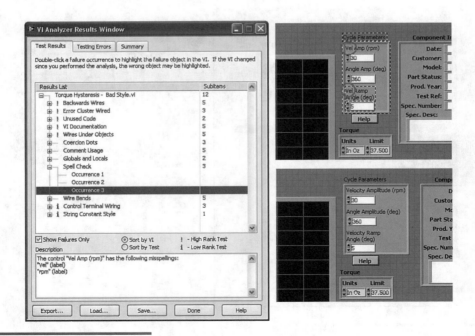

Figure 10-2I
Abbreviated control labels flag spelling errors. Expanding the term **velocity** eliminates two occurrences.

Chapter 10 • Code Reviews

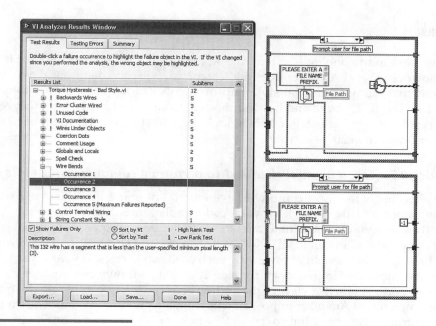

Figure 10-2J
The **Wire Bends** test identifies a wire segment with fewer than 3 pixels, known as a wire kink.

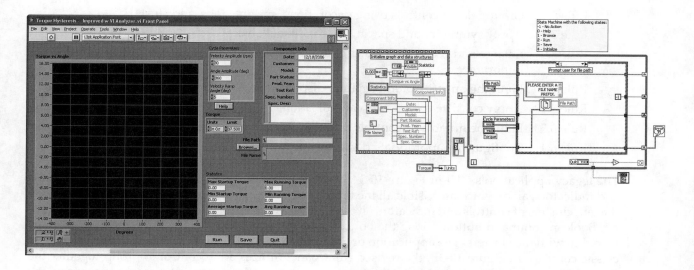

Figure 10-2K
The Torque Hysteresis VI is revised based on the recommendations provided by the VI Analyzer.

10.1.2 Manual Checklist

As demonstrated in the previous section, the VI Analyzer is an effective tool for fast, personal code reviews. However, the VI Analyzer does not test many style considerations. For example, you might have noticed that many additional style improvements can be made to the front panel and block diagram of Figure 10-2, above and beyond those recommended by the VI Analyzer. Indeed, the VI Analyzer performs approximately 60 tests, including 40 style-related tests, but *The LabVIEW Style Book* contains more than 200 style rules.

 Rule 10.7   Use a manual checklist to perform a comprehensive style review

A manual checklist is the most comprehensive method of going through every rule, reviewing the rule's significance and then checking your code's adherence. You can use the rules summary in Appendix B as the style checklist[1] or as a reference to derive your own. Simply not enough time exists during a peer review to evaluate all 200+ rules. The manual checklist approach is tedious but thorough and has the added benefit of reviewing each of the style rules. If your organization's style rules differ from the ones listed in the appendix, simply create your own checklist.

Let us now perform a self-review of the Torque Hysteresis VI by applying the manual checklist to the VI containing the VI Analyzer improvements from the previous section. The front panel and block diagram of Torque Hysteresis Improved w VI Analyzer VI are copied to Figures 10-3A and 10-4A, respectively. We begin with a self-review of the front panel. Reading through the checklist in Appendix B, the following style rules are violated by the front panel in Figure 10-3A:

Rule 3.1	Group related controls using decorations, spacing, tabs, and clusters
Rule 3.2	Apply symmetry and spacing to front panel objects
Rule 3.9	Enlarge and center the objects of an industrial GUI VI in proportion to their importance
Rule 3.10	Limit the quantity of information displayed on a GUI VI panel
Rule 3.19	Apply consistent fonts and capitalization
Rule 3.33	Limit the number of controls that are visible and enabled at any one time
Rule 6.4	Configure an appropriate default value for each control

The legacy application's GUI in Figure 10-3A consists of one large panel containing all the controls and indicators, all of which are visible all the time. The groupings alternate with a graph indicator on the left, clusters of controls and indicators on the right, a path control and indicator in the vertical center, Boolean command buttons along the bottom, and **Help** and **Browse** Boolean command buttons mixed in with the clusters. The application predates modern-style 3D controls and, therefore, utilizes classic controls. In Figure 10-3B, the revised front panel utilizes a tab control to group together the controls used most often on the **Main** tab, the controls that configure a measurement sequence on the **Configuration** tab, and the indicators that summarize the test results on the **Statistics** tab. Because it is an industrial application, the classic-style controls are replaced with modern-style 3D controls. The controls on the **Main** tab are enlarged for improved visibility and ease of use. In addition to grouping related controls, the tab control hides the controls until they are needed, thus limiting the quantity of information displayed at one time. On the **Configuration** tab, the control labels consistently use bold 13-point application font, with units in parentheses in normal font. Also, a default value has been added to the path control.

Chapter 10 • Code Reviews

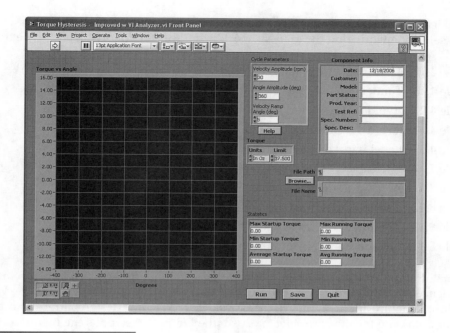

Figure 10-3A
Front panel window for Torque Hysteresis—Improved w VI Analyzer VI. Multiple style rules violations persist.

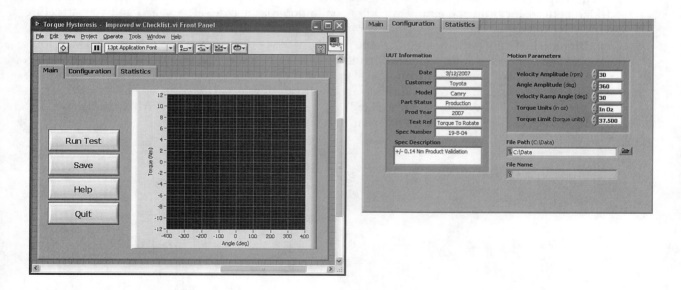

Figure 10-3B
The front panel is revised to incorporate additional improvements from the rules summary in Appendix B.

The block diagram for Torque Hysteresis—Improved with VI Analyzer VI is copied in Figure 10-4A. A self-review utilizing the manual checklist reveals the following additional rules violations:

Rule 4.11	Minimize wire bends; eliminate kinks and loops
Rule 4.13	Tunnel wires through left and right borders of structures
Rule 4.27	Avoid Sequence structures unless required
Rule 4.34	Avoid variables if wires are feasible
Rule 4.37	Label wires exiting the left shift register terminal
Rule 7.7	Use General Error Handler VI over Simple Error Handler VI
Rule 8.4	Avoid polling GUI objects
Rule 8.12	Use an enumerated type definition for the case selector
Rule 8.13	Minimize code external to the Case structure
Rule 8.14	Include states for Initialize, Idle, Shutdown, and Blank

The legacy diagram utilizes a design pattern that combines a Sequence structure and While Loop. The Sequence structure initializes several front panel controls and indicators. The While Loop contains a Case structure and is similar to the Classic State Machine design pattern, except that it utilizes an integer data type for the case selector. Also, the **Default** case polls the menu of Boolean controls and checks the **Units** element of the **Cycle** cluster for a value change. These features violate the rules for a proper state machine design pattern discussed in Chapter 8, "Design Patterns," Section 8.2, "State Machines."

The improved diagram shown in Figure 10-4B utilizes a proper Event-Driven State Machine design pattern, including an enumerated type definition for the case selector, and an Event structure within the **Idle** case. The state machine contains the standard states for **Initialize**, **Idle**, **Shutdown**, and **Blank**. The Event structure replaces the polling of the Boolean menu and **Torque** cluster. A local variable and shift register for propagating the **Units** are eliminated. The control initialization code is moved into the state machine's **Initialize** state, and the Sequence structure is eliminated. The state labels from the enumerated type definition are intuitive. All wires exiting the left shift register terminals are labeled. Finally, General Error Handler VI is used in place of Simple Error Handler VI. The resulting diagram is compact, readable, neat, and intuitive.

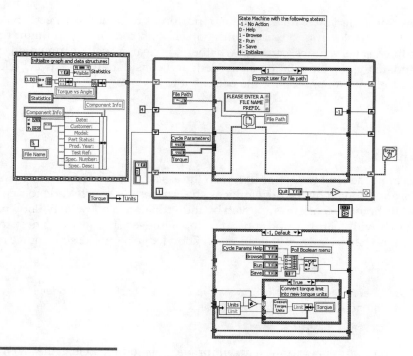

Figure 10-4A
Multiple style rule violations persist within the block diagram for Torque Hysteresis—Improved w VI Analyzer VI.

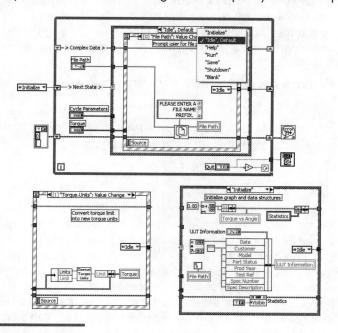

Figure 10-4B
The block diagram is revised to incorporate additional improvements from the rules summary in Appendix B, including a proper Event-Driven State Machine design pattern.

Self-reviews provide you with a specific opportunity to review, inspect, and refine your source code's style. This enables you to reacquaint yourself with code that you might have developed in a hurry or modified several times; consider whether it is clear, is well documented, and still makes logical sense; update the documentation as needed; and consider all the aspects related to style. Additionally, it is excellent preparation for the peer review.

10.2 Peer Reviews

Peer reviews are code reviews performed with the assistance of one or more peers. In practice, many coding issues are resolved by the developer as she *explains* her source code to a colleague. Also, some development techniques that are employed by individuals are not widely known. Indeed, LabVIEW is extremely rich with features, and most developers never stop learning new techniques. Both the developer and the participants can glean much knowledge by discussing source code with peers.

> ***Theorem 10.1:*** *Peer interaction is a powerful motivator. Consequently, peer reviews are an effective means of enforcing and evolving a style convention.*

Many developers are eager to learn and improve their coding style, either on their own or with the help of a mentor or peer group. It is common to reach a plateau at which a developer becomes comfortable with a particular design pattern or toolset; after that point, the developer is less likely to explore alternate styles. Peer interaction is a great motivational tool, and peer reviews are an excellent method of learning and improving development skills, including style. Moreover, peer reviews are an essential technique for enforcing a style convention within an organization.

 Rule 10.8   Perform at least one peer review per project

 Rule 10.9   Involve the following people in a peer review: project manager, lead developer, experienced peer, and an inexperienced peer

 Rule 10.10   Bring the requirements specification and style rules checklist to the peer review

The optimum number and frequency of peer reviews vary widely. Some factors to consider include the experience level of the developer or team, application size and complexity, and primary objectives of the peer review. If you are part of a multideveloper organization, always perform at least one peer review for each project. Involve the project manager, the lead developer, and experienced as well as inexperienced developers. Each brings different perspectives, and each contributes and benefits differently. The project manager understands the requirements from a functional perspective and can monitor how those requirements are being implemented. This can help identify any functional limitations of a specific software design. The lead developer is the software's primary architect and the direct beneficiary of the review. An experienced peer who is familiar with the organization's style convention provides constructive feedback. An inexperienced peer might contribute some new ideas and question existing methods, all while learning good style. Additionally, consider inviting a domain expert to critique any subsystems of your application that are highly specialized. For example, if a colleague is an expert in GUI design, ask her to participate in the review of your GUI. The same holds true for network communications, data acquisition, image analysis, real-time control, and more. Be sure to keep the requirements specification and style rules checklist on hand. These are important materials that are referenced throughout the meeting.

Begin with a review of the critical requirements, including priorities. Next, discuss the top-level architecture and justify the reason for it. This might include a presentation of a state diagram or other design method, followed by a top-level source code walk-through. Then discuss each of the application's subsystems, such as high-level software components and low-level device drivers. Every subsystem should be covered, including a discussion of the source code. Discuss the data structures and messaging or communication protocols used to connect each high-level component or subsystem. Also be sure to discuss the error handling strategy. Look for common mistakes, such as incomplete error trapping, missing VI descriptions, and unnecessary operations within loops.

Rule 10.11   Designate a neutral party to take notes on desired changes

Rule 10.12   Do not modify the functional source code during the peer review

It is best to designate a neutral party, not the lead developer or project manager, to take notes during the peer review. This way, the developers are free to concentrate on presenting their code and receiving feedback, and the meeting can progress quickly through the agenda. Also, a neutral party can maintain unbiased feedback and action items. For convenience, the developer might want to enter some free label comments within the source code, reminding her of recommended changes, similar to a bookmark. Include a search term within each comment, such as `review+<date>`, to quickly locate the comments after the review using the Find dialog. However, always avoid fixing or modifying your functional source code during the peer review. This slows down the meeting, while changing the level of perspective.

A peer review is performed on the Torque Hysteresis VI, subsequent to the automated and manual self-reviews. Referring to the front panel and block diagram in Figures 10-3B and 10-4B, the following additional rules violations were noticed by the peers that had gone undiscovered by the developer:

Rule 3.37	Set the Tabbing Order and use the Key Focus property to aid navigation
Rule 6.2	Choose memory efficient data types
Rule 6.27	Avoid clusters for interactive controls with dialog VIs
Rule 7.8	Implement an error log file for application deployment
Rule 7.9	Suppress dialog error reporting for unattended or remote operation
Rule 8.4	Avoid polling GUI objects

Additionally, several software components from the organization's software reuse library were recommended, including an industrial style two-button confirmation dialog, a close LabVIEW conditional VI, and an error logging routine. Diligent notes were taken, and the Torque Hysteresis VI was further improved after the code review, as shown in Figure 10-5.

Figure 10-5A contains the revised front panel. Program flow is improved by the addition of the **New UUT** Boolean control on the **Main** tab. This command button initiates a new test sequence, beginning with a dialog that prompts the user for the UUT Information. The Prompt UUT Information dialog VI uses large individual controls with tab key navigation, which simplifies ease of use compared to the corresponding cluster on the **Configuration** tab. Similarly, a Prompt Motion Params dialog VI prompts for the motion cycle parameters and updates the corresponding cluster on the **Configuration** tab.

Figure 10-5B contains the revised block diagram. Several data structures are propagated using shift registers near the top of the While Loop. Specifically, the UUT Information and Motion Parameters occupy separate shift registers instead of bundling the two together. This reduces the complexity of the data structure. Additionally, the Torque vs Angle and Statistics clusters are passed using shift registers. This reduces local variables and increases flexibility by having the data available in each case. The data structures and shift registers maximize data flow and memory efficiency.

Event cases have been added for **New UUT** and **Quit Value Change** events. The **New UUT Value Change** event case calls the dialogs for Prompt UUT Information and Prompt Motion Parameters. The **Quit Value Change** event replaces the polling of the **Quit** Boolean that was previously wired to the condition terminal at the bottom of the While Loop. Events are more efficient than polling. Also, a dialog VI for confirming before quitting is standard within the organization. Additionally, the General Error Handler VI is replaced with an error logging routine. This replaces the error dialog with an error log file that operates in the background. This error handling technique is preferred for a deployed application, to prevent an error dialog from confusing the user. Finally, the Quit LabVIEW function is replaced with a Close LabVIEW Conditional VI. This routine checks whether the VI is running as an executable before ending the LabVIEW session. The quit confirmation, error logging routine, and close LabVIEW conditional are standard VIs that are available from the organization's software reuse library. The experienced peer identified the opportunity for code reuse.

A brief visual comparison of the Torque Hysteresis VI before the code reviews, shown in Figure 10-2A, and the VI after improvements from all the three code reviews, shown in Figure 10-5, illustrates the substantial difference in software quality that can be achieved through code reviews. Additionally, code reviews increase developer skill level, depth of knowledgeable resources, and development productivity. Moreover, code reviews enforce the style conventions, which improve the ease of use, efficiency, readability, simplicity, performance, maintainability, and robustness of your LabVIEW applications.

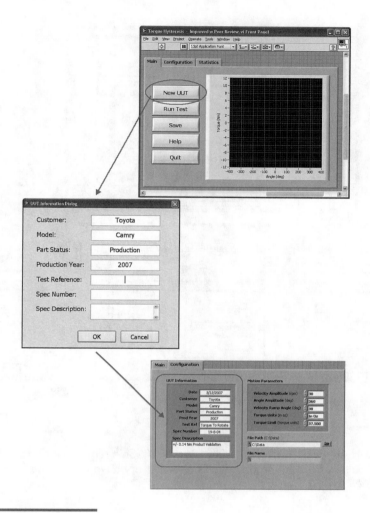

Figure 10-5A
Front panel window for Torque Hysteresis—Improved w Peer Review VI. Navigation of the GUI is improved via **New UUT** command button and dialogs that prompt the user for the configuration data, including the Prompt Information dialog.

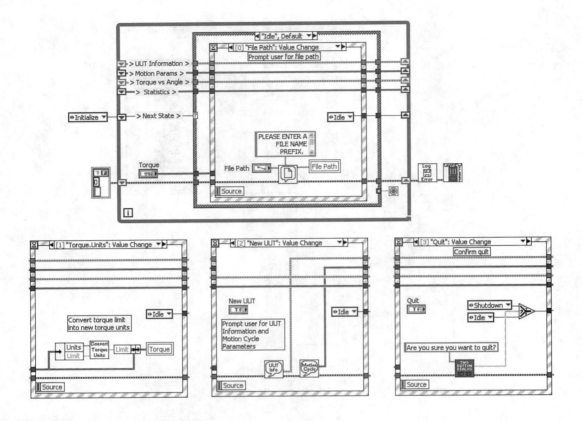

Figure 10-5B
Block diagram for Torque Hysteresis—Improved w Peer Review VI. Shift registers eliminate variables and promote data flow. Also, several routines from the organization's reuse library have been added for quit confirmation, error logging, and closing LabVIEW.

Endnotes

1. A custom test configuration file and a style checklist document in electronic form can be downloaded from www.bloomy.com/lvstyle.

   "Advice for VI-Based Code Reviews," NIWeek2006 presentation by Doug Norman of National Instruments.

Glossary

A

This glossary contains definitions of terms as applied within *The LabVIEW Style Book*. The terms are generally consistent with the LabVIEW documentation and appropriate software and virtual instrumentation industry standards. However, some of the terms are evolutionary, overlapping, or ambiguous. In some instances, I have taken the liberty to choose forms of the definitions that best support the relevant material in the book. The definitions remain consistent throughout the book.

A

application framework   An elaborate template filled in with multiple constructs, including design patterns, subVIs, data structures, and messaging.

application performance   The effectiveness with which an application or VI completes its intended mission.

artificial data dependency   A wiring assignment between two objects in which the receiving object does not actually utilize the data contained within the wire. Instead, the wire serves to specify the execution order of the objects via dataflow principles.

asynchronous loops   Two or more parallel looping structures that execute independently, without any timing or data-sharing relationship that would cause one loop to wait for another.

B

block diagram memory One of the four VI memory components, block diagram memory stores the graphical objects and images that comprise the block diagram. Block diagram memory usage is provided by the **Memory Usage** category of the VI Properties window selected by choosing **File»VI Properties**.

C

Call By Reference VI Server function that is used to execute a dynamically loaded subVI. The Dynamic Framework utilizes the Call By Reference to execute the high-level components of an application.

child panel The front panel of a component VI that is loaded into a subpanel control.

CHM See *Microsoft Compressed HTML Help*.

Classic State Machine Design pattern consisting of a Case structure within a While Loop that utilizes an enumerated data type for the case selector and a shift register for passing the next case selection between loop iterations. The cases are labeled intuitively based on the text items of the enumerated data type. The next case selection is determined programmatically.

Code and Fix Software lifecycle development model in which the source code is iteratively implemented, tested, and debugged, without any formal planning or design phase. This model is characterized by misunderstood requirements, inadequate architecture, sloppy source code, resulting in poor quality and prolonged development time and effort.

code memory or **code** One of the four VI memory components, code is the portion that contains the compiled source code. Code memory usage is provided by the **Memory Usage** category of the VI Properties window selected by choosing **File»VI Properties**.

code review Systematic method of reviewing source code quality for readability, maintainability, and robustness, utilizing one or more independent perspectives or reference points. Code reviews have three primary benefits: identifying and fixing problems, exchanging knowledge and ideas that improve developer skill level, and promoting and evolving standards across an organization.

cohesion A measure of how well the block diagram nodes within a module or subVI work together to perform a specific task. A subVI is considered cohesive if its purpose is clearly described using two or three sentences.

commercial subVI SubVI that is part of a product, such as a developer toolkit or instrument driver.

component VI High-level subVI or dynamically loaded plug-in VI that encapsulates a major portion or subsystem of the application.

compound design pattern Design pattern created using a combination of two or more single-loop design patterns and adding a messaging scheme to pass data between loops.

construct Any LabVIEW item that is constructed by the developer, such as data structures, subVIs, and design patterns. (See also *data construct*.)

Context Help LabVIEW Help window that displays the descriptions of controls and VIs as the user scrolls the mouse over them. The Context Help window opens via **Help»Show Context Help** or the **<Ctrl>+<H>** shortcut key combination.

Continuous Loop   Design pattern that consists of a single While Loop or Timed Loop, shift registers, loop timing, and error trapping. The design pattern is equally applicable to top-level VIs and subVIs.

custom control   Control that is customized using the Control Editor window, and saved to a CTL file. The instances of a custom control can be edited individually on the panel without affecting the appearance, properties, and data type of the source file. (See also *type definition* and *strict type definition*.)

D

data acquisition (DAQ)   Involves gathering signals from measurement sources and digitizing the signal for storage, analysis, and presentation. Data acquisition is often associated with the use of plug-in analog-to-digital converter boards.

data construct   A collection of one or more of the fundamental data types used to form a new data type. They include developer-defined constructs such as array, cluster, variant, variable, and queue, as well as built-in data types such as matrix, error, waveform, and dynamic.

data dependency   The use of the data flow principle to dictate the execution order of nodes and objects connected by one or more wires. Specifically, any node that receives data cannot execute until all nodes that source data have completed. The receiving node is data dependent on the sourcing nodes.

data flow   The fundamental principle of LabVIEW, in which data flows along wires from the output terminals of data sourcing nodes to the input terminals of data receiving nodes. A block diagram node executes when data is received at all wired input terminals. Upon completion, data is supplied to its output terminals and propagates to the next node in the dataflow path.

data memory   One of the four VI memory components, data memory contains all the data that flows through the diagram, as well as the diagram constants, default values for front panel controls, and the data that is copied when written to variables and front panel indicators. Data memory usage is provided by the **Memory Usage** category of the VI Properties window selected by choosing **File»VI Properties**. Additionally, the Profile Memory and Performance window, selected from **Tools»Profile»Performance and Memory**, gathers comprehensive memory statistics for the VI and its entire hierarchy.

data structure   Any LabVIEW data type or data construct utilized by an application for storing data. Data structures are defined by the developer's choice of controls, arrays, and clusters on the VI front panels, as well as the operations performed on the diagrams.

data type   The fundamental data elements depicted by unique terminal or wire styles. Data types define the memory size and functionality of the data.

datalog file   Binary file format that stores data as a sequence of identically structured records along with a data type descriptor. Use the functions on the **Datalog** palette to read and write to a datalog file. The record structure is defined by the data type of the cluster wired to the **record type** input terminal of the Open/Create/Replace Datalog function. Datalog files cannot be read or written outside of LabVIEW.

design patterns   Standard VI architectures that solve common software designs required in a wide variety of applications. They consist of an arrangement of structures, functions, controls, and error handling that forms a construct for common tasks such as looping, event handling, state transitioning, and data sharing and encapsulation.

desktop application   Any application that runs on a personal computer and is not intended for a real-time or embedded target.

desktop GUI VI A VI with a front panel designed for personal use in an office, private laboratory, or other individual computing environment. Desktop GUI VIs appear and behave similarly to other native applications on the computer's operating system.

developer Someone who creates custom software applications, which may entail any combination of low-level programming and high-level configuration.

development time The time required to develop, document, test, modify, and maintain an application throughout its entire life cycle.

Dialog function LabVIEW's native dialogs accessed from the **Dialog & User Interface** palette, including the One, Two, and Three Button Dialog functions, as well as Prompt for User Input and Display Message to User Express VIs. The LabVIEW dialog functions are simple to use and designed to resemble the dialogs native to the host computer's operating system.

dialog VIs GUI-related subVIs that open their front panels and prompt the user for information. Dialog VIs have far fewer subVIs and much less overall functionality than top-level VIs. Their purpose is to perform a specific transaction with the user.

Disk Operating System (DOS) The dominant operating system utilized by the IBM PC compatible throughout the 1980s, preceding Microsoft Windows.

documentation Any text-based description of the structure, components, or operation of a system, application, or source code. Documentation may include specifications, design documents, source code documentation, user manuals, and online help.

DOS See *Disk Operating System*.

dynamic A universal data type used with Express VIs. Designed for novice developers, dynamic is a very flexible data type that can store many types of data without learning about data types, data storage, and type conversion.

Dynamic Framework Application framework that utilizes VI server functions to dynamically load and run its high-level VIs, known as components.

E

ease of use The ease with which the end user operates an application to accomplish his or her objectives. This involves interacting with the application's graphical user interface (GUI).

efficiency The manner in which an application utilizes processor, memory, and input/output (I/O) resources. An efficient LabVIEW application executes quickly, without performing unnecessary operations, particularly ones that are performed repeatedly within looping structures. An efficient application also conserves memory by limiting the size of LabVIEW's four memory components: front panel, block diagram, data space, and code.

enumeration Also called an **enum**, a data type that provides a user-selectable list of strings that are associated with numeric values, similar to a text ring, except that the numeric values and string labels are part of the data type. Enums are very useful for presenting a discrete number of selections that are better described using text labels than numbers. Unlike ring controls, enums are always represented as unsigned integers, and the text selections are mapped to sequential numbers starting with 0.

error A failure of a function or VI to complete its programmed task. Most nodes in LabVIEW propagate error data using the **error in** and **error out** clusters.

Error Case Structure Case structure with an **error in** cluster terminal wired to the selector, resulting in cases for **Error** and **No Error**. The Error Case Structure is commonly used with subVIs.

error chain Method for trapping errors by propagating the error cluster among the nodes that have error terminals. A group of data-dependent nodes that execute sequentially comprises an error chain.

error cluster The **error in** and **error out** cluster controls, used extensively for error propagation.

error code The **code** control contained in an error cluster that is represented as a 32-bit signed integer. The **code** is used to uniquely identify the errors that functions and VIs generate. LabVIEW maintains an internal database for which error codes are associated to descriptions.

Error log file Error reporting technique that utilizes an ASCII or datalog file for logging error information to file.

Event-Driven State Machine Design pattern that combines a Classic State Machine, Event structure, and queue into a hybrid single-loop design pattern that performs GUI event handling as well as buffered state transitioning. The Event-Driven State Machine is appropriate for top-level and GUI VIs for applications of medium complexity.

Event-Handling Loop Design pattern that facilitates event-driven programming in LabVIEW. It consists of an Event structure within a While Loop. The Event structure contains an event subdiagram for each configured event and executes the subdiagram code when the event occurs. The Event-Handling Loop can process a wider variety of user interface activity in a more efficient manner than polling the controls within a continuous loop.

Event Machine Design pattern that consists of a single Event structure within a While Loop, with user events defined for each state, and subdiagrams configured for any combination of user and GUI events. The Event Machine is recommended for applications where multiple Event structures are required.

Explain Error A built-in utility that displays information about an error code in a dialog. It is invoked by selecting **Explain Error** from the shortcut menu from any error cluster or by selecting **Help»Explain Error**.

F

Field Programmable Gate Array (FPGA) A device that contains a matrix of reconfigurable gate array logic circuitry. When an FPGA is configured, the internal circuitry is connected in a way that creates a hardware implementation of the software application. Unlike processors, FPGAs use dedicated hardware for processing logic and do not have an operating system.

flexible sequencer A construct for executing a multiframe sequence implemented as a looped Case structure in which the execution order of the cases and stop condition are formed dynamically during execution. The Classic State Machine design pattern can be utilized to implement a flexible sequencer.

front panel memory One of the four VI memory components, front panel memory stores the graphical objects and images that comprise the front panel. Front panel memory usage is provided by the **Memory Usage** category of the VI Properties window selected by choosing **File»VI Properties**.

Functional Global Design pattern utilized for sharing data using subVIs instead of variables. It consists of a While Loop, a Case structure, an enumeration, and controls for reading and writing data. The While Loop contains one or more uninitialized shift registers, and the loop's conditional terminal is wired to a Boolean constant that forces it to stop after only one iteration. The loop's sole purpose is to store global data within the shift registers. The subVI is applied to each location of an application requiring access to the data, similar to a global variable.

G

glyph A recognizable graphical symbol, such as a public sign or traffic signal.

GUI VI Any VI with a user-viewable panel, including top-level and dialog VIs.

I

Immediate SubVI Design pattern that consists of the nodes that comprise the subVI and error trapping. There are no continuous loops, dialog windows, or GUI panels. Rather, an immediate subVI executes its code straight through completion, in the order in which the error cluster propagates through the nodes on the diagram.

industrial GUI VI VI with a front panel designed for multiple users to interface with industrial equipment. Industrial GUI VIs complement the industrial application and associated equipment, regardless of the computing platform upon which they are installed.

instrument prefix Very succinct text used to represent an instrument's manufacturer and model number in an instrument driver library. The first two or three characters of the prefix identify the manufacturer, and the following four or five characters identify the instrument model or family. The resulting six- to eight-character acronym is used in the VI or library name and appears on the icons throughout the driver.

I/O Input and output operations performed by nodes that make calls to device drivers, DLLs or shared libraries, the operating system, or any application or resource that is external to the LabVIEW environment, including remote instances of LabVIEW. I/O operations include all of the nodes available on the following palettes: **File I/O**, **Measurements I/O**, **Instrument I/O**, and **Data Communication**.

I/O names References to a specific instance of an open resource, such as an instrument or device. They function similarly to a pointer to a data structure describing the resource. I/O names provide an intuitive display of the current value. For example, VISA, IVI, and DAQmx Name Controls provide information about the specific hardware that they reference.

K

kink A very small bend in a wire.

L

LabVIEW 2 style globals Another name for **functional globals** because this was the only method of sharing data without data flow prior to the introduction of local and global variables in Version 3.0.

LabVIEW Advanced Virtual Architects (LAVA) Organization committed to the open and unbiased exchange of ideas on intermediate to advanced topics in LabVIEW. The URL is `www.lavausergroup.org`.

LabVIEW Tools Network An NI-sponsored site within the LabVIEW Zone that lists developer resources, including add-on products, books, tutorials, and more.

LAVA See *LabVIEW Advanced Virtual Architects*.

loop-subVI   Construct comprised of a loop that performs one of the application's principle tasks modularized into a subVI.

looped Case structure   Construct comprised of a Case structure embedded within a For Loop or While Loop. It functions similarly to a flexible alternative to a Sequence structure, or **Flexible Sequencer**, when the sequentially ordered code is placed in cases of the Case structure instead of the frames of the Sequence structure.

M

maintainable   The ease with which the source code is understood by other LabVIEW developers besides the author. The code can be edited to change or add new functionality.

matrix   Type definition consisting of two-dimensional array of either double-precision floating-point numbers (in a real matrix) or complex numbers (in a complex matrix). The matrix data types are used widely throughout the Linear Algebra VIs in the Mathematics library.

Microsoft Compressed HTML Help (CHM)   Standard help file format under Windows XP. CHM files are heavily indexed and include a hyperlinked table of contents that normally resides outside the main body text.

modal   GUI VI behavior in which the front panel window is locked on top of all open windows in the application until the interaction is completed and the window closes. Modal behavior is a default property setting with LabVIEW's **Dialog** window appearance, as selected from **File»VI Properties»Window Appearance**.

N

NaN   Abbreviation for Not a Number, a floating-point value that invalid operations produce, such as taking the square root of a negative number. The NaN constant is particularly useful for initializing floating-point numeric arrays because the NaN values are easily identified in a search algorithm and are not visible on a graph indicator.

nested data structure   Data construct containing multiple layers of arrays and clusters.

NI   National Instruments, the company that provides LabVIEW. The URL is www.ni.com.

NI Developer Zone   A forum for developers in need of LabVIEW resources, such as articles, example code, and support. Thousands of instrument drivers are available from **NI Developer Zone's** Instrument Driver Network (www.ni.com/idnet).

node   Any object on the block diagram that has inputs and/or outputs and performs operations when a VI runs. Nodes include functions, subVIs, and structures.

O

OpenG   An organized community committed to the development and use of open source LabVIEW tools. The URL is www.openg.org.

P

PDA See *Personal Digital Assistant*.

PDF See *Portable Document Format*.

peer review Code review performed with the assistance of one or more peers. Many bugs and coding issues are discovered by the developer as she explains her challenges to a colleague.

performance Very broad term that describes software effectiveness (see also *application performance*) or execution speed (see also *subVI performance*).

Personal computer (PC) Computer that runs a non-deterministic operating system, such as Microsoft Windows.

Personal Digital Assistant (PDA) An electronic device that can include some of the functionality of a computer, a cell phone, a music player, or a camera. PDAs are an alternative to bound notebooks for recording project specifications during a meeting.

plug-in VIs that are dynamically loaded by a Dynamic Framework, including both high-level components and low level subVIs.

polling Condition in which a loop continuously monitors a resource until it reaches a specific value or state. Avoid polling front panel terminals and variables, as there are more efficient alternatives.

polymorphism Term that describes a function or VI with one or more terminals that can accept more than one data type. Polymorphic nodes adapt to the input data type instead of breaking the wire or forcing a coercion to occur on the input terminal.

Portable Document Format (PDF) Format created by Adobe® Systems for representing two- and three-dimensional documents in a fixed-layout, cross-platform compatible file format. PDF files encode the exact look of a document in a device-independent manner.

Profile window Tool that monitors the data memory use and execution speed for a set of VIs loaded in memory. The Profile window is accessed from **Tools»Profile»Performance and Memory**.

programmer Someone who develops custom software applications using a conventional text-based language.

Q

queue Similar to a first-in, first-out buffer used to pass data between parallel loops without variables.

R

readability The ease with which developers can comprehend source code. Readability pertains to both the front panel and the block diagram.

real time An application or operating system that runs with precise and reliable timing, also known as deterministic.

Refnum Reference to a specific instance of an open resource, such as a file, instrument or device, network connection, image, LabVIEW application, VI, or control. Refnums function similarly to a pointer to a data structure describing the resource.

ring controls   Controls that map text selections to numeric values. They are very useful for presenting a discrete number of selections that are better described using text labels than numbers.

robust   Characteristic of a LabVIEW application if it performs reliable error handling and shuts down gracefully.

S

shift registers   Terminals on looping structure borders that shift data between loop iterations. They are functionally and conceptually similar to terminals that extend wires from the end of one iteration to the beginning of the next.

simple controls   Controls that are intuitive and easy to operate, and that represent simple data types. Simple controls possess properties that can be configured to help validate user and programmatic input.

simple data structures   Data structures that store data in contiguous memory addresses. They include all scalar data types, such as Boolean, numeric, and string; an array of Boolean and numeric; and clusters containing only the aforementioned simple data types.

simplicity   Inversely relates to the quantity of nodes and terminals that comprise an application. It is the opposite of complexity. The fewer front panel objects and block diagram nodes, the simpler the application.

standard subVI   User-developed subVI that is not part of a commercial product or toolkit.

state machine   The most popular LabVIEW design pattern of all time, consisting of a Case structure within a While Loop, with a shift register or messaging construct wired to the Case structure's selector terminal.

strict type definition   Custom control defined using the Control Editor window that maintains the exact appearance, properties, and data type for each instance of the strict type definition. (See also *type definition*.)

subVI   A VI that is called from the diagram of a higher level VI. In Chapters 3 and 8, subVIs are also assumed not to open their panels during program execution.

subVI performance   The execution speed of a subVI.

T

target   Any computing device upon which a LabVIEW VI can execute. Targets include personal computers, embedded controllers, PDAs, and FPGAs.

task   A set of operations normally encapsulated by a subVI or While Loop.

TBD   To be determined—Pertains to undefined requirements in a specification.

tight loop   A loop that executes at the highest possible rate, without any loop timing inefficiently utilizing CPU resources. A tight loop may unnecessarily monopolize the processor. Each call to a loop timing function—such as Wait (ms), Wait Until Next ms Multiple, and the Time Delay Express VI—forces the execution thread to yield control of the processor to allow other tasks to run.

time stamp   Data type that stores an absolute time with very high precision using separate 8-byte fields for seconds and fractions of a second.

top-level VI VI that resides at the highest level of the application hierarchy, having a front panel that comprises the primary display screen.

type definition Also know as **typedef**, a custom control defined using the Control Editor window that maintains its data type information in a CTL file. When the data type of a type definition changes, all instances of the type definition automatically update. See also *custom control* and *strict type definition*.

U

unit under test (UUT) The objective of many LabVIEW applications is to validate or test UUTs. UUTs are also commonly referred to as devices under test, or DUTs.

user event Custom developer-defined event that is processed by an Event structure's corresponding Event case. User events are created, registered, and fired programmatically using the functions on the **Events** palette, located under **Programming»Dialog&User Interface**.

utility VI SubVI that performs a low-level task that complements or extends the capabilities of the built-in LabVIEW functions.

UUT See *unit under test*.

V

variant Self-describing data type that encodes the data name, data type, data, and attributes or information about the data into a generic format. Conceptually, variant can be thought of as a wrapper that converts any data into a new format that is described in a universal manner.

vi.lib Folder containing all of LabVIEW's standard shipping VIs provided on the palettes.

VI Analyzer Toolkit An add-on product from NI that automatically inspects VIs for style and performance. The VI Analyzer contains more than sixty tests that can be configured and run interactively or programmatically. The VI Analyzer Toolkit is very useful for performing self-reviews.

W

warning An error cluster value consisting of a nonzero **code** and FALSE **status**. Warnings are considered to be similar to, but less severe than, an error.

waveform data type (WDT) A special type of cluster, consisting of three elements and attributes. The elements include a start time stamp (t_0), a time interval between data points (**dt**), and a one-dimensional array of numbers representing the samples of an analog or digital waveform (**Y**). The attributes can contain any number of developer-specified name and value pairs.

white space Empty space on a GUI window.

Style Rules Summary

B

This appendix lists all the style rules presented throughout *The LabVIEW Style Book*. Use this as a manual checklist when performing code reviews, as discussed in section 10.1.2, "Manual Checklist," of Chapter 10, "Code Reviews." You can also use it as a quick look-up reference when discussing style with peers. An electronic copy is available from www.blooomy.com/lvstyle.

Chapter 2

	2.1	Maintain a LabVIEW project journal
	2.2	Write a requirements specification document
	2.3	Maintain good LabVIEW style throughout the proof of concepts
	2.4	Document your LabVIEW options and back up the LabVIEW.ini file
	2.5	Develop reusable SubVIs
	2.6	Make reusable libraries accessible from the LabVIEW palettes
	2.7	Place reusable templates in the LabVIEW\templates folder
	2.8	Maintain an organized repository on disk
	2.9	Create an LabVIEW source folder hierarchy that reflects your application's architecture
	2.10	Create the folder hierarchy before you begin coding
	2.11	Organize LabVIEW source files into cohesive project libraries, where appropriate
	2.12	Create unique and intuitive source filenames
	2.13	Do not abbreviate filenames

2.14　Never use LabVIEW's default filenames
2.15　Identify the top-level VIs
2.16　Follow your organization's CM Rules
2.17　Avoid moving source files on disk

Chapter 3

3.1　Group related controls using decorations, spacing, tabs, and clusters
3.2　Apply symmetry and spacing to front panel objects
3.3　Size similar objects the same
3.4　Maximize the top-level VI panels for industrial applications
3.5　Size dialog VI panels much less than full screen
3.6　Center dialog VI panels
3.7　Use LabVIEW's dialogs for desktop applications, avoid them for industrial applications
3.8　Use system controls for desktop dialogs; use 3D controls for industrial dialogs
3.9　Enlarge and center the objects of an industrial GUI VI in proportion to their importance
3.10　Limit the quantity of information displayed on a GUI VI panel
3.11　Avoid overlapping visible objects
3.12　Hide the toolbar
3.13　Include your company logo for a professional appearance
3.14　Use default appearance for subVI panels, objects, and most text
3.15　Arrange controls to resemble the connector assignments
3.16　Resize the panel for a snug fit
3.17　Minimize front panel text
3.18　Delete template instructions immediately after edits are performed
3.19　Apply consistent fonts and capitalization
3.20　Choose only one font, and vary the size, boldness, and color to obtain multiple styles
3.21　Use succinct, intuitive control labels and embedded text
3.22　Apply 13-point black Application font for most subVI panels
3.23　Provide default values and units in parentheses at the end of owned labels
3.24　Combine bold text labels with plain text parentheses, for control labels of commercial subVIs
3.25　Maximize the contrast between text color and background color
3.26　Use large text size for command buttons and critical data
3.27　Allow extra space between labels and objects for multiplatform applications
3.28　Apply color judiciously
3.29　Create a color theme
3.30　Follow universal conventions for green, yellow, and red
3.31　Leave the panel and objects of subVIs gray
3.32　Keep the color schemes simple and time limited
3.33　Limit the number of controls that are visible and enabled at any one time
3.34　Restrict the range of all controls to values that are relevant to the application
3.35　Set the Data Range property of numeric controls
3.36　Use ring or enumerated controls over string controls when feasible
3.37　Set the Tabbing Order and use the Key Focus property to aid navigation
3.38　Customize the runtime menus for top-level VIs
3.39　Always include a Help menu or button
3.40　Be consistent!

Chapter 4

4.1	Use 1280 × 1024 display resolution
4.2	Leave the background color white
4.3	Use a high object density
4.4	Limit the diagram size to one visible screen, or limit scrolling to one direction
4.5	Create a multilayer hierarchy of subVIs
4.6	Modularize top-level diagrams with subVIs
4.7	Modularize the high-level subVIs with lower level subVIs
4.8	Do not create subVIs just to save space
4.9	Avoid trivial subVIs containing few nodes
4.10	Create a meaningful icon and cohesive description for every subVI!
4.11	Minimize wire bends; eliminate kinks and loops
4.12	Maintain even spacing of parallel wires
4.13	Tunnel wires through left and right borders of structures
4.14	Do not wire through structures unnecessarily
4.15	Never obstruct the view of wires and nodes
4.16	Limit wire lengths such that source and destination are visible on one screen
4.17	Never use local and global variables for wiring convenience
4.18	Label long wires and wires from hidden source terminals
4.19	Place unwired front panel terminals in a consistent location
4.20	Modularize wires of related data into clusters
4.21	Save clusters as type definitions
4.22	Always flow data from left to right
4.23	Propagate the error cluster
4.24	Avoid array and cluster coercions
4.25	Create controls and constants from a terminal's context menu
4.26	Disable dots at wire junctions
4.27	Avoid Sequence structures unless required
4.28	Avoid nesting beyond three layers
4.29	Use write local variables for initializing control values
4.30	Use global variables for simple data sharing between parallel loops or VIs
4.31	Use a Sequence structure to order operations if no data dependency exists
4.32	Use only Flat Sequence structures when required
4.33	Avoid polling variables within continuous loops
4.34	Avoid variables if wires are feasible
4.35	Use shift registers over local and global variables
4.36	Group most shift registers near the top of the loop
4.37	Label wires exiting the left shift register terminal
4.38	Use looped Case structures over Sequence structures

Chapter 5

5.1	Have fun creating icons
5.2	Create a unique and meaningful icon for every VI
5.3	Never use LabVIEW's default icons

	5.4	Save VIs with subVI icons visible instead of terminals
	5.5	Use a black border
	5.6	Combine a glyph with color and text for best style
	5.7	Choose universally recognized glyphs
	5.8	Use 8- or 10-point small fonts for most text
	5.9	Choose a unified style for related VIs
	5.10	Budget your time in proportion to intended VI reuse
	5.11	Use textual foreground, colored background, and a black border for fastest results
	5.12	Contrast the text and background colors
	5.13	Choose a color convention for the type of VI
	5.14	Create an icon template for related VIs
	5.15	Reuse one glyph, color scheme, and font for related VIs
	5.16	Copy graphics
	5.17	Avoid text and graphics not universally understood on international icons
	5.18	Choose connectors and terminal assignments that promote clear wiring and proper data flow
	5.19	Select a pattern with extra terminals
	5.20	Choose a unified pattern for related VIs
	5.21	Use the 4×2×2×4 connector pattern for most VIs
	5.22	Assign controls to left terminals, indicators to right terminals
	5.23	Never cross wire stubs in the Context Help window
	5.24	Specify terminal assignments resembling the panel layout
	5.25	Assign error clusters to bottom left and right terminals
	5.26	Assign references and I/O names to top left and right terminals
	5.27	Choose left and right vertical edge connector terminals for high priority inputs and outputs
	5.28	Choose top and bottom horizontal edge connector terminals for lower priority inputs and outputs
	5.29	Specify required priority for critical inputs and outputs
	5.30	Specify optional priority for inputs and outputs that are normally not used

Chapter 6

	6.1	Choose controls that simplify the operation of the panel
	6.2	Choose memory efficient data types
	6.3	Choose controls and data types that facilitate consistent data structures throughout an application
	6.4	Configure an appropriate default value for each control
	6.5	Enter control descriptions
	6.6	Save custom controls as strict type definitions
	6.7	Create arrays and clusters that associate related data
	6.8	Use Booleans if two states are logical opposites
	6.9	Assign names that identify the TRUE and FALSE value behavior
	6.10	Use command buttons for action, slide switches for parameter settings
	6.11	Label the TRUE and FALSE states of slide and toggle switches
	6.12	Avoid using buttons or switches as indicators, and LEDs as controls
	6.13	Use I32 representation for integers and DBL for floating-point numbers
	6.14	Use automatic formatting unless a specific format is required

	6.15	Show radix for hex, octal, or binary data
	6.16	Use enums liberally throughout your applications
	6.17	Save enums as type definitions
	6.18	Avoid string controls on GUI VI panels unless required
	6.19	Use enum, ring, and path controls in place of string controls where possible
	6.20	Keep the Browse button visible for path controls on GUI VI panels
	6.21	Use arrays for multivalued data items; use clusters for grouping multiple distinct items
	6.22	Use arrays to store large or dynamic length data sets
	6.23	Enter descriptions for array and cluster shells and control elements
	6.24	Use alignment tools to keep clusters neat and compact
	6.25	Save all clusters as type definitions
	6.26	Always use Bundle and Unbundle By Name
	6.27	Avoid clusters for interactive controls with Dialog VIs
	6.28	Organize complex data using nested data structures
	6.29	Avoid manipulating nested data during critical tasks
	6.30	Limit the size of arrays by initializing to maximum length

Chapter 7

	7.1	All VIs must `trap` and `report` the errors returned from error terminals
	7.2	Trap errors by propagating the error cluster among the error terminals
	7.3	Trap errors from all iterations of loops
	7.4	Disable indexing of errors with continuous loops
	7.5	Trap all errors from all nodes that have error terminals
	7.6	Report errors using a dialog and/or log file
	7.7	Use General Error Handler VI over Simple Error Handler VI
	7.8	Implement an error log file for application deployment
	7.9	Suppress dialog error reporting for unattended or remote operation
	7.10	Avoid subVIs with built-in error reporting
	7.11	Maintain user-defined error codes within an XML file
	7.12	Use negative codes for I/O device errors, and positive codes for warnings
	7.13	Skip most subVI diagrams on error using an Error Case Structure
	7.14	Use unwired defaults over constants for output tunnels of Error case
	7.15	Use the SubVI with Error Handling template
	7.16	Error trapping is **required** for nodes that perform I/O operations, **recommended** for nodes that contain error terminals, and **optional** for diagrams that do not contain nodes with error terminals
	7.17	Tunnel the error cluster near the bottom of structures
	7.18	Leave automatic error handling disabled

Chapter 8

	8.1	Use multiple criteria for the loop condition
	8.2	Use a Timed Loop for highly precise or complex timing, and use a While Loop otherwise
	8.3	Include a delay within continuous While Loops

	8.4	Avoid polling GUI objects
	8.5	Use the Value Change event for most GUI controls
	8.6	Place control terminals within their Value Change event case
	8.7	Resize the Event Data Node to hide unused terminals
	8.8	Avoid continuous timeout events
	8.9	Use a state machine design pattern in most VIs of medium or greater complexity
	8.10	Derive the application's primary states from the specification or design document
	8.11	Divide the primary states into additional states
	8.12	Use an enumerated type definition for the case selector
	8.13	Minimize code external to the Case structure
	8.14	Include states for Initialize, Idle, Shutdown, and Blank
	8.15	Avoid timeout with the Enqueue and Dequeue Element
	8.16	Use queues, shared variables, or RT FIFOs for parallel loop messaging
	8.17	Prioritize loops using delays or thread priorities
	8.18	Size parallel loops to the same width and align vertically
	8.19	Minimize space between loops
	8.20	Label each loop in the top left corner
	8.21	Use the Call By Reference Node over the Run method
	8.22	Choose standard connector terminal assignments
	8.23	Maintain all components within a dedicated directory
	8.24	Pass data between components with shift registers
	8.25	Assign one input and one output terminal as type variant
	8.26	Display plug-in panels within subpanels
	8.27	Create a parallel loop for each cohesive parallel task
	8.28	Include loops for Event Handling, Main State Machine, Hardware I/O, and Error Handling
	8.29	Pass control references into subVIs via a type-defined cluster
	8.30	Avoid subVI from selection with continuous loops
	8.31	Keep the Event-Handling Loop at the top level
	8.32	Keep high speed display updates at the top level

Chapter 9

	9.1	Keep front panel terminal labels visible on the diagram
	9.2	Apply free label comments in select locations
	9.3	Label every subdiagram of every multidiagram structure
	9.4	Label algorithms, constants, and Call Library Function Nodes
	9.5	Use default 13-point plain black application font for all diagram text
	9.6	Leave notes for the development team
	9.7	Use enumerated data types with Case structures
	9.8	Create the icons and descriptions as you develop your source code
	9.9	Provide online documentation with deployed applications

Chapter 10

10.1	Enforce your organization's style convention using code reviews
10.2	Utilize a combination of self-reviews and peer reviews for best results
10.3	Perform a self-review prior to every peer review
10.4	Use the VI Analyzer to automate tedious inspections
10.5	Customize the VI Analyzer's test criteria
10.6	Customize the rank for each test according to the priority of each rule
10.7	Use a manual checklist to perform a comprehensive style review
10.8	Perform at least one peer review per project
10.9	Involve the following people in a peer review: project manager, lead developer, experienced peer, and an inexperienced peer
10.10	Bring the requirements specification and style rules checklist to the peer review
10.11	Designate a neutral party to take notes on desired changes
10.12	Do not modify the functional source code during the peer review

Index

A

abbreviated filenames, 44
Account buttons, 66
Acquire global variables, 105, 109
acquisition section (requirements
 specifications), 26
algorithms, labels, 305
Align Objects menu, 50, 94
aligning
 dialog VI panels, 51
 objects, 50
Although Generate Report VI, 171
AML (Assistance Markup Language), 312
analysis section (requirements specifications), 26
Analyze Data events, 267
applications
 Capacitor Test & Sort, 79
 data types, 165-169
 dialogs, 51-52
 enums, 306

frameworks, 239, 272
 Dynamic Framework, 272-278
 examples, 287-297
 Modular Multiple-Loop Framework, 283-286
 Multiple-Loop Framework, 278-282
Left to Right VI, 121-122
online documentation, 314
optical filter test, 127-128
performance, 17
Right to Left VI, 120
Spectralyzer, 82-83
applying
 controls, 52
 dialogs, 51-52
 fonts, 62
arrays
 descriptions, 301
 simple, 181-184
assigning
 error clusters, 144
 references, 144
 terminals, 145-154

357

Assistance Markup Language (AML), 312
automatic error handling, 230
avoiding overlapping objects, 57-58

B

background icons, 135
backups, LabVIEW.ini file, 33
Backwards Wire failures, 323-325
banners, 148
best practice requirement specifications, 24-25
binary notation, 176
black borders
  applying, 132
  selecting, 135
Blank state, 256
block diagrams, 87
  Centrifuge DAQ VIs, 122-123
  data flow, 101-111, 114
  documentation, 304-308
  Excessively Nested VIs, 117
  Haphazard VIs, 119-120
  LabVIEW environment options, 32
  layout, 88-93
  maintainability, 11
  readability, 9-11
  Screw Inspection VIs, 124-126
  wiring, 95-101
Bloomy Controls, 147
Boolean data types, 173-174
borders, applying, 132
Bottom Edges alignment tools, 95
Browse buttons, 180
bugs. *See* error handling
Bundle By Name function, 183, 308
business objectives, 22
buttons
  Account, 66
  Help, 301-304

C

Call By Reference Nodes, 274
Call Library Function Nodes, labeling, 305
Capacitor Test & Sort applications, 79
capitalization, applying consistent, 62
centering dialog VI panels, 51
Centrifuge DAQ System VIs, 199-200
Centrifuge DAQ VIs, 122-123, 292
centrifuge data acquisition (DAQ), 81
CHM (Microsoft Compressed HTML Help), 312
Classic State Machine design pattern, 257-258
Clear Mode, 228
clearing errors, 228
Cluster Wired tests, 323
clusters
  errors, 144
  modularization, 97-101
  multiple distinct items, grouping, 308
  shells, 301
  simple arrays and, 181-184
CM (configuration management), 21
CM (content management), 46. *See also* source control
Code and Fix software lifecycle development model, 23
code
  errors, 214, 217
  reuse, 34-39
  reviews, 18, 319
    peer reviews, 334-336
    self reviews, 320-329
  source. *See* source code
Coercion Dots test, 322
colors
  front panels, 69, 71
  icons, selecting, 135
  reusing, 136
Comment Usage errors, 328
comments, applying, 305
commercial subVIs, 48

complex application frameworks, 272
  Dynamic Framework, 272-278
  Modular Multiple-Loop Framework, 283-286
  Multiple-Loop Framework, 278-282
Complexity, 16. *See also* simplicity
components, 90, 272
compound design patterns, 267-272
Compute Statistics VI, 171
conditions, race, 244
configuration management (CM), 21
Configuration tab, 336
configuring
  data constructs, 170, 172
  Data Range property, 72
  data structure properties, 169
  LabVIEW environment, 32
    code reuse, 34-39
    LabVIEW Options dialog box, 32-33
  resolution, 88
  tabbing order, 73
connector panes, 140-148, 150-154
consistency, 74, 165-169
constants, labels, 305
Cont Acq&Graph Voltage-To File(Binary) VI, 230-231
content management. *See* CM (content management)
continuous acquire to file, 230-231
Continuous Loop design patterns, 246-250
Control Alignment tests, 323
Control Editor, 169
controls
  code errors, 214, 217
  data types, 158-172
  descriptions, 301
  grouping, 49
  Interval, 59
  labels, 66-67
  managing, 59
  Mode, 59
  Pause, 59
  scope, 71-73

Controls palettes, 36
converting LLBs to project libraries, 44
copying graphics, 138-139
Curve local variables, 108
custom palettes, 33
customizing run-time menus. *See also* configuring

D

DAQmx Task, 146
Data Communication palette, 223
data constructs, 157, 180-184
  creating, 170-172
  nested, 187-192
  special, 186
data dependency, 106
data flow, 101, 103-114
Data Range property, configuring, 72
data structures, 157
  design methodology, 158, 160-172
  examples of, 193-197
  properties, 169
data types, 157
  simple, 172
    Boolean, 173-174
    numeric, 175-176
    path, 179-180
    picture, 179-180
    special numeric, 177-179
    strings, 179-180
decimal notation, 176
default appearance, 59
default filenames, 44-45
default values, labels, 301
Definition 1.1 (development time), 18
deleting template instructions, 62, 301
Dequeue Element function, 262
descriptions, VIs, 311

design, 29
  data structures, 158, 160-172
  patterns, 239, 241
    Classic State Machine, 257-258
    compound, 267-272
    Continuous Loop, 246-250
    Event Machine, 265-267
    Event-Driven State Machine, 262-264
    Event-Handling Loop, 250-254
    examples, 287-297
    Functional Global, 244-245
    Immediate SubVI, 241-244
    Queued State Machine, 260-262
    state machines, 254-256
  proof of concept, 30-31
  resources for, 30
  revision of specifications, 31
  templates. *See* templates
desktop applications, applying dialogs, 51-52
desktop GUI VIs, 48
development style. *See* style
development time. *See* time
diagrams. *See* block diagrams
Dialog & User Interface palettes, 51
Dialog Controls tests, 323
Dialog Using Events templates, 37
dialog VIs, 47, 51
Dialog Window Appearance, 53
dialogs
  applying, 51-52
  controls, 52
Digital Inputs, 146
disk organization, 40
  file repositories, 40
  folder hierarchies, 40-42
Distribute Objects menu, 50
documentation
  block diagrams, 304-308
  examples of, 315, 318
  front panels, 301-304
  icons, 311
  online, 311-314
  tests, 321
  VI descriptions, 19, 311

double precision (DBL) representation, 175
dynamic data constructs, 187
dynamic event terminals, 266
Dynamic Framework, 272-278
dynamic loaders, 272

E

ease of use, 6-7
efficiency, 7-9, 160-164
Elapsed Time VI, 287
embedding text, 301
enlarging objects, 52-57
Enqueue Element function, 262
entering control descriptions, 301
enums
  applying, 306
  saving, 307
Equation 1.1 (modularity index), 13
Error case, 218
Error Cluster Wired failures, 323-326
errors
  clusters, 144, 218
  error handling, 204
    automatic, 230
    clearing errors, 228
    error codes, 214-217
    examples, 230-238
    merging errors, 226-227
    prioritizing errors, 222-225
    reporting errors, 211-214
    robustness, 12-15
    structure wiring, 226
    subVI, 217-222
    trapping errors, 205-209
    terminals, 109, 218-222, 323-326
evaluation tools for style, 18
Event Machine design patterns, 265-267
Event-Driven State Machine design patterns, 262-264
Event-Handling Loop design patterns, 250-254

Index

events
  efficiency, 8
  registration, 266
  source terminals, 266
  Timeout, 251
  users, 266
  Value Change, 250
Excessively Nested VIs, 117
Execution category, 230
Exit Boolean
  controls, 109
  text, 63
Exit Value Change events, 105-109
Express Input subpalettes, 223

F

Fetch Waveform VI, 167
File Dialog functions, 326
File I/O palette, 36
File Path indicators, 120
files
  naming conventions, 44-45
  repositories, 40
Filled Rectangle tool, 135-136
Flexible Sequencers, 112
folder hierarchies, 40-42
fonts
  applying, 62
  icons, 133
  reusing, 136
  selecting, 63-65
foreground icons, 135
formatting. *See also* design
  block diagrams, 88-93
  data constructs, 170-172
  front panels, 48
    aligning objects, 50
    applying controls, 52
    applying dialogs, 51-52
    avoiding overlapping objects, 57-58
    centering dialog VI panels, 51
    enlarging objects, 52-57
    grouping controls, 49
    importing images, 59
    maximizing top-level VIs, 51
    navigating GUIs, 71-73
    sizing objects, 50-51
    text, 61-71
  icons, 132
frameworks
  applications, 239, 272
    Dynamic Framework, 272-278
    examples, 287-297
    Modular Multiple-Loop Framework, 283-286
    Multiple-Loop Framework, 278-282
front panels
  colors, 69, 71
  documentation, 301-304
  GUI navigation, 71-73
  LabVIEW environment options, 32
  layout, 48
    aligning objects, 50
    applying controls, 52
    applying dialogs, 51-52
    avoiding overlapping objects, 57-58
    centering dialog VI panels, 51
    enlarging objects, 52-57
    grouping controls, 49
    importing images, 59
    maximizing top-level VIs, 51
    sizing objects, 50-51
  maintainability, 11
  readability, 9-11
  tests, 321
  text, 61-69
Function parameters, 177
Functional Global design pattern, 244-245
functions
  Bundle By Name, 308-310
  Dequeue Element, 262
  Enqueue Element, 262
  error handling, 204
    automatic, 230
    clearing errors, 228

error codes, 214, 217
  examples, 230-238
  merging errors, 226-227
  prioritizing errors, 222-225
  reporting errors, 211-214
  structure wiring, 226
  subVIs, 217, 220-222
  trapping errors, 205-209
Two Button Dialog, 52
Unbundle By Name, 308-310
Functions palette, 35, 150

G–H

General Error Handler VI, 210-211
Get Raw I16 Waveform VI, 166
global design patterns, 244-245
glyphs
  reusing, 136
  selecting, 133
graphical user interfaces. *See* GUIs
graphics, copying, 138-139
grouping controls, 49
GUIs (graphical user interfaces), 6, 47
  ease of use, 6-7
  front panels, navigating, 71-73
guidelines. *See* rules

Hamburger, Bob, 147
Haphazard VIs, 119-120
Help menus, 301-304
hexadecimal notation, 176
hiding
  labels, 307
  toolbars, 58
HTML (HyperText Markup Language), 312

I–J

I/O (input/output), 146
  Datalog palettes, 212
  resource efficiency, 7-9

icons, 132-135, 311
  creating, 311
  examples, 145-154
  international, 139-140
  shortcuts, 135-139
ID Query controls, 148
Idle state, 256-257
images, importing, 59
Immediate SubVI design patterns, 241-244
importing images, 59
industrial applications, 51-52
industrial GUI VIs, 48, 69
Initialize case, 113, 236
Initialize state, 256
input/output. *See* I/O
instrument drivers
  code reuse, 34
  file-naming conventions, 45
instrument prefixes, 148
interfaces
  connector panes, 140-145
  GUIs. *See* GUIs
  Parafoil Guidance Interfaces, 83-84
international icons, 139-140
Interval controls, 59
intuitive control labels, applying, 301

K–L

labels
  algorithms, 305
  comments, 305
  constants, 305
  controls, 66-67
  default values, 301
  hiding, 307
  loops, 305
  subdiagrams, 305
  terminals, 304
  wiring, 305

LabVIEW environments
  configuring, 32
    code reuse, 34-39
    LabVIEW Options dialog box, 32-33
  GUI Essential Techniques, 48
  .ini file backups, 33
  Options dialog box, 32-33
  project requirements specifications, 25-28
  Tools Network website, 30
LAVA website, 30
layouts
  block diagrams, 88-93
  front panels, 48
    aligning objects, 50
    applying controls, 52
    applying dialogs, 51-52
    avoiding overlapping objects, 57-58
    centering dialog VI panels, 51
    enlarging objects, 52-57
    grouping controls, 49
    importing images, 59
    maximizing top-level VIs, 51
    sizing objects, 50-51
Left Audio Output indicator, 53
Left to Right VI, 121-122
libraries
  code reuse, 35-37
  LLBs
    converting to project libraries, 44
    project libraries versus, 44
  project libraries, 43-44
Load Script case, 236
Log Data case, 120
Log Interval labels, 63
looped Case structures, 111-114
loops
  Continuous Loop design patterns, 246-250
  efficiency, 7-8
  Event-Handling Loop design patterns, 250-254
  labels, 305
  Modular Multiple-Loop Framework, 283-286
  Multiple-Loop Framework, 278-282

  parallel, 269-272
  tight, 279

M

Main tab, 336
maintainability, 11
managing controls, 59
manual checklists, 330-334
Manual Resolution parameters, 177
maximizing top-level VIs, 51
memory
  data types, 160-164
  efficiency, 7-9
Memory Usage VI properties page, 18
menus, Help, 301-304
Merge Errors VI, 232
merging errors, 226-227
message output terminals, 212
Meticulous VIs, 2, 6
  ease of use, 6
  efficiency, 8
  error handling, 13
  maintainability, 11
  modularity index, 13
  performance, 17
  readability, 9
  simplicity, 16
Microsoft Assistance Markup Language (AML), 312
Microsoft Compressed HTML Help (CHM), 312
minimizing text, 62
Mode controls, 59
modifying colors, 69
Modular Multiple-Loop Framework, 283-286
modularization
  clusters, 97, 100-101
  indexes, 13
  subVIs, 90
moving source files, 46
Multiple-Loop Framework, 278-282
multivalued data types, applying, 308

N

naming conventions. *See* file naming conventions
navigation, GUI, 71-74
nested data constructs, 187-192
nested structures, efficiency, 9
Nested VIs, 4
　ease of use, 6
　efficiency, 7-9
　error handling, 14
　maintainability, 12
　modularity indexes, 13
　performance, 17
　readability, 10
　simplicity, 16
New UUT Boolean controls, 336
New UUT command buttons, 337
New UUT events, 336
NI Developer Zone website, 30
NI Example Finder, 30
NI Instrument Driver Guidelines, 148-151
No Error case, 218
nodes, simplicity, 15-17
notification output terminals, 109
numeric data types, 175-176
Numeric palettes, 223

O

objects
　aligning, 50
　default appearance, 59
　enlarging, 52-57
　overlapping, 57-58
　sizing, 50-51
online documentation, 311-314
OpenG
　variants, 194-197
　website, 30
optical filter test application, 127-128
optimizing data flow, 111, 114

optional priority, 144
output. *See* I/O
Output subpalettes, 223
overlapping objects, 57-59
Overview tab, 56

P

palettes
　code reuse, 35-37
　custom palettes, 33
　Dialog & User Interface, 51
　System, 53
panels
　default appearance, 59
　front. *See* front panels
　resizing, 59
　subVI, 48, 68
Parafoil Guidance Interfaces, 83-84
parallel loops, 269-272
Parse and Transmit Dialog VIs, 78
Parse and Transmit VIs, 77
path data types, 179-180
patterns
　design, 239, 241
　　Classic State Machines, 257-258
　　compound, 267-272
　　Continuous Loop, 246-250
　　Event Machine, 265-267
　　Event-Driven State Machines, 262-264
　　Event-Handling Loop, 250-254
　　examples, 287-297
　　Functional Global, 244-245
　　Immediate SubVI, 241-244
　　Queued State Machines, 260-262
　　state machines, 254-256
　terminals, 140
　VIs, 141
Pause controls, 59
PDF (Portable Document Format), 311
peer reviews, 334-336
performance, 17-18

Index

picture data types, 179-180
Pixel Tolerance parameter, 323
plug-ins, 272
Poll Instrument Response VI, 288
polling, 108
polymorphism, 175
Portable Document Format (PDF), 312
Power Monitor System VI, 54-57
prefixes, instruments, 148
preparation
  design documentation, 29
    proof of concept, 30-31
    resources for, 30
    revision of specifications, 31
  disk organization, 40
    file repository, 40
    folder hierarchy, 40-42
  file-naming conventions, 39-45
  project organization, 39-44
  requirements specifications, 22-24
    best practices, 24-25
    projects, 25-28
  source controls, 39-46
preparations, 32
  code reuse, 34-39
  LabVIEW Options dialog box, 32-33
presentation section (requirements specifications), 28
priority
  errors, 222-225
  inputs/outputs, 144
  requirements specifications, 28
processor resources, efficiency, 7-9
Profile Performance and Memory window, 7, 17-18
Project Explorer, 42
projects
  journals, 24-25
  libraries, 43-44
  organization, 39-43
  requirements specifications, 25-28
proof of concept, 30-31

properties
  data structures, 169
  Visible, 222

Q-R

Queued State Machine design pattern, 113, 260-262
Quit Value Change event, 114, 336

race conditions, 244
radix, visible, 176
rapid code development, 22
Read Waveform VIs, 167-168
readability, 9-11
recommended priority, 145
record type input terminals, 212
references, assigning, 144
reporting errors, 211, 213-214
required priority, 144
requirements specifications, 22-24
  best practices, 24-25
  projects, 25-28
  revision of, 31
  Theorem 2.1, 22
Reset controls, 148
Reset Energy Boolean text, 63
Resize Objects tool, 50
resizing panels, 59
resolution, configuring, 88
resources for design documentation, 30
reuse. See code reuse
revision
  numbers, 45
  of requirements specifications, 31
Revision Number labels, 63
Right Audio Output indicators, 53
Right Edges tool, 50
Right to Left VIs, 120
Ritter, David, 48
robustness, 12-15

rules
  applications, providing online documentation, 314
  applying Timed Loops, 246
  arrays
    applying for multivalued data items, 181
    entering descriptions, 182, 301
    storing large or dynamic length data sets, 181-182
  block diagrams
    applying high object density, 89
    configuring resolution, 88
    creating multilayer hierarchy of subVIs, 90
    leaving backgrounds white, 89
    limiting, 89
  Boolean data types
    applying if two states are logical opposites, 173
    assigning names, 173
    configuring parameters, 173
    formatting controls, 174
    labeling, 173
  Call By Reference Node, 274
  clusters
    applying alignment tools, 182
    avoiding for interactive controls, 184
    bundling/unbundling by name, 183
    entering descriptions, 301
    grouping multiple distinct items, 308
    saving, 183
  CM (content management) Rules, 46
  code reviews
    applying enumerated type definitions, 332
    applying symmetry and spacing, 330
    automating inspections, 321
    avoiding clusters, 335
    avoiding polling GUI objects, 332, 335
    avoiding Sequence structures, 332
    avoiding variables, 332
    configuring controls, 330
    configuring Tabbing Order, 335
    consistent fonts and capitalization, 330
    customizing test criteria, 321
    designating neutral parties to take notes, 335
    enlarging objects, 330
    General Error Handler VI over Simple Error Handler VIs, 332
    grouping controls, 330
    implementing error log files, 335
    including states for Initialize, Idle, Shutdown and Blank, 332
    labeling wires, 332
    limiting controls, 330
    limiting information on GUIs, 330
    minimizing wire bends, eliminating kinks and loops, 332
    modifying source code, 335
    peer reviews, 320, 334
    prioritizing rules, 323
    selecting memory efficient data types, 335
    style conventions, 320
    suppressing dialog error reporting, 335
    tunneling wires, 332
  colors, 69
    creating themes, 70
    following universal conventions, 70
    leaving panels/objects subVI gray, 70
    modifying, 69
    simplifying color schemes, 71
  comments
    applying, 305
    leaving notes for development teams, 306
  components
    maintaining, 274
    passing data between shift registers, 275
  connector panes
    applying 4×2×2×4 patterns, 142
    assigning controls, 142
    assigning error clusters, 144
    assigning references, 144
    avoiding crossing wire stubs, 142
    selecting left and right vertical edge connector terminals, 144
    selecting patterns with extra terminals, 140
    selecting top and bottom horizontal edge connector terminals, 144

selecting unified patterns for VIs, 141
specifying priority for inputs/outputs, 144
specifying terminal assignments, 144
controls, 52
  applying, enum, ring, and path controls, 180
  applying ring or enumerated, 72
  arranging, 59
  avoiding strings, 179
  combining labels, 68
  configuring default values, 169
  configuring tabbing order, 73
  Data Range property, 72
  entering descriptions, 169, 301
  grouping, 49
  including Help buttons/menus, 73
  labels, 66-67
  limiting, 71
  passing references, 283
  placing within Value Change event cases, 251
  providing default vales, 68
  restricting ranges, 71
  saving custom, 169
data constructs, creating arrays and clusters, 170, 172
data flow
  always flow left to right, 102
  applying Flat Sequence structures, 105
  applying global variables for simple data sharing, 105
  applying looped Case structures, 111-114
  applying shift registers, 111
  avoiding array and cluster coercions, 103
  avoiding nesting beyond 3 layers, 104
  avoiding sequence structures, 103-104
  avoiding variables, 108
  creating controls and constants, 103
  disabling dots at wire junctions, 103
  grouping shift registers, 111
  initializing control values, 105
  labeling wires exiting left shift register terminals, 111
  ordering operations, 105
  propagating error clusters, 102

data structures
  selecting consistent data types, 165-169
  selecting controls, 158
  selecting memory efficient data types, 160-161, 164
data types, applying multivalued, 308
dialog VI panels
  centering, 51
  sizing, 51
enums, applying, 306
error handling
  applying Error Case Structure, 218-220
  applying Error Handling templates, 222
  applying negative codes of I/O device errors, 217
  applying unwired defaults over constants, 220-222
  avoiding subVIs with built-in error reporting, 213
  disabling indexing, 208-209
  General Error Handler VI, 210-211
  implementing error logs, 211-213
  maintaining user-defined error codes, 217
  propagating error clusters, 205
  reporting, 205, 210
  suppressing dialog error reporting, 213-214
  trapping errors, 224-226, 230
  trapping from all iterations of loops, 206
  trapping from nodes, 209
events
  avoiding continuous timeouts, 251
  resizing data nodes, 251
  Value Change, 250
files
  naming conventions, 44-45
  repositories, 40
folder hierarchies, 40-42
fonts
  applying 13 point black, 68
  applying consistent, 62
  selecting, 63, 65
functions, Bundle and UnBundle By Name, 308-310

GUIs
  limiting information, 54
  navigation, 71-73
Help menus, 301-304
icons
  applying black borders, 132
  avoiding default icons, 132
  avoiding text and graphics, 139
  budgeting time for intended VI reuse, 135
  combining glyphs with colors and text, 132
  copying graphics, 138-139
  creating, 132, 311
  creating templates for, 136
  formatting fonts, 133
  reusing glyphs, 136
  saving VIs, 132
  selecting colors, 135
  selecting glyphs, 133
  selecting unified styles, 134
images, importing, 59
industrial GUI VIs
  large text size for command buttons, 69
  maximizing text colors, 69
labels
  algorithms, 305
  constants, 305
  hiding, 307
  loops, 305
  providing default values, 301
  subdiagrams, 305
  viewing terminal, 304
  wiring, 305
LabVIEW.ini backups, 33
layers, avoiding nesting beyond 3, 241
loops
  applying multiple criteria, 246
  applying shift registers, 246
  avoiding subVI selection for continuous, 283
  creating parallel, 278
  grouping shift registers, 246
  including, 279
  including delays, 246
  labeling, 270
  maintaining Event-Handling at top-level, 284

  messaging between, 269
  minimizing space, 270
  prioritizing, 269
  sizing, 270
modularize high-level subVIs into lower level subVIs, 90
nested data structures
  avoiding manipulating, 188
  limiting size of arrays, 188-192
  organizing complex data, 188
numeric data types
  applying automatic formatting, 176
  applying I32 representations, 175
  formatting, 176
objects
  aligning, 50
  avoiding overlapping, 57
  avoiding polling GUI, 250
  default appearance, 59
  enlarging and centering, 52
  sizing, 50
panels
  resizing, 59
  text, 61-69
projects
  journals, 24-25
  libraries, 43-44
proof of concept, style and, 31
reusable libraries, 35-37
reusable subVIs, 35
reusable templates, 38-39
run-time menus, customizing, 73
sequences, avoiding structures, 241
source file movement, 46
special numeric data types
  applying enums and rings, 178
  saving enums and rings, 179
state machines
  applying, 255
  applying enumerated type definitions, 255
  avoiding timeout, 262
  deriving application primary states, 255
  dividing primary states, 255

including Initialize, Idle, Shutdown, and Blank, 256
minimizing external code, 256
subVIs
  avoiding trivial containing few nodes, 91
  creating meaningful icons, 93
  creating to save space, 91
  modularizing top-level diagrams, 241
templates, deleting instructions, 62, 301
terminals
  assigning, 275
  selecting connectors, 274
text
  applying default 13-point plain application fonts, 305-306
  embedding, 301
  minimizing, 62
  minimizing front panels, 301
toolbars, hiding, 58
top-level VIs, 45, 51
updates, maintaining high speed display at top-level, 284
wiring
  avoiding local and global variables, 96
  avoiding structures, 96
  labeling, 96, 246
  limiting length, 96
  locating unwired front panel terminals, 97
  maintaining spacing of parallel wires, 94
  minimizing bends, 94
  modularizing into clusters, 97, 100
  obstructing view of, 96
  saving clusters as type definitions, 101
  tunneling, 96
written requirements specifications, 25
Run Test Boolean controls, 105
Run Test events, 109, 113, 267
run-time menus, customizing, 73

S

saving enums, 307
schemes, colors, 71

scope
  controls, 71-73
  creep, 22
  Theorem 2.1, 23
Screw Inspection VIs, 124, 126, 234-235
Select Tests dialog, 322
selecting
  controls, 158, 160-172
  fonts, 63-65
  icons, 133
  subVIs, 75-76
Selection utility, subVIs from, 115-117
self reviews, 320
  manual checklists, 330-334
  VI Analyzer Toolkit, 320-323, 329
Sensor Scaling clusters, 105
Sensor Scaling write and read local variables, 108
Sensor Selection listbox controls, 199
Sequence structures, 104-109
Serial Port Settings cluster, 148-149
Serial Ports, 148
shift registers, efficiency, 8
shortcuts icons, 135-136, 139
Shutdown state, 114, 256, 267
simple arrays, 181-184
simple controls, 158
simple data structures, 158
simple data types, 172
  Boolean, 173-174
  numeric, 175-176
  path, 179-180
  picture, 179-180
  special numeric, 177-179
  strings, 179-180
Simple Error Handler VIs, 210-211
simplicity, 15-17
single-precision (SGL), 175
sizing objects, 50-51
SMTP Email palette, 213
source code
  maintainability, 11
  peer reviews, 334-336
  readability, 9-11
  self reviews, 320

manual checklists, 330-334
VI Analyzer Toolkit, 320-323, 329
source control, 39-46. *See also* CM (content management)
source files, moving, 46
Spaghetti VIs, 5
  ease of use, 6
  efficiency, 7
  error handling, 14
  maintainability, 12
  modularity index, 13
  performance, 17
  readability, 11
  scope creep, 23
  simplicity, 16
special data constructs, 186
special numeric data types, 177-179
specifications. *See* requirements specifications
specifying terminal assignments, 144
Spectralyzer, 82-83
Standard State Machine design patterns, 37
standard subVIs, 48
state machines, 254-256
  Classic State Machine design patterns, 257-258
  Event Machine design patterns, 265-267
  Event-Driven State Machine design patterns, 262-264
  Queued State Machine design patterns, 260-262
Statistics clusters, 324
Stop global variables, 105
strict type definitions, 101
string data types, 179-180
structures
  data. *See* data structures
  Sequence, 104-109
  wiring, 226
style
  ease of use, 6-7
  efficiency, 7-9
  evaluation tools, 18
  icons, 132-135
  importance of, 2-6
  maintainability, 11

  performance, 17-18
  proof of concept and, 31
  readability, 9-11
  robustness, 12-15
  simplicity, 15-17
  Theorem 1.0, 2
  Theorem 1.1, 19
  time versus, 18-19
subdiagrams, labels, 305
subVIs
  code reuse, 35
  efficiency, 9
  error handling, 37, 217, 220-222
  from Selection utility, 115-117
  icons, 311
  modularization, 90
  panels, 48, 68
  performance, 17
  robustness, 12-15
  selecting, 75-76
  utility VIs, 36
Suss Interface Toolkit, 137, 231-232
symbolic mappings, 63
Synchronization palettes, 223
System palettes, 53

T

tabbing order, configuring, 73
tabs, overview, 56
task ID out 2, 146-147
templates
  code reuse, 36-39
  Error handling, 222
  icons, 136
  instructions, 62, 301
terminal labels, viewing, 304
terminals
  assigning, 144-154
  interfaces, 140-145
  top and bottom horizontal edge connectors, 144
  vertical edge connectors, 144

Index

Test Executive VIs, error handling, 235-238
Test to Run numeric control, 306
tests
  documentation, 321
  Front Panels, 321
  methodology section (requirements specifications), 28
text
  default
    13-point plain black, 305-306
    appearance, 59
  embedding, 301
  front panels, 61-69, 301
  industrial GUI VIs, 69
  minimizing, 62
  subVIs, 68
themes, creating, 70
Theorem 1.1 (style elements), 2
Theorem 1.2 (style and development time), 19
Theorem 2.1 (requirements specifications), 22
Theorem 2.2 (scope creep), 23
Thermometer VI, 193-194
tight loops, 279
time
  style versus, 18-19
  Theorem 1.2, 19
Timeout events, 251
Timing palettes, 212, 223
toolbars, hiding, 58
tools
  Bottom Edges alignment, 95
  Filled Rectangle, 135-136
  Resize Objects, 50
  Right Edges, 50
  Vertical Compress, 50
top and bottom horizontal edge connector terminals, 144
Top level application window, 58
top-level VIs, 47
  arrays, bundling, 167
  identifying, 45
  maximizing, 51
Torque Hysteresis VI, 171-172, 189-191, 214, 330, 336

transducers, 293, 295
trapping errors, 205-209
Two Button Dialog function, 52
type definitions, 101
type of dialog input terminals, 212

U

Unbundle By Name function, 183, 308-310
unified patterns, selecting for related VIs, 141
unit under test (UUT), 170
Unused Code failures, 323, 326
Use transparent name labels, 67
User Controls palette, 36
user event data type input terminals, 266
user events, 266
User Interface category, 323
User Libraries palette, 36
utility VIs, 36
UUT (unit under test), 170

V

Value Change events, 250
variables, data flow, 104-109
variants, 187, 194-197
velocity, 324, 328
Vertical Compress tool, 50
vertical edge connector terminals, 144
VI Analyzer Toolkit, 320-323, 329
VI descriptions, style versus time, 19
VI Documentation failures, 327
VI Hierarchy windows, 166
VI History, style versus time, 19
VI Metrics window, 12, 15, 18
VI Properties dialog, 230
VI Properties window, 8
View as Icon (shortcut menu), 152
VIs
  descriptions, 311
  error handling, 204

automatic, 230
   clearing errors, 228
   error codes, 214, 217
   examples, 230-238
   merging errors, 226-227
   prioritizing errors, 222-225
   reporting errors, 211-214
   structure wiring, 226
   trapping errors, 205-209
templates. *See* templates
top-level VIs
   arrays, bundling, 167
   identifying, 45
   maximizing, 51
Visible property, 222
visible radix, 176

W–Z

Wait n mSec VIs, 225
WDT (waveform data type), 186
Window Appearance property pages, 58
Wire Bends tests, 324, 329
Wires Under Objects errors, 323, 327
wiring, 93-101
  connector pane, 140-145
  labels, 305
  structures, 226
written specifications. *See* requirements
    specifications

THIS BOOK IS SAFARI ENABLED

INCLUDES FREE 45-DAY ACCESS TO THE ONLINE EDITION

The Safari® Enabled icon on the cover of your favorite technology book means the book is available through Safari Bookshelf. When you buy this book, you get free access to the online edition for 45 days.

Safari Bookshelf is an electronic reference library that lets you easily search thousands of technical books, find code samples, download chapters, and access technical information whenever and wherever you need it.

TO GAIN 45-DAY SAFARI ENABLED ACCESS TO THIS BOOK:

- Go to **http://www.prenhallprofessional.com/safarienabled**
- Complete the brief registration form
- Enter the coupon code found in the front of this book on the "Copyright" page

If you have difficulty registering on Safari Bookshelf or accessing the online edition, please e-mail customer-service@safaribooksonline.com.

Also Available from Prentice Hall

LabVIEW for Everyone
Graphical Programming Made Easy and Fun

Jeffrey Travis and Jim Kring

Master LabVIEW 8 with the industry's friendliest and most intuitive tutorial. Top LabVIEW experts Jeffrey Travis and Jim Kring teach LabVIEW the easy way: through carefully explained, step-by-step examples that give you reusable code for your own projects.

This third edition has been fully revamped and expanded to reflect new features and techniques introduced in LabVIEW 8. You'll find two new chapters, plus dozens of new topics, including Project Explorer, AutoTool, XML, event-driven programming, error handling, regular expressions, polymorphic VIs, timed structures, advanced reporting, and much more. Certified LabVIEW developer (CLD) candidates will find callouts linking to key objectives on NI's newest exam, making this book a more valuable study tool than ever.

Jeffrey Travis provides expert consulting and creates books, courses, and products for remote Internet controls and monitoring, virtual instrumentation, and Web applications through his company, Jeffrey Travis Studios. He has more than fifteen years of experience developing software, teaching, and consulting on LabVIEW and related technologies

Jim Kring is president of James Kring, Inc., a leader in LabVIEW development, system integration consulting, and custom software design. He is founder of OpenG, a foundation promting open-source LabVIEW tools, applications, frameworks, and documentation.

0-13-185672-3 • © 2007

For additional information, please visit www.prenhallprofessional.com/title/0131856723.